The Second

JEREMY M. NORMAN

THE SECOND PRINTING REVOLUTION

INVENTION OF MASS MEDIA

The Grolier Club • New York • 2026

Contents

Acknowledgements ix
Prologue xi

1.
Book Production During the First Century of the Industrial Revolution 1

2.
Louis-Nicolas Robert and the Development of the Papermaking Machine in the First Half of the Nineteenth Century. . . 25

3.
Friedrich Koenig Invents the Steam-Powered Printing Machine 43

4.
Railroads, Power Looms, and Bibles: Innovation Versus Tradition 63

5.
Henry Brougham, Charles Knight, and the Society for the Diffusion of Useful Knowledge Use the Printing Machine to Reach Large Audiences 77

6.
Image-Reproduction Methods Appropriate for Rotary Printing Machines: Wood Engraving, Lithography, Steel Engraving, Electrotyping 97

7.
Developments in Mechanized Book and Newspaper Production, 1800–1850 . . . 109

8.
Charles Dickens and His Imitators Exploit the New Technologies with Great Success 125

9.
The Development of Mechanized Printing in America: Daniel Treadwell, the American Bible Society, R. Hoe & Co., Isaac Adams, the Harper Brothers 135

10.
The Role of Women in Eighteenth- and Nineteenth-Century Book Production: Emily Faithfull 151

11.
Mechanizing Typesetting & Type Distribution from William Church to Young and Delcambre's Pianotyp, and to Linotype and Monotype 165

12.
The Mechanization of Bookbinding: William Burn, William Pickering, Archibald Leighton, and Followers . . . 193

13.
The Mechanization of Book Production in the United States and Europe, 1851–1904: The Great Exhibition, George Baxter, Manuals and Promotional Books on Mechanized Printing, the Caxton Celebration and Bible 213

14.
The Mechanization of Newspaper Production in the United States and Europe, 1826–1900: The Hoe Family and Hippolyte Marinoni . . 237

CODA
William Morris, Theodore Low De Vinne, and Robert Hoe III Reflect on Nineteenth-Century Developments in Book Production 257

Notes. 263
Bibliography 271
Index 293

Hippolyte Marinoni (front left with arm outstretched) in black coat and top hat, demonstrates the high-speed web press that he invented, printing one million copies of the illustrated color supplement to *Le Petit Journal* in 1890. As the inventor and builder of the press and the owner of *Le Petit Journal*, then the largest circulation newspaper in Europe, Marinoni placed himself in historic company. The tableau above the image includes Marinoni's name on the far right next to famous hand press printers who had preceded him: Elzevier, Estienne, Gutenberg, and Dolet.

Acknowledgements

The enthusiasm for this project expressed by Nancy Boehm, President of The Grolier Club, when she and I met at the London bookfair in 2023 set the wheels in motion for what became this book and the exhibition at The Grolier Club from January 14 to April 11, 2026. Club staff essential for the project were Shira Buchsbaum and Oscar Salguero (exhibitions), Ann Donahue (publications) Amanda Domizio (communications), April Rose (Programs), and Misha Beletsky (book design).

Diana Hook, my professional colleague for the past 40 years, contributed significantly in many ways. Martha Rose Noble took all the photographs. Patricia Fogarty edited my manuscript.

Michael Suarez read an early draft and provided valuable advice.

Stephen Galbraith and Amelia Hugill-Fontel of the RIT Cary Graphic Arts Collection granted permission for the reproduction of the paintings from the *Graphic Communications through the Ages* series.

Eric Frank, U.S. sales manager for Koenig & Bauer AG, supplied images.

The essence of the history of the second printing revolution is in the patents, which are often the only way to understand how inventions worked. They are not easy to identify or approach. Three essential books that were of invaluable help in providing background on some of the patents, and how printing, papermaking, and typesetting technologies evolved, are James Moran, *Printing Presses: History and Development from the Fifteenth Century to Modern Times* (1973), R. H. Clapperton, *The Paper-Making Machine: Its Invention, Evolution and Development* (1967), and Richard Huss, *The Development of Printers' Mechanical Typesetting Methods 1822–1925* (1973). They sent me on the process of collecting the 208 patents that begin the bibliography starting on page 271, and underly my research for this book. All the books by Michael Twyman were essential for understanding the history of graphic processes.

Thanks also to Alexandra Norman and Max Norman for their support over the twenty years that I devoted to this project.

Jeremy M. Norman
November 22, 2025

Theodore Low De Vinne at the age of thirty-six.

Prologue

At the beginning of the twentieth century, in 1901, Theodore Low De Vinne, a leading American printer and one of the founders of The Grolier Club, published an article titled "Printing in the Nineteenth Century."[1] De Vinne was a master of all aspects of the printing trades, including the design and production of large-edition books, high-circulation magazines such as *The Century Illustrated Monthly Magazine*, and fine limited editions; he was also a scholar of the history of printing and typography. In his article, De Vinne reviewed the enormous advances during the nineteenth century that had taken book production from a handcraft to a modern industrial process.

The nineteenth has been a century of wonderful achievement in every branch of printing. The Fourdrinier paper-making machine, the Bruce type-caster, the Linotype type-casting and type-setting machine, and other mechanical type-setters of merit; composition inking-rollers, the cylinder press, the web press, and mechanisms of many kinds for the rapid printing of the smallest label or the largest sheet in black or many colours; machines for folding, sewing, and binding books; the arts of stereotype, electrotype, and photo-engraving—all these are its outgrowth, and the more important have been invented or made practicable within the memory of men now living.[2]

Radical changes in book production in the nineteenth century led to the invention of mass media—very large editions of newspapers, periodicals, and books sold to increasingly large readerships at low costs. These changes amounted to a second printing revolution. At the beginning of the nineteenth century, *The Times* of London daily newspaper, printed on iron handpresses, could only publish 4,000 copies of a four-page edition. In 1863, decades after printing was mechanized, the French daily newspaper *Le Petit Journal* began publication with a circulation of 83,000; it reached a circulation of one million copies in 1890 and claimed two million by 1895. This transition from 4,000 to two million copies per day was revolutionary, but it was a revolution that did not affect all elements of printing. Many printing projects did not require high speed, and while dramatic increases in printing output occurred during the nineteenth century, traditional handpress printing and smaller editions of books continued to be produced. Explaining how and why high-speed innovation occurred in parallel with traditional handpress book production is one of the purposes of this book.

At the beginning of the nineteenth century, the Industrial Revolution had been underway in England for nearly a century, yet book production had hardly changed since Johannes Gutenberg's invention of printing by movable type in the mid-fifteenth century. Printing in England was still being done on wooden handpresses, only incrementally improved from those Gutenberg built and used. Paper was still

"Mr. Rutt's Printing Machine"
From Thomas Curson Hansard, *Typographia* (London, 1825)

Invented by Thomas Rutt in 1819, this was one of several smaller early printing machines designed to be powered by hand crank rather than a steam engine. Rutt based its design on that of Friedrich Koenig's first cylindrical steam-powered press, which was finished in 1812.

Jan van der Straet, called Stradanus
Impressio librorum (Antwerp, c. 1600)
8¼ × 10½ in. (21 × 26.5 cm)

A view of workers in a typical sixteenth-century printing office, engraved by Jan Collaert I, after Stradanus, and published by Philips Galle in the series of prints entitled *Nova reperta*. The arrangement of the printing shop shown here had not changed significantly since Gutenberg's day and remained little changed, except for incremental improvements to the wooden handpress, until the invention of printing machines in the early nineteenth century.

being made by hand. The process of typecasting and typesetting had not significantly changed since the fifteenth century; it was still done by hand, as was bookbinding. The relatively static nature of English book-production technology at the time contrasted with the growing impact of steam power in England that began with Thomas Newcomen's invention of the Atmospheric Steam Pumping Engine for pumping water out of mines in 1710–1712, and the mechanization of English textile production that began in the 1760s with James Hargreaves's invention of the spinning jenny (1764) and Richard Arkwright's invention of his spinning machine (1769–1775). These key inventions led to some of the first developments in mass production and the factory system, and they ushered in a period of profound change in the socio-economic relationship between workers and employers. Inevitably, there was resistance to mechanization from workers who feared, often with justification, that machines would eliminate their jobs.

By the end of the eighteenth century, England was at the forefront of the Industrial Revolution, especially in the production of steam engines and power-loom textile production, but those technological innovations and their social impact did not reflect the kind of sudden upheaval that we might associate with a political revolution. Instead, what we call the Industrial Revolution describes the gradual development over about 200 years of technologies that would reach their full social and economic impact by the middle and end of the nineteenth century. It was in the second half of the Industrial Revolution, the period from around 1800 to around 1890, that enterprising people with new inventions caused the transition of printing and book production processes from traditional hand processes to mechanized production.

In the early nineteenth century, evolving technologies contributed to social and economic change. These included the new railroads carrying passengers at previously unheard-of speeds—twelve to thirty miles per hour. In this changing and sometimes disruptive environment, newspapers in growing cities in Britain, such as *The Times* of London, could not print enough copies to satisfy demand. To meet this demand, the processes of printing and papermaking were gradually mechanized in Britain (and somewhat later in France, Germany, and the United States), increasing the size of sheets of paper beyond the limitations of handmade paper, speeding the rate of paper production, speeding up printing, and enabling newspapers and magazines to print and sell far more copies. A few decades later, starting in the 1830s and 1840s, the more complex processes of typesetting and type distribution (returning lead type to the case after printing) were gradually mechanized in many imperfect or impractical attempts—over 200 different attempts during the nineteenth century. The savings and efficiencies of mechanized production invented for newspapers were also gradually applied to book production, and elements of the bookbinding process were also mechanized. The availability of cheaper books, magazines, and newspapers printed in larger editions at lower costs, resulting from mechanization, stimulated the growth of literacy, which in turn further increased the demand for more books, magazines, and newspapers at lower cost.

To trace the histories of these developments, we benefit from the aspect that many nineteenth-century advances in printing and other elements of book production were regarded as newsworthy, often reported in considerable detail in the newspapers and magazines as they adopted the new technology. For example, in November 1890, Theodore Low De Vinne published "The Printing of 'The Century'" in *The Century Illustrated Monthly Magazine*.[3] In that article, De Vinne, whose company printed the magazine, described and illustrated advances in printing machines, paper, ink, the reproduction of illustrations, and advances in bindery machinery that had improved the quality of magazine production within the previous twenty years and enabled the magazine's publishers to produce 200,000 copies every month. We also benefit from publications that numerous printers and manufacturers of printing equipment issued to promote their businesses, especially after mechanization resulted in the growth of large industrial printing concerns during the second half of the nineteenth century; these are often fine examples of the printing art. Less frequently, printers stated within publications they wrote or printed that those publications were the result of technological innovations.

Chapters in this book concern the mechanization of printing, papermaking, reproduction of illustrations, typesetting, and bookbinding, with emphasis on these developments in England, France, Germany, and the United States from 1800 to 1900. Whenever possible, this book cites and illustrates specific examples of the incunabula of the second printing revolution, particularly the first examples of new technologies when they were introduced. These examples are analogous to the traditionally collected incunabula from the so-called "cradle of printing" from c. 1455 to 1500.

Like the first printing revolution, in which printing from movable type only gradually replaced the medieval process of manuscript copying as the dominant means of information storage and distribution in Europe, mass media produced by mechanized printing, papermaking, typesetting, and bookbinding in the second printing revolution did not suddenly replace the production of handcrafted books. Instead, once established, it served new and expanding readerships and assumed the leading role in the efficient distribution of information to ever-larger audiences. Smaller editions of books continued to be printed on iron handpresses, sometimes using hand-set type, sometimes on handmade paper, and sometimes hand bound, throughout the nineteenth century and into the twentieth. In the twenty-first century smaller editions of books, such as the one you are presently reading, continue to be printed using the latest printing technology, and makers of handcrafted books continue their crafts in the present day.

Wilhelm Haas, the younger
Beschreibung einer neuen Buchdruckerpresse 1772
(Basel, 1790), detail. See p. 17.

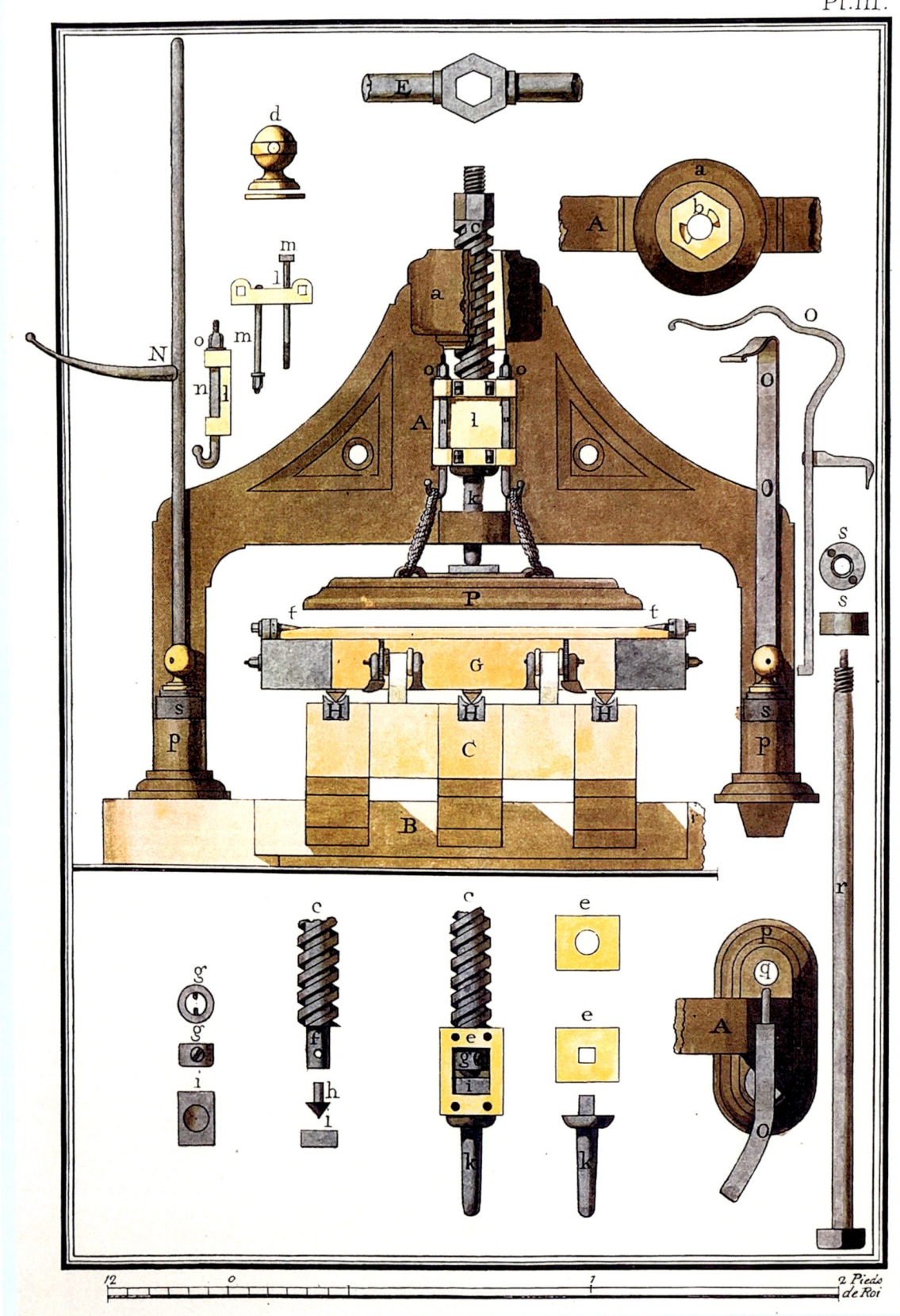

CHAPTER

Book Production During the First Century of the Industrial Revolution

During the eighteenth century, given enough time, and a lot of hand labor, remarkably large numbers of books and pamphlets could be printed on handmade paper with handpresses and distributed. A dramatic and exceptional example were the *Cheap Repository Tracts*, a series of more than 200 sixteen-page pamphlets on moral, religious, and occasionally political subjects devised by the English religious writer Hannah More that sold for one penny each. These efforts to provide "religious and useful knowledge" were intended for sale or distribution by "Hawkers in Town and Country" as well as by booksellers. More than two million copies (about 10,000 copies of each pamphlet) were sold during 1795, the first year of their production.[4] Series of these pamphlets continued to be issued until 1817. Each *Cheap Repository Tract* was a single printed sheet folded and stitched to create a sixteen-page pamphlet without wrappers. Printing two million copies of both sides of a single printed sheet in one year would have challenged a large group of handpress printers, just as producing that many sheets of handmade paper would have challenged many hand paper mills. But they were able to meet the challenge with traditional methods. Except for some incremental improvements in the handpress in the seventeenth century, and incremental improvements in the efficiency of book production toward the end of the eighteenth century and during the first decade of the nineteenth century, book production during the first century of the Industrial Revolution, from about 1708 to about 1800, remained essentially unchanged from Gutenberg's time. Printing was still done on wooden handpresses, paper was still made by hand, type casting and typesetting were done by hand, and binding remained a handcraft. One reason that the traditional system worked was that the number of copies printed of most publications tended to be relatively small and therefore manageable with the existing handpress technologies. For example, *The Gentleman's Magazine*, one of the more popular general-interest magazines in England, had a monthly circulation of 3,000 copies in 1746.[5]

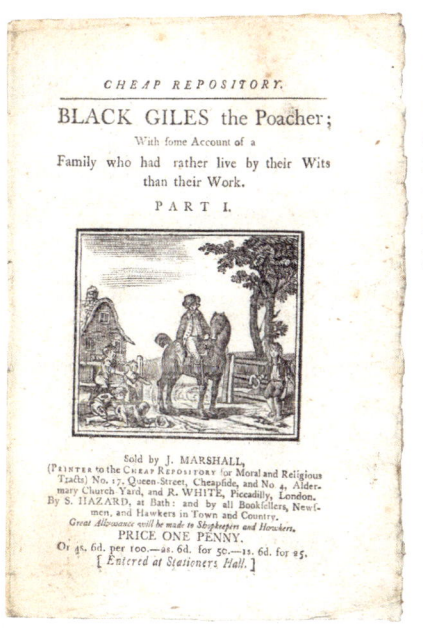

[Hannah More,] *Black Giles the Poacher* (London, 1795)

One example of the two million copies of pamphlets in More's series *Cheap Repository Tracts* that were printed by handpress and sold during the year 1795.

CHAPTER 1

Book Production Retains Traditional Technology as Mining, Textiles, and Agriculture Mechanize

For the first 200 years after Gutenberg, the technique of printing was passed down from generation to generation, secretively, through guild apprenticeship, without any detailed, printed technical description of the process or the equipment. The first writer to provide a thorough technical description of the equipment involved in printing, and the process of printing from movable type, was the English hydrographer, printer, punch cutter, globe maker, and instrument maker Joseph Moxon, who published his *Mechanick Exercises on the Whole Art of Printing* in 1683 and 1684; it formed the second volume of his *Mechanick Exercises, or the Doctrine of Handy-Works* (1677-1684), a survey of the chief trades of his day. This was the first comprehensive manual in any language on printing, and the first printing manual published in English. Moxon had worked for years as a master printer; he had also cut steel punches for typefaces, made typecasting moulds and matrices, and cast and sold type. As a printer, Moxon specialized in mathematical books and maps; he also made globes and mathematical instruments, and he produced the first English-language dictionary of mathematics. He was the first tradesman elected a fellow of the Royal Society of London for Improving Natural Knowledge, the equivalent of our National Academy of Sciences.

The first volume of *Mechanick Exercises*, devoted to blacksmithing, joining, carpentry, and related arts, was also the first book in England published in parts, or fascicules. Fourteen numbers were issued between 1677 and 1680, before possible disappointment with sales and the "breaking out of the Popish Plot"—which "took off the minds of my few customers from buying . . ." (Moxon's "Advertisement," vol. II)—forced Moxon temporarily to cease production. Moxon resumed the series in 1683 with *Mechanick Exercises on the Whole Art of Printing*, issued in twenty-four parts during 1683 and 1684. "Although 500 copies were printed, very few complete sets have been preserved, the work being, perhaps, the most difficult to obtain in the whole range of typographical literature."[6]

Frederick Henrik van Hove
Joseph Moxon (London, 1699)
7 × 4¾ in. (17.5 × 12 cm)

Portrait of English hydrographer, printer, punch cutter, globe maker, and instrument maker Joseph Moxon, who wrote and published *Mechanick Exercises on the Whole Art of Printing*, the first comprehensive manual on printing in any language.

In the second volume of *Mechanick Exercises*, Moxon provided detailed technical accounts of the tools of the compositor and pressman, the art of typefounding, and the work of the compositor, corrector, pressman, and other members of the printing trades as they had come down to his day. Most of these tools and the skills to operate them had not changed significantly for the previous 200 years and would remain essentially unaltered until the mechanization of printing in the nineteenth century. For the next 200 years after its publication, elements of Moxon's manual were copied or adapted by writers of printing manuals. Regarding the speed of output on a wooden handpress,

Joseph Moxon
From *Mechanick Exercises on the Whole Art of Printing* (London, 1684)

👉 The type case as illustrated by Joseph Moxon in *Mechanick Exercises*.

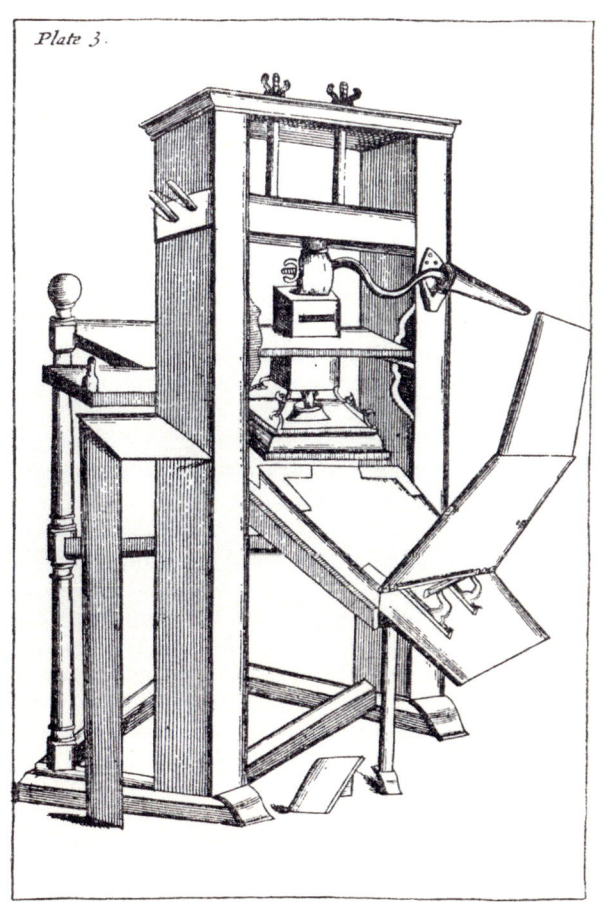
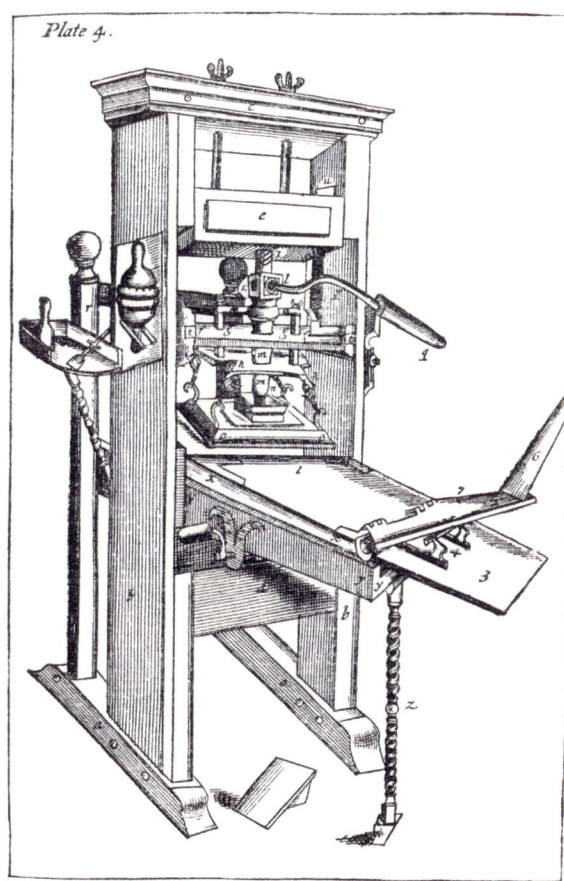

👉👉 The traditional wooden handpress.

👉 The latest seventeenth-century version of the wooden handpress as improved by the Dutch printer and publisher Willem Janszoon Blaeu.

Moxon provided the first measure of handpress printing efficiency: 250 sheets per hour, printed on one side of a sheet by two pressmen, which Moxon called the "token." In thinking about that number, we might assume that, toward the end of a twelve-hour working day of such intense manual labor, the rate might have dropped, and a more reasonable average production figure would have been around 200 sheets per hour.[7] However, Moxon's estimate of 250 sheets per hour remained an average production limit on handpress printing until printing was first mechanized during the second decade of the nineteenth century. To increase production output of a single title by handpress within a set period, it was necessary to use more handpresses and more pressmen printing the same title. This generally meant dividing up the gatherings so that each press printed only portions of a book, or it could require the time-consuming and expensive process of resetting the type for the second or other presses if a printer owned enough type for that purpose.

For centuries, manual typesetting was the most complex and labor-intensive, and one of the costliest parts of the printing process. It was also challenging and difficult for typesetters, who were usually paid by piecework. The print-shop environment could also pose serious risks to a worker's health. Though Moxon did not concern himself with the occupational diseases that could affect people in the printing trades, the Italian physician Bernardino Ramazzini devoted a chapter to the diseases of printers and typesetters in his pioneering book on the diseases of workers, *De morbis artificum diatriba* (1700, 1713). Ramazzini noted that typesetters could injure their eyesight from staring at small type for hours on end, often in poor light.[8] Lead poisoning from type metal was also a hazard. In a letter famous in the history of occupational medicine but little known in the history of printing, Benjamin Franklin wrote to British physician and political radical Benjamin Vaughan on 31 July 1786 of his early experiences of symptoms of lead palsy as an apprentice compositor in London at the printing house of Samuel Palmer in 1724.

In 1724, being in London, I went to work in the Printing-House of Mr. Palmer, Bartholomew Close, as a Compositor. I then found a Practice I had never seen before, of drying a Case of Types, (which are wet in Distribution) by placing it sloping before the Fire. I found this had the additional Advantage, when the Types were not only dry'd but heated, of being comfortable to the Hands working over them in cold weather. I therefore sometimes heated my Case when the Types did not want drying. But an old Workman observing it, advis'd me not to do so, telling me I might lose the Use of my Hands by it, as two of our Companions had nearly done, one of whom that us'd to earn his Guinea a Week could not then make more than ten Shillings and the other, who had the Dangles, but Seven and sixpence. This, with a kind of obscure Pain that I had sometimes felt as it were in the Bones of my Hand when working over the Types made very hot, induc'd me to omit the Practice.[9]

Franklin's letter is one of the earliest accounts of the symptoms of lead poisoning experienced by typesetters, and it is of special significance for being a firsthand account by a skilled observer of the actual symptoms of lead palsy from handling lead type.

The first development in book production that reflected significant innovation after Gutenberg's invention of printing by movable type was stereotyping—a process of creating printing plates from moulds taken from the surface of typeset pages, which made it possible to reprint a text without having to reset the type. Multiple copies of stereotype plates could be made, so that a text could be printed on multiple presses at the same time, and the stereotype process also allowed printers to store typeset pages for future reprints without tying up any of their movable type. It is now understood that stereotype printing has an extensive but undocumented history,[10] and several printers during the eighteenth century are thought to have employed the process without mentioning it in their publications. The first printer to state in print that he had printed from stereotype plates was the Scottish goldsmith and printer William Ged, whose 1739 edition of Sallust's *Belli Catilinarii et Jugurthini Historiae* included the following imprint:

Edinburghi, Gulielmus Ged, Aurifaber Edinensis, non Typis mobilis, ut vulgo fieri solet, sed Tabellis seu Laminis fusis, excudebat, MDCCXXXIX. (Edinburgh: Printed by William Ged, Goldsmith of Edinburgh, not from movable type, as is commonly done, but from cast plates, 1739).

Robert Thom
Benjamin Franklin (c. 1960)
Oil on canvas, 32 × 24 in. (81 × 61 cm)
From Kimberly-Clark, *Graphic Communications Through the Ages*
Courtesy of RIT Cary Graphic Arts Collection

The young Benjamin Franklin is shown pulling his handpress in Philadelphia.

Ged's process appears not to have been adopted by other printers, so his invention did not significantly advance book production at the time.

Consistent processes and procedures in book production were conducive to stability. One positive result of this stability was that Luddite-style machine-breaking appears undocumented in book production trades during the eighteenth century. This stability contrasted with the examples of Luddite machine-breaking directed toward steam engines in mining, power looms and the factory system for textiles, and threshing machines in agriculture. The new mechanization caused disruption of the traditional socio-economic order. For workers enduring periodic economic downturns and excessive exploitation during the Industrial Revolution, machine-breaking was a dramatic means of protest. By 1768, enough instances of machine-breaking of Newcomen engines

Sallust *Belli Catilinarii et Jugurthini historiae* (Edinburgh: William Ged, 1739 and 1744)

The two issues of William Ged's pocket-size edition of Sallust were the first books to state on their title pages that they were printed from "cast plates," later known as stereotypes, rather than from individual pieces of lead type.

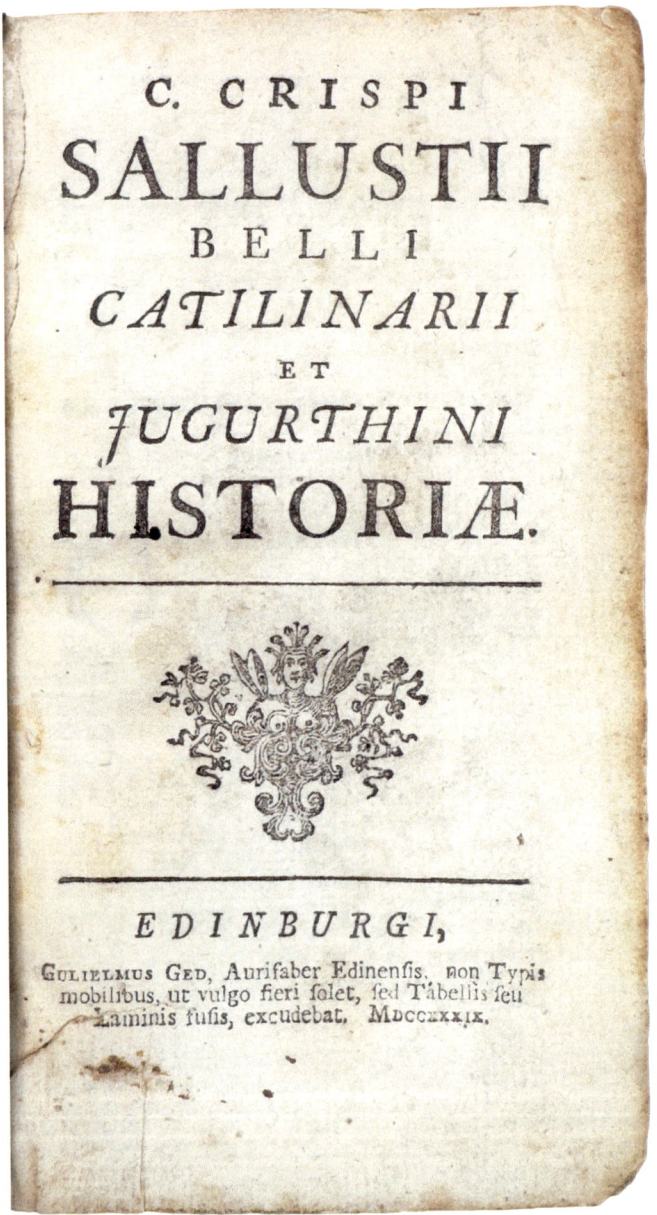
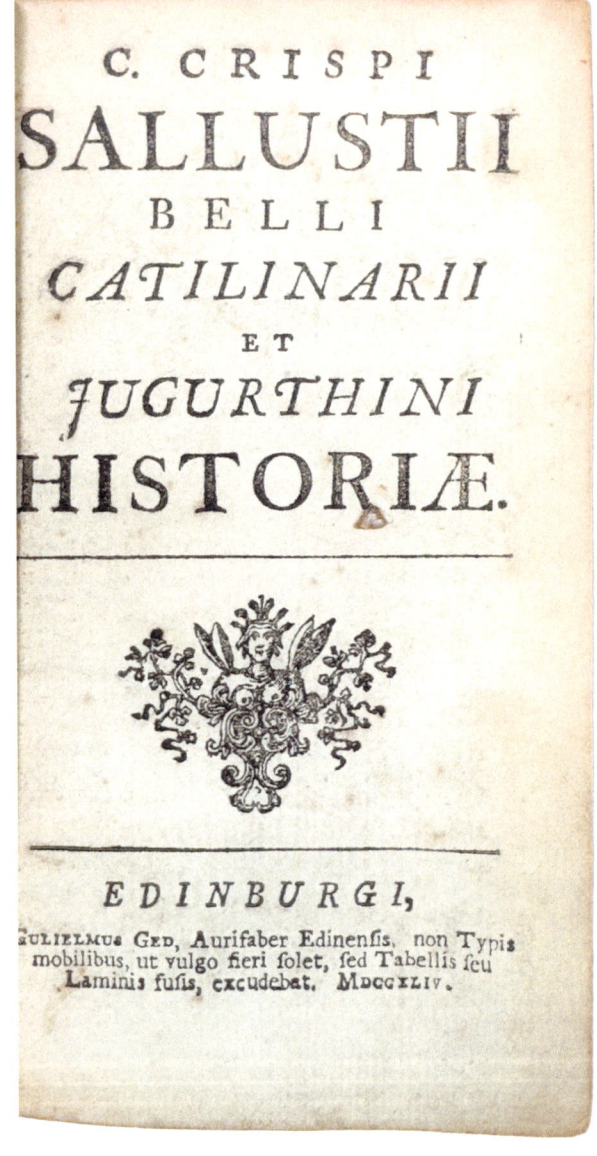

pumping water out of mines had occurred that Britain passed laws against it. In 1789, England faced a severely cold winter, disappointing spring weather, and poor harvests. Facing bitter cold and low pay, miners rioted against mine owners, destroying several steam engines in different mines. In response, George III issued a proclamation, a copy of which is illustrated in this book. After a very long-winded and verbose explanation of the context, the king offered a fifty-pound bounty for the apprehension of colliers who destroyed steam engines and other mine company property.[11]

But since book-production technologies and working conditions had changed little over the centuries, printing presses and related machines were not a target for frustration in the printing trades during the eighteenth century, though inevitably there were conflicts between printers and their employers. Except for newspapers in larger cities, which could not print enough daily copies to satisfy demand, traditional handcrafted means of printing remained satisfactory.

By the KING.

A PROCLAMATION

For suppressing RIOTS and TUMULTS committed by COLLIERS, and Others, in the Counties of Northumberland and Durham, and for apprehending and bringing to Justice the Persons who have committed or shall commit the same.

GEORGE R.

WHEREAS it has been represented to Us, that a great Number of Pitmen, to the Number of Two Hundred and upwards, did, in a tumultuous, disorderly, and riotous Manner, assemble themselves, upon the Tenth Day of March last, at Long Benton Colliery, in the County of Northumberland, and there wilfully and maliciously damaged and destroyed several Engines erected for drawing Coals out of several of the Pits of the said Colliery, and did also wilfully and maliciously set on Fire one Pit belonging to the said Colliery, and the same continued on Fire for two whole Days, to the great Damage of the Proprietors thereof; and also that, on the said Tenth Day of March, a Number of Pitmen, to the Number of Two Hundred and upwards, assembled themselves in like Manner at a certain Pit, called the Robert Pit, belonging to Shire Moor Colliery, in the County aforesaid, and there wilfully and maliciously destroyed and damaged various Implements, Materials, and Engines belonging to the same: And also, that on the said Tenth Day of March, a Number of Pitmen, to the Number of One Hundred and upwards, did, in like Manner assemble themselves at Walls-End Colliery in the County aforesaid, and wilfully and maliciously did considerable Damage to the Gins and Machines belonging to the said Colliery: And also that on the said Tenth Day of March, a Number of Pitmen, to the Number of One Hundred and upwards, in like Manner assembled, wilfully and maliciously broke and damaged the Engine for drawing Coal, at the Engine Pit of East Benton Colliery in the County aforesaid: Whereas it has also been represented unto Us, that a Number of Persons, to the Number of One Hundred and upwards, assembled themselves together upon the said Tenth Day of March last, at a Colliery called Walker Colliery, situate in the said County, and by Force and Violence prevented the working of a certain Coal Pit belonging to the said Colliery, and broke and damaged a certain Engine or Engines belonging to the same, and violently assaulted some of the Workmen of the said Colliery, and threw down Heap Lamps into the said Pit with Fire in them: And whereas it has also been represented to Us, that, upon the Eleventh or Twelfth Days of the said Month of March, divers Persons assembled themselves together in a like riotous, tumultuous, and illegal Manner at Gosforth Pit, belonging to the said Colliery, and threw Lamps and other Materials down the same: And whereas it has also been represented to Us, that on the several Days aforesaid, divers Persons being riotously, tumultuously, and illegally assembled, did various other Outrages and Mischiefs to several of the Coal Mines in the said Counties of Northumberland and Durham, in open Defiance of Justice and the Laws, by which the Peace of the said Counties, and the Lives and Properties of Our good Subjects are greatly endangered; We, therefore, being deeply sensible of the many mischievous Consequences that must inevitably ensue to the Peace of Our Kingdom, and the Lives and Properties of Our loving Subjects from such wicked and illegal Practices, if they shall go unpunished; and being firmly resolved to cause the Laws of this Our Kingdom to be put in Execution for the Punishment of all such Offenders, have thought fit, by the Advice of Our Privy Council, to issue this Our Royal Proclamation, hereby strictly commanding and requiring all Our Justices of the Peace, Sheriffs, Under Sheriffs, and all other Our Civil Officers whatsoever, that they do use their utmost Endeavours for discovering and apprehending the Persons who We are informed are concerned in the said Outrages and Riots, and who shall be concerned in any such riotous and dangerous Practices, to the End that they may be brought to Justice. And We do hereby promise and declare, that if any Person or Persons concerned in the said Outrages and Felonies, or in any of the said Crimes, shall, on or before the Third Day of June next, discover and apprehend, or cause to be discovered and apprehended, any other Person or Persons, who hath or have or shall have been concerned in any of the said Offences that have or shall have been committed on or since the said Tenth Day of March last, so as he or they may be convicted thereof, such Discoverer or Discoverers thereof shall have Our most Gracious Pardon. And, as a further Encouragement to such as shall discover the said Offenders, We do hereby further promise and declare, that any Person or Persons who shall, on or before the said Third Day of June next, discover and apprehend, or cause to be discovered and apprehended, any of the Persons concerned in the said Outrages and Riots, so as he or they may be convicted thereof, shall have and receive, for every Person so discovered, apprehended, and convicted, the Sum of FIFTY POUNDS, which said Sum the Commissioners of Our Treasury, or Our Lord High Treasurer for the Time being, are hereby directed and required to pay accordingly, without any further Warrant in that Behalf. And we do hereby further strictly charge and command all Our Justices of the Peace, Sheriffs, Under Sheriffs, and other Officers aforesaid, that they do use the most effectual Means for suppressing all Riots and Tumults, and to that End effectually put in Execution an Act of Parliament made in the First Year of the Reign of Our late Royal Ancestor, King George the First, of glorious Memory, intituled, "An Act for preventing Tumults and riotous Assemblies, and for the more speedy and effectual punishing the Rioters," and all other Laws and Statutes made against Riots, Routs, and unlawful Assemblies; and that the said Justices, and all other Our Civil Officers to whom it doth appertain, do give the necessary Directions that sufficient Watch and Ward be duly kept at such Times and Places as they shall judge necessary for the preventing and suppressing the like Disorders. And We do hereby further strictly charge and command all Our Officers, Civil and Military, and all other Our loving Subjects, that they be aiding and assisting in the Execution of Our Commands herein, and in the apprehending and taking the said Persons, and all other Persons who have offended, or shall hereafter offend in Manner aforesaid. And We do hereby charge and command, that the said Offenders be prosecuted with the utmost Severity and Rigour of the Law, We being resolved to suppress such Riots and Tumults by a most strict and exemplary Punishment of all such Offenders. And We do hereby command the respective Justices of the Peace, and other Magistrates aforesaid, that they do from Time to Time transmit an exact Account of what they shall do pursuant to this our Royal Proclamation, to one of Our principal Secretaries of State.

Given at Our Court at Windsor, the Third Day of April, in the Twenty-ninth Year of Our Reign. 1789.

GOD save the KING.

By the King, "A Proclamation" ([London,] 1789)
20 × 14 in. (51 × 36 cm)

Luddite behavior began relatively early in the Industrial Revolution. This proclamation, issued in 1789, offered a bounty of 50 pounds sterling for the apprehension of coal miners who destroyed steam engines and other mine-company property.

CHAPTER 1

Origins of the Industrial Revolution

The Industrial Revolution began in Britain at the beginning of the eighteenth century in the mining and iron industries with Thomas Newcomen's invention in 1712 of the "atmospheric" or steam engine for pumping water out of mines. Newcomen engines were "house-built," using the masonry structure in which they were constructed as an integral part of the engine. Newcomen engines made previously inoperable mines operable, greatly stimulating the production of coal, though the working conditions of coal miners in England remained among the most difficult, unhealthy, and dangerous of all occupations for men and boys throughout the Industrial Revolution.

Almost at the same time as Newcomen, in 1709, British ironmaster and foundryman Abraham Darby developed a method of producing pig iron in a blast furnace fueled by coke—made by heating coal in the absence of air—rather than charcoal. Invention of the coke-fired blast furnace was a significant advance in the production of iron as a raw material for the Industrial Revolution, especially since coal was plentiful in England, while the firewood needed to make charcoal was becoming scarce. Increased coal mining enabled by the Newcomen engines and improved iron production enabled by coke-fired blast furnaces fueled the Industrial Revolution—first in Britain, and later in France, Germany, and the United States.

Machines in the Industrial Revolution were typically built out of iron until the production of steel became economical via the Bessemer process in the

🔖 Jean-Théophile Désaguliers *Cours de physique expérimentale*, traduit de l'Anglois par le R. P. Pezenas, vol. 2, plate 37 (Paris, 1751)

The brick structure, or "engine house," shown in this image was an integral part of Newcomen engines. This engine would most likely have been 20 or 30 feet (6 × 9 m) high.

🔖 🔖 From R. Ackermann's *Repository of Arts* (London, 1799) Hand-colored mezzotint, 11¼ × 8⅔ in. (28.5 × 22 cm)

This copy was hand-colored on both sides so that it resembles a transparency.

mid-nineteenth century; they were powered by water, by wind, by horse, or by steam. Some early printing machines were powered by men turning large cranks. The textile factories developed by Richard Arkwright, who invented the spinning frame for manufacturing thread or yarn in the period 1769–1775, were initially powered by water, and many cotton mills based on Arkwright's factory model remained driven by waterpower well into the nineteenth century. The use of hydropower was not new: water-powered mills had been used at least since the third century BCE (as recorded by Philo of Byzantium), and the technology remained in use with variations through the Middle Ages up to the eighteenth century. What was new about Arkwright's application was his weaving machinery, along with the scale of his operations. Arkwright's water-frame spinning machine, initially

P. J. de Loutherbourg
"Iron Works, Colebrook Dale"
From *The Romantic and Picturesque Scenery of England and Wales* (1805)
Hand-colored aquatint reproduction, 9 × 12¾ in. (23 × 32.5 cm)

Blast furnaces and iron foundries were among the least glamorous or least romantic of all places in the Industrial Revolution, but that did not deter artist P. J. de Loutherbourg from including this reproduction of his romanticized painting in his book.

John Raphael Smith, after the painting by Joseph Wright of Derby
Sir Richard Arkwright (London, 1801)
Mezzotint, 25½ × 18 in. (65 × 46 cm)

On the table, to Arkwright's left, is a model of his water-frame spinning machine.

intended to be powered by a horse, was one of the first developments of mass production, which eventually caused the disruptive economic and social changes characteristic of the Industrial Revolution.[12]

Prior to Arkwright, the first major stage in the mechanization of the British textile industry had occurred in the sixteenth century, when in 1589 William Lee of Calverton (near Nottingham) invented the frame knitting machine for the production of stockings. The stocking frame, as improved by one of Lee's assistants, John Ashton, was responsible for the growth of the cotton-stocking industry that developed in Nottingham in the seventeenth century. An economic problem with the invention was that individual stocking knitters typically could not afford to buy the machines; they were instead forced to rent them from wealthy machine owners, creating an environment for conflict between masters and workers that resulted in early examples of Luddite-style machine-breaking early in the eighteenth century. The machine-breaking behavior typically associated with Luddism in the early nineteenth century was a revival of this earlier behavior, as documented by William Felkin in his *A History of Machine-Wrought Hosiery and Lace Manufactures*.[13] Felkin pointed out that the war of 1803 resulted in an increase of taxation, and this, combined with bad harvests, caused working people in England to suffer, especially in the Midlands district, resulting in a revival of Luddism—a practice that had begun in England around 1710 when unemployed journeyman stocking makers broke about 100 stocking frames and physically beat their masters and apprentices. The system of apprenticeship in force at the time had resulted in a surplus of unemployed workers who wandered about the country and disrupted various manufacturing trades. At the time, the masters acceded to the demands of the workers; however, rather than reform the apprenticeship system, in 1727 a committee of the House of Commons passed an act punishing with *death* those who destroyed the machinery used in making cloth or hosiery of woolen materials. Whether it was from the death penalty or just the growth of the textile industry, which caused more employment, Luddism practically ceased for the next forty years.

Large-scale mechanization of the textile industry toward the end of the eighteenth century originated with the near-simultaneous inventions of Arkwright and James Hargreaves. About 1764, Hargreaves invented the spinning jenny, which allowed a single operator to spin multiple threads at one time, greatly increasing the production of thread or yarn for weaving. Rather than patenting the spinning jenny, Hargreaves kept it secret for some time, but he sold some of his machines to workers in the growing textile industry in Blackburn, England. The use of spinning jennies drove down the price of yarn, angering the large hand-spinning community in Blackburn, which had become a center for yarn production. Eventually, spinners broke into Hargreaves's house and smashed his machines, forcing Hargreaves to flee to Nottingham in 1768. In 1770, Hargreaves received a patent for the spinning jenny;[14] however, because he had sold some of his machines before the patent was issued, the patent was invalidated.

In 1771, Arkwright installed his water-powered spinning frames in his cotton mill at Cromford, Der-

"The Spinning Jenny"
From Edward Baines, *History of the Cotton Manufacture in Great Britain* (London, 1835)

Invented in 1764, the spinning jenny was a key invention in the Industrial Revolution that allowed one person to spin multiple threads simultaneously.

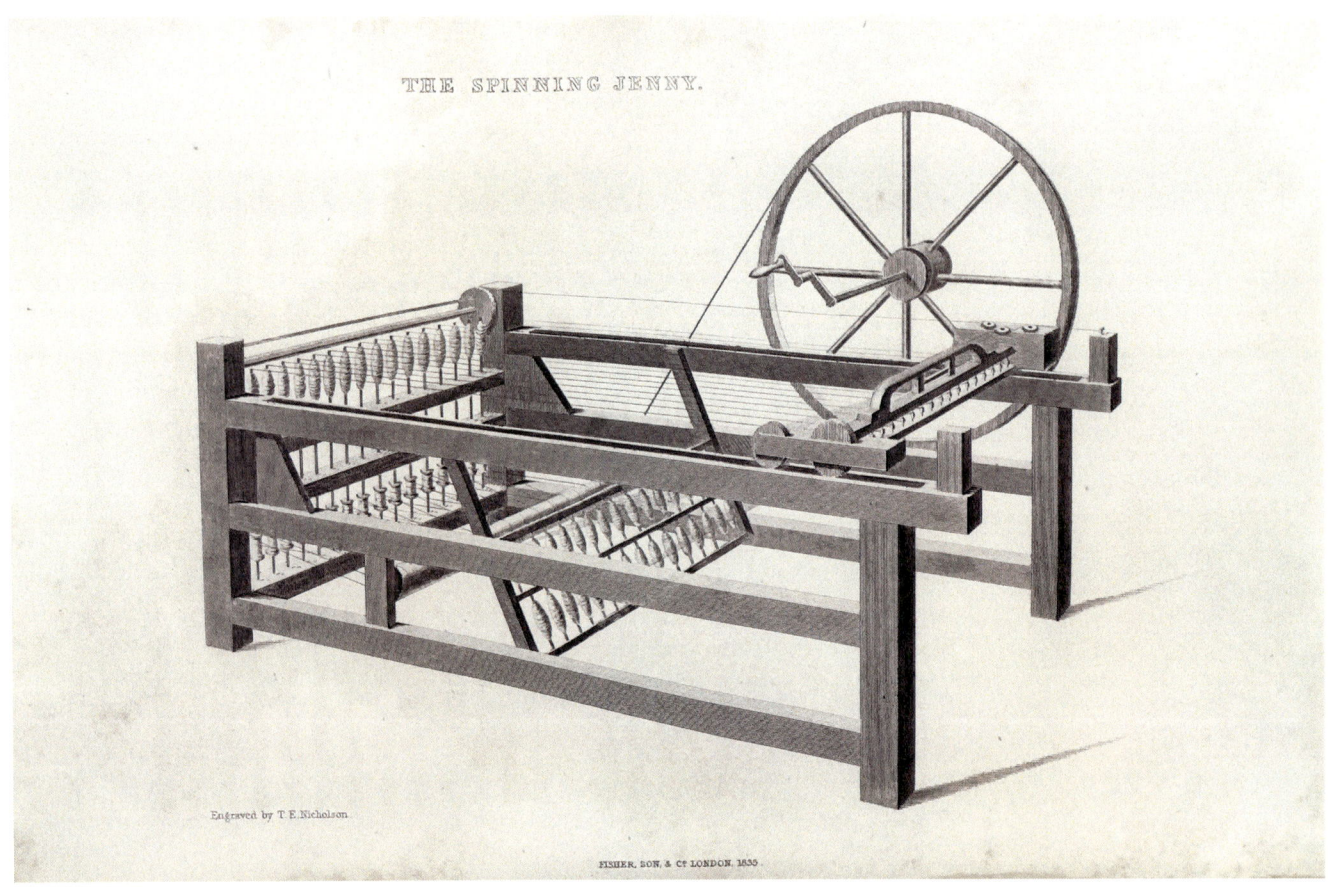

byshire, on the River Derwent. This mill was one of the first factories built specifically to house machinery rather than to create a workplace for manual labor. The factory that Arkwright established and systematized became one of the first places where the workday was set by the clock instead of the daylight hours, and where people were employed rather than contracted to do piecework. In its final form, Arkwright's factory, equipped with both his spinning frames and his patented rotary carding machine, was the first manufacturing facility designed to follow a series of operations in a continuous process from raw material to finished product.

Arkwright's extraordinary series of inventions, and their commercial development, necessarily resulted in competition from other textile manufacturers. In 1781, Arkwright successfully sued nine firms for patent infringement, and in 1785, he won a further case validating his 1775 patent.[15] Arkwright's patents were so fundamental to the industry that, by this time, there were about 30,000 infringements, and out of necessity, several manufacturers grouped together and obtained a writ of *scire facias* for a new trial to repeal the 1775 patent. The trial of the case known as *Rex vs. Arkwright*,[16] which occurred in June 1785, hinged on the necessity to overturn Arkwright's claim to be the inventor of each separate machine. To give evidence, a procession of millwrights, frame makers, mechanics, smiths, and manufacturers, some who had been employed by Arkwright, came forward; they included the widow and son of James Hargreaves, inventor of the spinning jenny. But the central witnesses were John Kay, a watchmaker who had worked closely with Arkwright and had made models for him, and Thomas Highs, who claimed to be the real creator of the roller drawing process. The trial judgment went against Arkwright, and his patent was invalidated; the cotton industry expanded rapidly in England, and Lancashire eventually superseded Arkwright's native Derbyshire as the center of cotton weaving.

25th Congress, 3rd Session, House of Representatives, Treasury Dept., Doc. no. 21
Steam-Engines (Washington, D.C., 1838)

Embellished with pen-and-ink drawings by an unknown artist on both covers for presentation to a friend. The ribbon held in the beak of the bird flying above the steamboat carries the names of Robert Fulton, inventor of the first steamboat, and of the entrepreneur Robert Livingston, Fulton's financial backer.

The first steam engines to provide smooth, uninterrupted power were the reciprocal rotative engines developed from 1769 to 1784 by engineers James Watt and Matthew Boulton. Watt and Boulton engines could be of equal size, larger, or somewhat smaller than Newcomen engines, but all early rotative Watt and Boulton steam engines were too large to drive small factories or smaller machines.

By 1800, Britain led the world in steam technology with around 2,000 steam engines. France had few, if any,[17] and the United States and Germany were in similar circumstances. The first American steam engine builder, Oliver Evans of Philadelphia, started building his first stationary steam engine around 1800. This was about ninety years after Newcomen invented the atmospheric steam engine in England and about twenty-four years after James Watt introduced his fundamental improvements to the steam engine in 1776. By 1801 or 1803, Evans had a working high-pressure steam engine in Philadelphia. In 1805, he published the obscurely titled first American book on steam technology, *The Abortion of the Young Steam Engineer's Guide*; this had a stimulating effect on the nascent American steam-engine industry, which grew rapidly. In 1838, the U.S. government published its first survey of the use of steam engines in America.[18] That 472-page report identified 3,010 steam engines in the United States, of which 2,653 were confirmed and the remainder estimated. The report stated that 800 steam engines were on steamboats and 350 in railroad locomotives; the remainder were in manufacturing. Most of the steam engines employed in printing were powering presses that printed textiles, rather than books or newspapers. According to the report, in Cambridge, Massachusetts, there was one steam engine powering a printing press; in Philadelphia, there were six steam engines (typically of 4 horsepower) powering printing operations, including the *Philadelphia Public Ledger* newspaper. Other engines powered a few papermaking plants. The report, which was undoubtedly not a complete survey, did not identify any steam engines powering printing presses in New York City. Overall, the report was concerned with injuries resulting from explosions of steam engines, focusing primarily on steamboats and railroads rather than industries like printing or papermaking.

In 1800, most steam engines in Britain were used

Most eighteenth-century textile mills were water-powered, but steam engines were used throughout the eighteenth century in mining and to power some large factories. Steam engines of the Newcomen type, which were typically about 20 feet (6 meters) tall and could weigh twenty tons, continued to operate for seventy-five years in the United Kingdom and on the Continent of Europe. However, those engines, designed for pumping, could not generate the smooth, uninterrupted power needed for factory production.

in large-scale industries such as mines, cotton mills, and larger-scale manufacturing. Almost 500 of these engines were built by Watt and Boulton, who built both pumping engines for mines and reciprocal rotative engines for other industries. Even factories built next to rivers, which used traditional waterwheels for power, started to use the Watt engine as a backup when local water levels were too low for efficient production. In the United States, Robert Fulton built the first proper steamboat, the *North River Steamboat* (later known as the *Clermont*) in 1807 after obtaining a Boulton and Watt rotative steam engine from England. By 1812, the first steamboat in England, Henry Bell's PS *Comet*, was providing the first commercial steamboat passenger service in England, operating along the River Clyde in Scotland.

As the Industrial Revolution advanced, in 1786 English inventor Edmund Cartwright patented the first power loom. His first of several patents for this invention was specification no. 1565 for what he called a Newly Invented Weaving Machine. Cartwright's first attempts, and his later attempts too, to improve the power loom, were problematic, but the technical obstacles he encountered were gradually solved by other inventors, and by the end of the eighteenth century, there were 2,400 wooden power looms in England, mostly powered by water. As the nineteenth century advanced, steam power began replacing hydropower in England's cotton mills; this enabled the construction of mills away from rivers, in more populous districts where more workers were available and close to supplies of coal to fuel steam engines. Steam power also increased the speed of the power looms and allowed them to be built larger, appropriate for operation by older children or adults, rather than young children.

Steam engines gradually made their way into other industries. Albion Mills in Southwark, London, engineered and built by Matthew Boulton, was the first steam-powered commercial flour mill in the United Kingdom. Completed in 1786, it became a symbol of the changes brought by the Industrial Revolution. The mill was powered by two huge, double-acting, rotative steam engines by Watt and Boulton that produced 50 horsepower and drove 20 pairs of millstones, each grinding nine bushels of wheat per hour and produc-

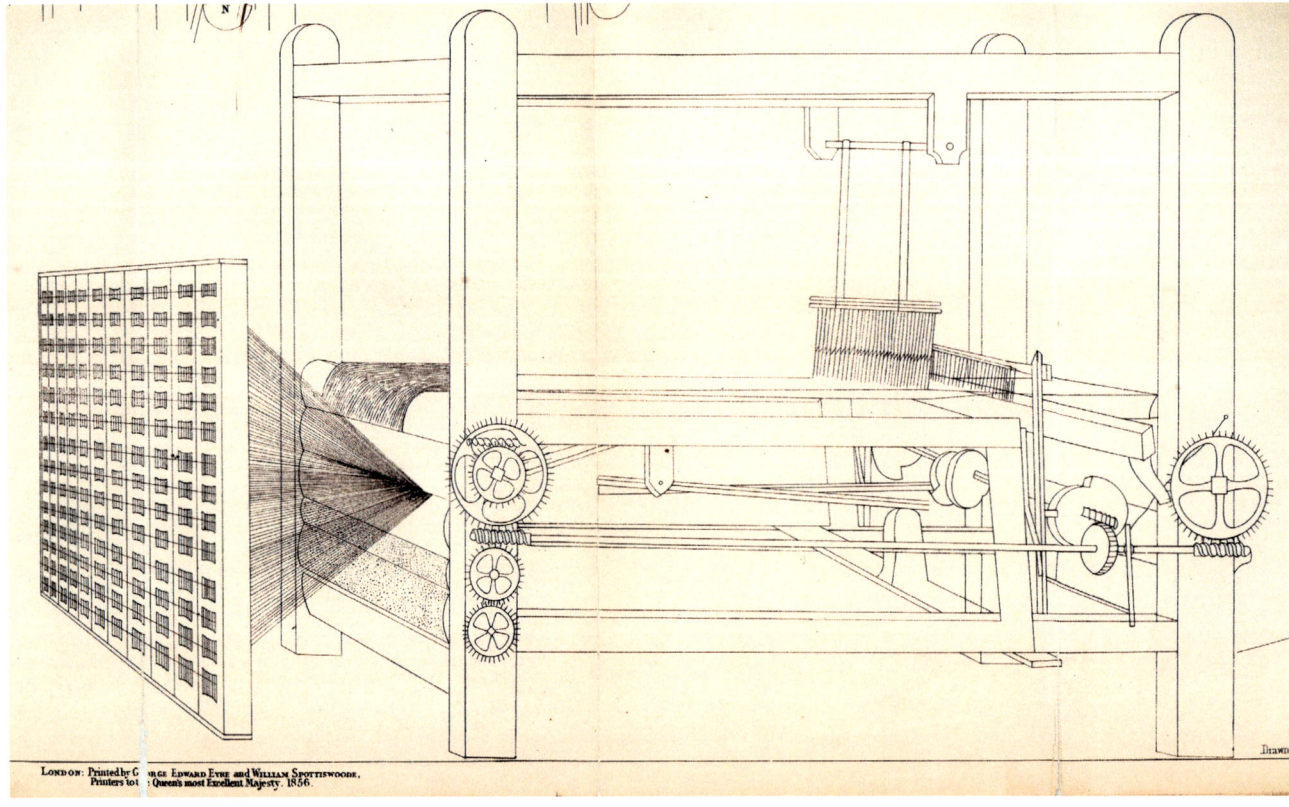

🕯 Drawn by Augustus Charles Pugin,
with figures by Thomas Rowlandson
"Fire in London"
Plate no. 35 from *The Microcosm of London* (London, 1808)
Hand-colored aquatint, 9½ × 11½ in. (24 × 29 cm)

The burning of Albion Mills in 1791, a landmark event in the Industrial Revolution. *The Microcosm of London* was printed by Thomas Bensley, who, just a few years later, became the primary financier of the first steam-powered printing machine, invented by Friedrich Koenig.

🕯 Edmund Cartwright
Looms for Weaving, patent no. 1565 (1786) (London, 1856)

A drawing of the first power loom, invented by Cartwright, and patented in 1786. The patent was first printed in 1856.

ing 6,000 bushels of flour per week. The engine also provided power for fanning the wheat to free it of impurities, for sifting and packaging the flour, and lowering the packaged product into barges on the Thames. These factors made the mill more efficient than traditional flour mills in London and drove some of its competition out of business, reportedly giving the mill a virtual monopoly on the London flour market. This had the effect of raising the cost of bread.

On 2 March 1791, fire destroyed Albion Mills. Damage was estimated at £150,000, and over 500 people working there became unemployed. Arson was suspected. London's independent millers celebrated with placards reading "Success to the mills of Albion but no Albion Mills." Opponents referred to the fac-

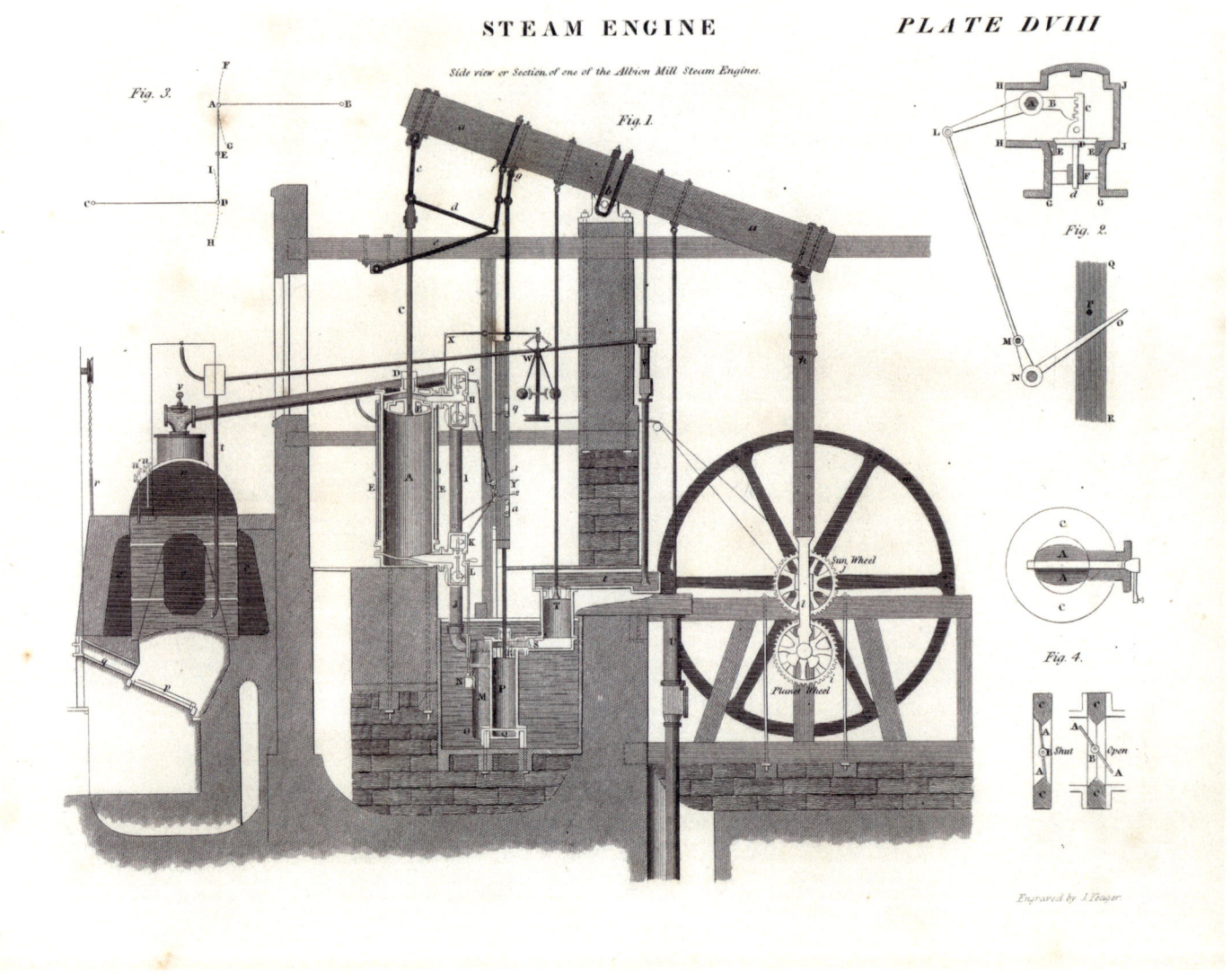

tory as satanic; they accused its owners of adulterating flour and using cheap imports at the expense of British producers. Albion Mills was a short distance from the home of the poet and artist William Blake, and it is believed that the shell of the Albion Mills building was the inspiration for the reference to "dark satanic mills" in his poem "And did those feet in ancient time," also known as "Jerusalem":

> *And did those feet in ancient time,*
> *Walk upon Englands mountains green:*
> *And was the holy Lamb of God*
> *On Englands pleasant pastures seen!*

> *And did the Countenance Divine*
> *Shine forth upon our clouded hills?*
> *And was Jerusalem builded here,*
> *Among these dark Satanic Mills?*

> *Bring me my Bow of burning gold:*
> *Bring me my Arrows of desire:*
> *Bring me my Spear: O clouds unfold:*
> *Bring me my Chariot of fire!*

> *I will not cease from Mental Fight,*
> *Nor shall my Sword sleep in my hand:*
> *Till we have built Jerusalem,*
> *In Englands green & pleasant Land.*

Incremental Advances in Printing Toward the End of the Eighteenth Century

Toward the end of the eighteenth century, independent of the advances of mechanization of larger industries, some incremental improvements began to be made to the traditional wooden handpress invented by Gutenberg three centuries earlier. In 1772, Wilhelm Haas, a typefounder in Basel, Switzerland, built a new type of printing press for printing books in which all parts of the press subject to stress during the printing process—including the bed and the platen—were made of iron. This significantly improved the power and efficiency of the press. To publicize these advances to printing, in 1790 Haas's son, Wilhelm Haas (the younger), issued a pamphlet in Basel titled *Beschreibung einer neuen Buchdruckerpresse 1772* (Description of a new printing press 1772) describing the new press invented by his father.[19] According to James Moran,

Despite the fact that the staple was made of iron, it tended to break, and a model with a strengthened frame was constructed in 1784. This fracturing of the frame, which later was to concern the inventor of the first all-metal press, Lord Stanhope, indicated that the simple replacement of wood by iron in itself was not sufficient to counterbalance the stress provided by more powerful printing mechanisms. Attention had to be given to the precise design of the iron frame as well.[20]

Haas's innovations were not immediately applied by other press builders, who continued to refine wooden handpresses. In 1783, Étienne-Alexandre-Jacques Anisson, known as Anisson-Dupéron, direc-

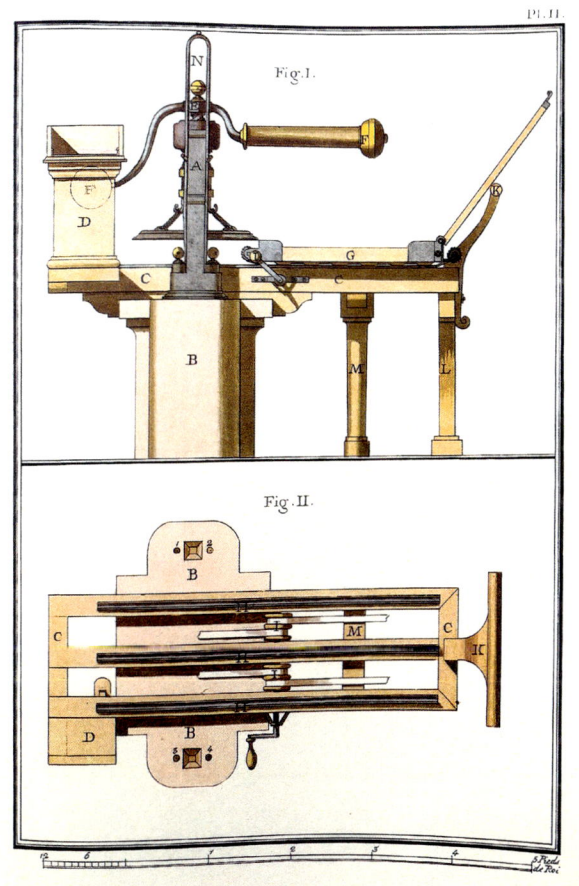

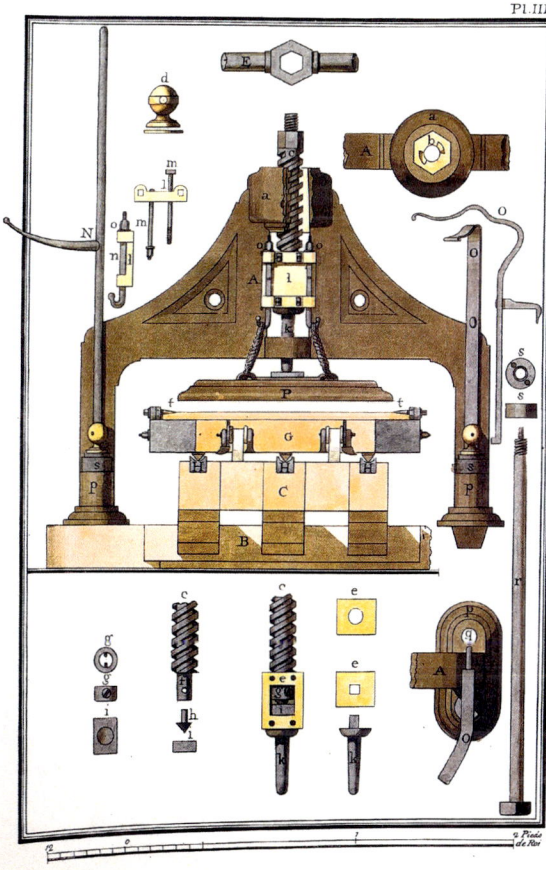

🐦 Wilhelm Haas, the younger
Beschreibung einer neuen Buchdruckerpresse 1772 (Basel, 1790)

A drawing of Wilhelm Haas's improved handpress in which the parts subject to stress during the printing process were made of iron, but the remainder was made of wood. From the booklet describing the press published by Wilhelm Haas, the younger.

🐦 🐦 Schematic engraving from an unidentified encyclopedia

One of the two double-acting rotative steam engines built by Watt and Boulton for Albion Mills. The engines produced 50 horsepower and drove 20 pairs of millstones, each grinding 9 bushels of wheat per hour.

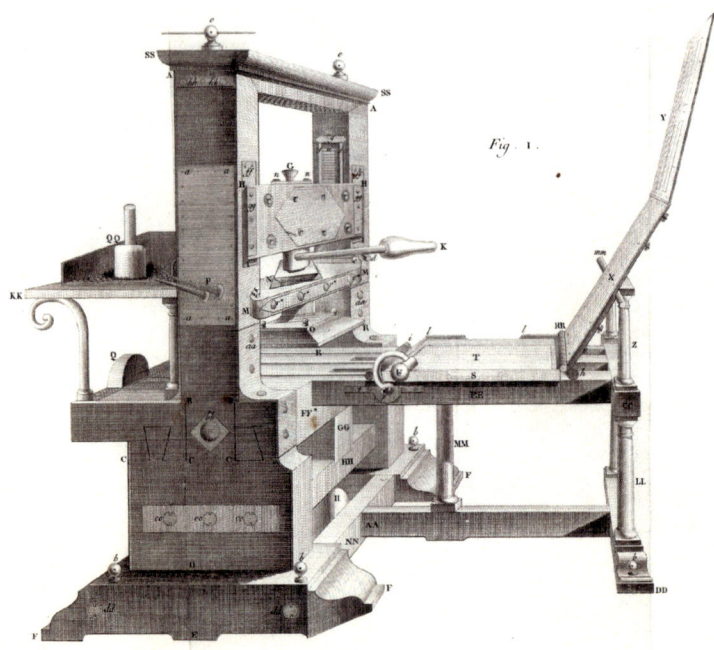

Étienne-Alexandrte-Jacques Anisson (known as Anisson-Dupéron)
"Premier mémoire sur l'impression en lettres, suivi de la description d'une nouvelle presse"
From *Mémoires de mathématiques et de physique des Scavants Etrangers*, tome X (1785)

Improvements to the screw mechanism in Anisson-Dupéron's mostly wooden handpress, shown in the engraving, allowed the press to be operated with a single pull.

Engraving by T. Watson, after a painting by Antoine Prud'homme (London, 1775)
Charles Stanhope, Viscount Mahon (Third Earl Stanhope)
Mezzotint, 15 × 11½ in. (38 × 29 cm)

Lord Stanhope invented the first entirely iron handpress, called the Stanhope Press, as well as an improved method of stereotyping.

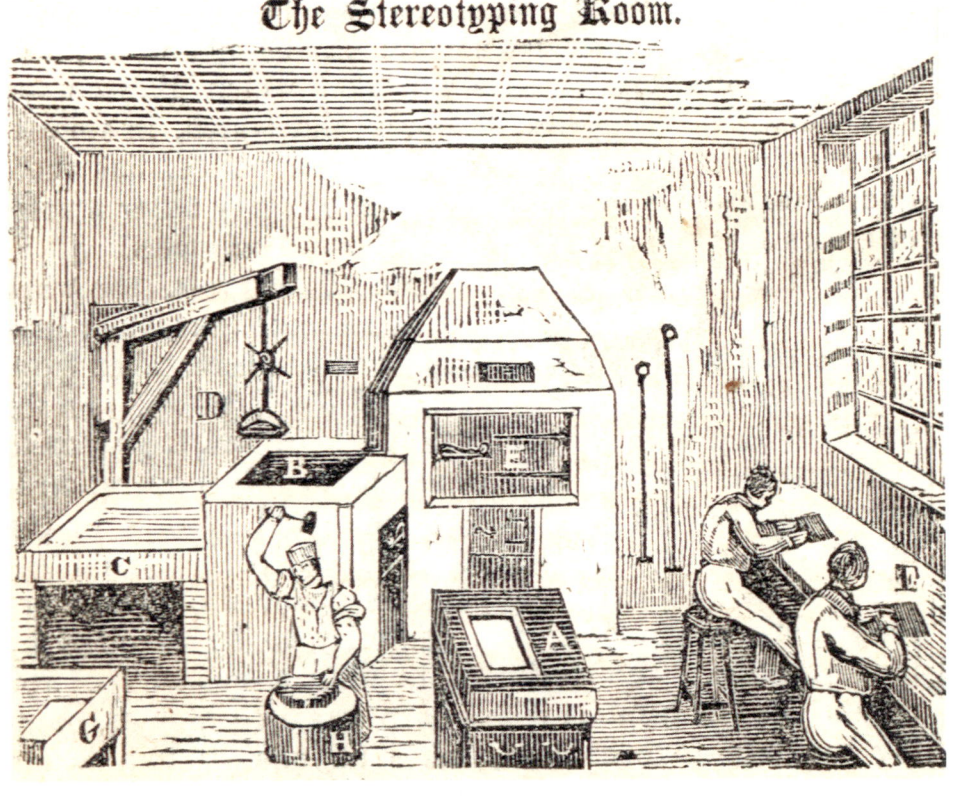

"The Stereotyping Room"
Parley's Visit to the Printing Office (London, 1841)

The caption to the small woodcut reads:

A. Table with stone top, on which the form is placed to be moulded.

B. The melting pot, made of passive cast-iron.

C. A stone trough for cooling the casting pots.

D. A crane with casting-pot attached, for immersing the mould in the boiling metal, and carrying it to the cooling trough.

E. The oven, fitted with shelves, for baking the gypsum moulds.

F. A bench at which a picker and an engraver are at work.

G. A trough filled with water for "white-washing" the forms previous to moulding; that is, filling up such interstices as would otherwise cause the plate to be cast with holes in various parts.

H. A workman separating the plaster mould from the plate, after being taken from the casting-pot.

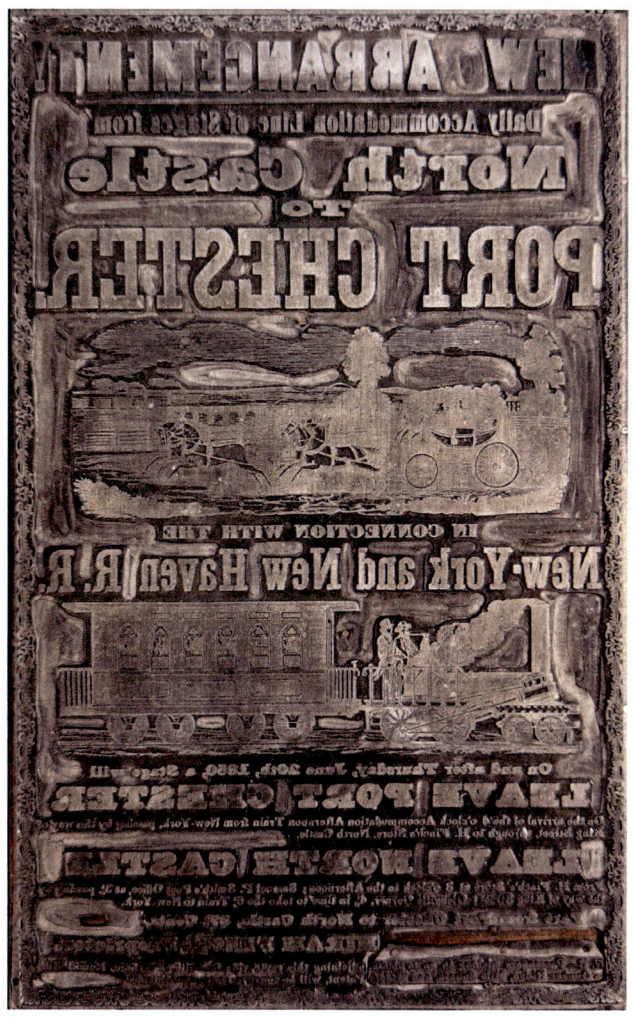 

"New Arrangement. Daily Accommodation Line of Stages from North Castle to Port Chester in Connection with the New York and New Haven R.R." (c. 1850)
Stereotype plate, 10¼ × 6¼ in. (26 × 16 cm)

A stereotype plate with the type set backward for printing (left), together with the stereotype plate reversed for legibility (right).

tor of the Imprimerie Royale in Paris, improved the strength of the traditional handpress but, more significantly, made refinements to its screw mechanism, allowing the press to be operated with a single pull. He described and illustrated his new press in fine detail in a journal article published in 1785.[21]

Anisson-Dupéron had his press installed at the Imprimerie Royale. During the French Revolution, he was initially appointed as head of the same operation, renamed the Imprimerie Nationale; however, he lost favor and was guillotined on 25 April 1794. The Didot family of printers, typographers, and papermakers later claimed that they had invented the single-pull screw mechanism prior to Anisson.

In 1784, Andrew Foulis, printer to the University of Glasgow, and Alexander Tilloch, a printer in Glasgow, were awarded British patent no. 1431 for an improved method of stereotyping.[22] Tilloch claimed to have invented their process in 1781 without knowledge of his fellow Scotsman William Ged's prior work. In their brief patent specification, Foulis and Tilloch provided a description of their process, stating that their

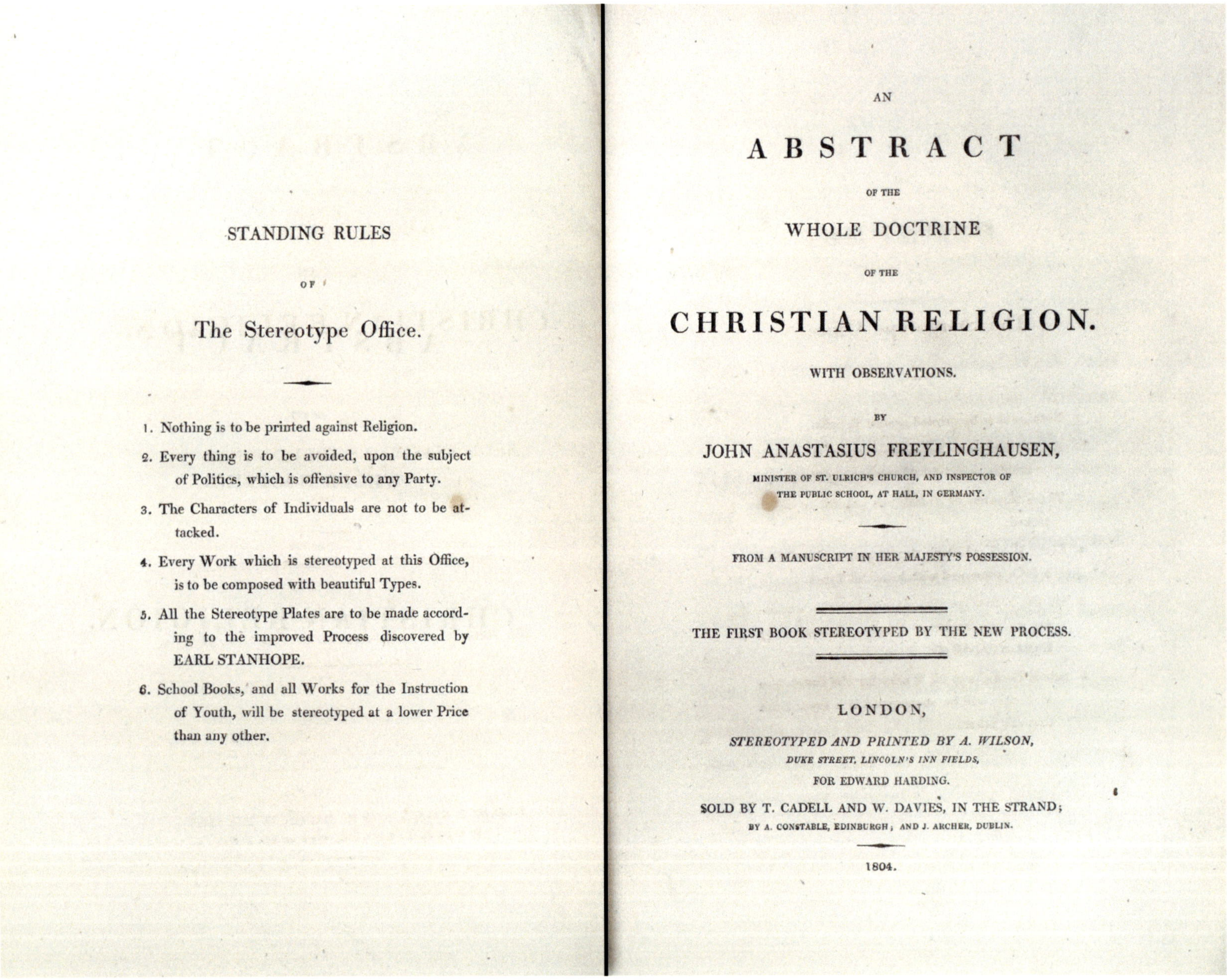

CHAPTER 1

STANDING RULES

OF

The Stereotype Office.

1. Nothing is to be printed against Religion.
2. Every thing is to be avoided, upon the subject of Politics, which is offensive to any Party.
3. The Characters of Individuals are not to be attacked.
4. Every Work which is stereotyped at this Office, is to be composed with beautiful Types.
5. All the Stereotype Plates are to be made according to the improved Process discovered by EARL STANHOPE.
6. School Books, and all Works for the Instruction of Youth, will be stereotyped at a lower Price than any other.

AN

ABSTRACT

OF THE

WHOLE DOCTRINE

OF THE

CHRISTIAN RELIGION.

WITH OBSERVATIONS.

BY

JOHN ANASTASIUS FREYLINGHAUSEN,

MINISTER OF ST. ULRICH'S CHURCH, AND INSPECTOR OF
THE PUBLIC SCHOOL, AT HALL, IN GERMANY.

FROM A MANUSCRIPT IN HER MAJESTY'S POSSESSION.

THE FIRST BOOK STEREOTYPED BY THE NEW PROCESS.

LONDON,
STEREOTYPED AND PRINTED BY A. WILSON,
DUKE STREET, LINCOLN'S INN FIELDS,
FOR EDWARD HARDING.
SOLD BY T. CADELL AND W. DAVIES, IN THE STRAND;
BY A. CONSTABLE, EDINBURGH; AND J. ARCHER, DUBLIN.

1804.

method of making plates for the purpose of printing by or with such plates, instead of the moveable types commonly used, which is performed by making a plate or plates for their page or pages of any book or other publication, and in printing off such book or other publication at the press; the plates of the pages to be arranged in their proper order, and the number of copies wanted thrown off, instead of throwing the impressions wanted from moveable types locked together in the common method; and such plates are made either by forming moulds or matrices for the page or pages of the books or other publications to be printed by or with plates, and filling such moulds or matrices with metal or with clay, or with a mixture of clay and earth, or by stamping or striking with these moulds or matrices the metal, clay, earth or mixture of clay and earth.

In spite of the apparent labor-saving advantages of stereotyping, which eliminated the need to reset type when books were reprinted, the Foulis and Tilloch

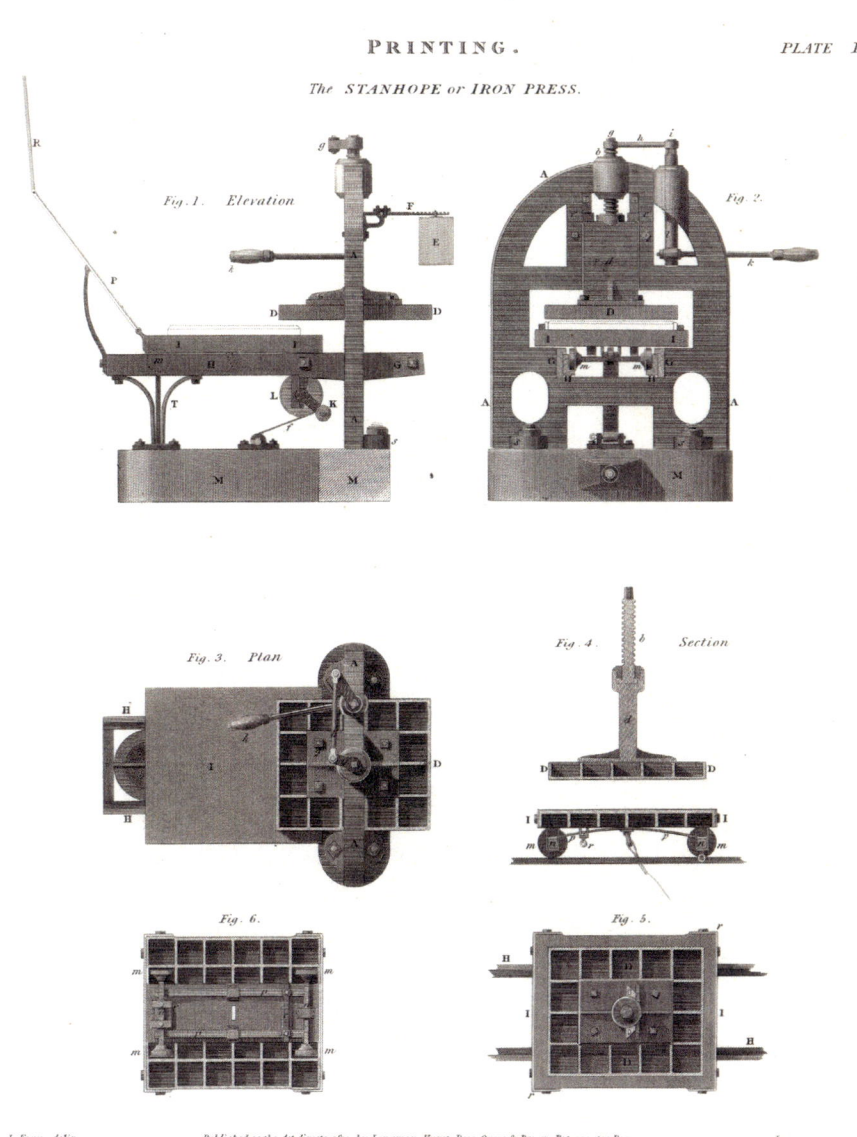

The Stanhope Press, or Iron Press
From an unidentified encyclopedia (c. 1820)

The Stanhope Press was the first entirely iron handpress.

John Anastasius Freylinghausen
An Abstract of the Whole Doctrine of the Christian Religion (London, 1804)

As the title page suggests, this was the first book printed from stereotype plates made by Earl Stanhope's improved process. Standing Rules for the printer, Andrew Wilson's, Stereotype Office were printed facing the title page.

process seems to have met with resistance from printers until Charles Stanhope, third Earl Stanhope, purchased the Foulis and Tilloch stereotype patent and perfected the method. A prolific inventor, Stanhope was idealistically motivated to improve the printing process. Having improved the stereotype printing process, which he did not patent, in 1803 Stanhope sponsored London printer Andrew Wilson to promote the new method through the production of numerous new stereotyped editions. In 1804, Wilson printed John Anastasius Freylinghausen's *An Abstract of the Whole Doctrine of the Christian Religion. With Observations. From a Manuscript in Her Majesty's Possession: The First Book Stereotyped by the New Process.**

---

*Lowndes, Bibliographer's Manual of English Literature II (1865), p. 841, states that "the translator was Queen Charlotte, consort of George III, and the editor of the volume was Beilby Porteus, Bishop of London. The Rev. Philip Bliss states that he had seen the original MS. in Her Majesty's handwriting."

CHAPTER 1

Drawn by George Walker, engraved by R. & D. Havell
"The Collier"
Hand-colored aquatint, 11¾ × 8 in. (30 × 20 cm)
Plate 3 in *The Costume of Yorkshire* (London, 1814)

Behind the coal miner we see the first published illustration of a steam-powered railroad hauling coal. Far behind that, to the left of the tall chimney, a Newcomen engine is shown pumping water out of the coal mine.

Stanhope's stereotype process was intended to reduce printing costs, since type did not have to be reset when books were reprinted; however, it was opposed by some printers on the grounds that the quality of printing from the stereotype plates remained inferior to printing from conventionally set type. Wilson's edition of Freylinghausen's *Abstract* was intended to counter these objections, reflecting the high quality of printing that could be obtained from stereotype plates. Cambridge University Press seems to have quickly recognized the economies that could be gained from the stereotype process, as they underwrote some of the cost of development of the process and employed Wilson to print stereotyped Bibles and prayer books in 1805.[23] Oxford University Press quickly followed suit in 1806.[24]

Lord Stanhope's second major improvement to the printing process was the invention of the first completely iron handpress, which became known as the Stanhope Press. The greatly increased rigidity resulting from the iron, rather than wood construction, and Stanhope's innovative combination of levers turning the screw caused the platen to descend with decreasing rapidity and consequently with increasing force, until it reached the type, when a much-increased pressure was obtained. These features significantly improved the efficiency of the handpress; however, output increased only modestly, from an average of 200 sheets per hour on a wooden handpress to around 250 sheets per hour with two men working the Stanhope Press. An additional advantage provided by the press was that the platen was made the full size of the bed, enabling impression to be done in one pull compared with two pulls on traditional presses.

Just as Stanhope did not patent his stereotype process, so too he did not patent his iron printing press, so the precise year of its origin is unknown; the earliest surviving example is dated 1804. Early models had straight side frames, which were prone to breaking due to the immense pressure that could be exerted. These castings were changed in about 1806 to a heavier "rounded" style, and in this form, the press continued to be manufactured into the mid-nineteenth century. It remained in use to a limited to extent into the late nineteenth century, though by around 1880, it had been superseded by the Albion and Columbian iron handpresses.[25]

This was the "state of the art" in printing and book production during the first decade of the nineteenth century. In England, where larger industries, such as textiles and agriculture, were mechanizing, book production remained hardly touched by the Industrial Revolution. Lord Stanhope had improved the stereotype process, and his new, improved iron handpress was available. However, printing output on the completely iron Stanhope Press remained essentially the same as Moxon had stated in 1683: 200 to 250 sheets per hour. The quality of the printing was, as it had always been, highly variable, from mediocre to superb, depending upon the equipment and skills of the printers involved. Handmade paper of the full range of quality was available, as were reproductions of images by woodcuts or copperplate engraving, mezzotint or aquatint. Bookbinders, working by hand, could produce bindings of whatever quality was desired.

During the second decade of the nineteenth century, the prominent London printer Thomas Bensley played a dual role as a quality handpress printer and as a sponsor of the new technology that led to the second printing revolution. In 1814, Bensley issued *The Costume of Yorkshire*, a superbly printed book in deluxe, large-quarto format on Whatman art paper with forty beautifully hand-colored aquatint engravings. This book, which included numerous illustrations of members of the working class, captioned in English and French, was obviously intended for a wealthy clientele, who might not be accustomed to visiting workmen and women holding those jobs. Among the hand-colored plates is an image of a coal miner (plate 3); behind him is a coal mine with a Newcomen pumping engine in its engine house and a steam-powered train hauling coal—the first railroad to be depicted in a book illustration. Plate 36 depicts children who work in a textile mill, their faces and hands gray with dirt. Ironically, the same Thomas Bensley whose handpress printed this traditionally produced deluxe volume was also the primary financier of the world's first steam-powered printing press, invented by Friedrich Koenig and built in London by Koenig and his business partner, engineer Andreas Bauer. In 1814, the same year that Bensley's staff printed *The Costume of Yorkshire* on a handpress, Koenig and Bauer's revolutionary new printing machine, financed by Bensley and partners, became operational. This double-cylinder, steam-powered machine printed the 29 November issue of *The Times* of London newspaper at the rate of 1,100 sheets per hour, more than four times the speed of a handpress. This issue was the first ever printed on the first high-speed printing machine. As John Walter II, the owner of *The Times*, stated in that issue, the 29 November issue was "the practical result of the greatest improvement connected with printing, since the discovery of the art itself." Its publication opened the door to new possibilities for printing and publishing, which will be the subject of chapter 3.

Drawn by Emile Bourdelin and engraved by H. Linton (detail). See p. 38.

CHAPTER 2

Louis-Nicolas Robert and the Development of the Papermaking Machine in the First Half of the Nineteenth Century

Papermaking was the first of the book-production technologies to be invented, and coincidentally perhaps, many centuries later, it was the first of the book-production technologies to be mechanized. Originally developed in China around the year 100 CE, papermaking technology gradually spread westward over the next thousand years; it was introduced to Europe in the twelfth century, with the first documented European paper mill established in Xàtiva, Spain, in 1150. By the early seventeenth century, papermaking had spread to many countries in Europe, and at the end of the eighteenth century, with Louis-Nicolas Robert's invention of the papermaking machine, it became the first book-production technology to be mechanized.

Robert, a trained engineer, was an inspector of personnel at Pierre-François Didot's paper mill in Corbeil-Essones, which supplied paper to the French Ministry of Finance for the printing of currency. In the mid-1790s, exasperated by "constant strife and quarrelling among the workers of the handmade papermakers' guild,"[26] Robert set out to design a machine that would replace hand labor in the industry. His first prototype, completed in 1797, was a failure, but with Didot's financial backing and encouragement, he was able to build a successful model the following year. Robert's machine had a wide, moving wire-mesh belt that received a continuous flow of pulp, shook it free of water, and deposited an unbroken sheet of wet paper onto a pair of squeeze rolls. As

Louis-Nicolas Robert
Watercolor
From Henry Morris, *Nicolas Louis Robert and His Endless Wire Papermaking Machine* (Newton, Pennsylvania, 2000)

The only portrait of Louis-Nicolas Robert, painted by his sister.

the continuous sheet of wet paper came off the machine, it was manually hung over a series of cables or bars to dry, then cut by hand into smaller sheets.

Robert's original papermaking machine, once it was improved by others, mechanized a process that for centuries had been the monopoly of skilled men and women; it greatly increased the production of paper. In a day, a skilled handmade-paper crew could produce up to five reams, or about 2,400 sheets of paper, the size of which depended upon the measurements of the hand mold they were using, while the output of Robert's machine, which produced a continuous sheet of paper that would later be cut into smaller pieces, could be measured in feet per minute.

Papermaking Prior to Robert's Invention

Prior to Robert's invention, paper was made one sheet at a time, a laborious and time-consuming process. A worker known as a vatman would dip a rectangular frame or mould with a wire-mesh bottom into a vat of pulp, or stock, made from fermented and beaten linen rags mixed with water. After shaking the frame to evenly distribute the fibers, the vatman would give the frame to a coucher, whose job was to remove the damp sheets of pulp and stack them between absorbent layers of felt. When the stack reached a certain height, it was passed to a layer, who would remove the paper sheets from the felts and restack them; this new stack was then compressed in a screw press in order to extract more water. The sheets were then hung in a drying loft to dry completely.

The only major advance in papermaking technology before Robert's machine was the water-powered Hollander beater for processing rag pulp, developed in the Netherlands around 1680. In papermaking, this eventually replaced the stamp mill, a machine using a water wheel, cams, and hammers that had been in use since the Hellenistic period (fourth century CE). The Hollander beater, which used metal blades and a chopping action to cut the raw material, could produce in one day the same amount of pulp that would take a traditional stamp mill eight days to make. However, the resulting pulp contained shorter cellulose fibers that were less easily hydrated and less fibrous than those produced by stamp mills, and the paper made from it was weaker and more prone to foxing (discoloration). Apart from the Hollander beater, papermaking remained a manual process until the end of the eighteenth century.

As with printing, the techniques of papermaking were chiefly passed down as trade secrets from master to apprentice. No comprehensive treatise on papermaking existed until 1761, when Jérôme de Lalande's *L'Art de faire le papier* was published in the fourth volume of the Académie royale des sciences' encyclopedic *Descriptions des arts et métiers*. Lalande was an astronomer, not a papermaker, but he consulted with numerous professional papermakers in different regions of France and made careful and accurate observations of the process. His work, comprising 150 pages illustrated with fourteen engraved plates, covers all aspects of the papermaking trade, including the design and construction of buildings, the design of machinery and equipment, the various manufacturing processes, and the economics of the business; he also included a glossary of terms of the trade and noted the many state regulations governing the papermaking industry in France. According to Cohen and Wakeman,[27] the fourteen plates reproduced by de Lalande "date from 1698 and were originally prepared for a text, now lost, by Gilles Filleau des Billetes completed in 1706." Lalande's reproduction of the plates sixty years later indicates that essentially nothing had changed in the papermaking process between the completion of the illustrations and Lalande's writing. Three of the plates depict women sorting rags and drying and folding paper. If these plates were engraved in 1698, they are among the earliest images of women working in any aspect of book production.

The publication of Lalande's work in a handsomely and expensively printed scientific series would suggest that it was intended not necessarily for papermakers themselves, but for students of technology or entrepreneurs who might enter the papermaking industry. Certainly, the luxurious, expensive folio format of the *Descriptions des arts et métiers*, with its large copperplate illustrations, was not conducive to hands-on practical use within a paper mill. A quarto edition of the *Arts et métiers*, published in Neuchatel in 1766, might have been more affordable and useful to the trade.

Reflecting an international demand for information on papermaking, Lalande's work was translated into German, along with the rest of the *Descriptions des arts et métiers* series, between 1762 and 1765. An Italian translation of *L'Art de faire le papier* appeared in 1762, a Spanish edition in 1778, and a Dutch edition in 1792. An anonymous English translation of roughly one-quarter of Lalande's text, credited as "from a late Treatise, in French" was serialized in London in *The Universal Magazine* issues for March, May, and June of 1762 and February and April of 1763, with reproductions of five of the plates.[28] Additional informative plates and text, especially on the design of water-driven Hollander beaters, was included in the section on "Papeterie" in Diderot and d'Alembert's *Encyclopédie des sciences*.[29]

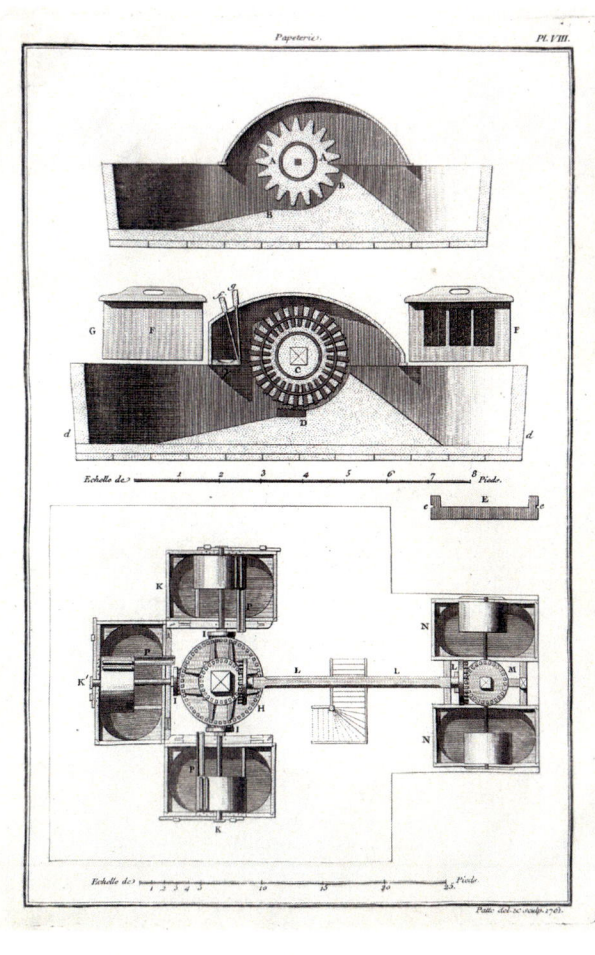

From Jérôme de Lalande's *L'Art de faire le papier* (Paris, 1761)

🐛🐛 Plate 8 illustrates the mechanism of a Hollander beater.

🐛 The upper image shows women sorting rags. This plate, designated as plate 1 in the book, is dated 1698 at the bottom of the plate. That Lalande would use a plate engraved sixty-three years before his publication confirms that little had changed in papermaking technology during that interval.

🐛🐛 Women drying and folding paper. This image is also dated 1698 at the bottom of the engraved plate.

🐛 Women folding and packing paper. This group of three images of women working in a paper mill is probably the earliest series of published images showing women working in any aspect of book production. Note that all three images are dated 1698 at the base of their plates.

CHAPTER 2

Development of Mechanized Papermaking in England

Robert obtained a French patent (*brevet*) for his papermaking machine in 1799, but his employer, Pierre-François Didot, believed that they would have better success developing and marketing the machine in England. Didot made the decision partly because of disagreements between himself, Robert, and his brother Saint-Leger Didot, and also because he recognized that France lagged far behind England in industrial development after the French Revolution. The Industrial Revolution occurred in France significantly later than in England; although dates for the beginning of the Industrial Revolution in France vary, one date used by scholars is the mid-1830s. The massive economic disruption caused by the French Revolution was one factor causing the delay; another was that France did not possess large and accessible supplies of coal and iron ore like Great Britain and Belgium. On the other hand, measures introduced by the French government after the 1789 revolution encouraged industrialization. The introduction of the Code Civil (the Napoleonic Code) occurred simultaneously with the abolition of the old guild restrictions and internal customs tariffs. The state created the Bank of France and stabilized the currency, and the state became involved in the construction of roads and canals. Ironically, this may have indirectly discouraged industrialization in France. The 1789 revolution freed farmers and peasants from debts and taxes, thereby guaranteeing them a relatively secure existence in agriculture, and preventing the creation in France of the large supply of unemployed workers that had occurred in England as a result of agricultural mechanization. In 1789, Great Britain had more than 20,000 spinning jennies

John Gamble
Patent 2487, awarded in 1801, for an "invention of Making Paper in single Sheets, without Seam or Joining, from One to Twelve Feet and upwards Wide...."

Gamble's patent was essentially an English translation of the *brevet* granted to Robert in 1799.

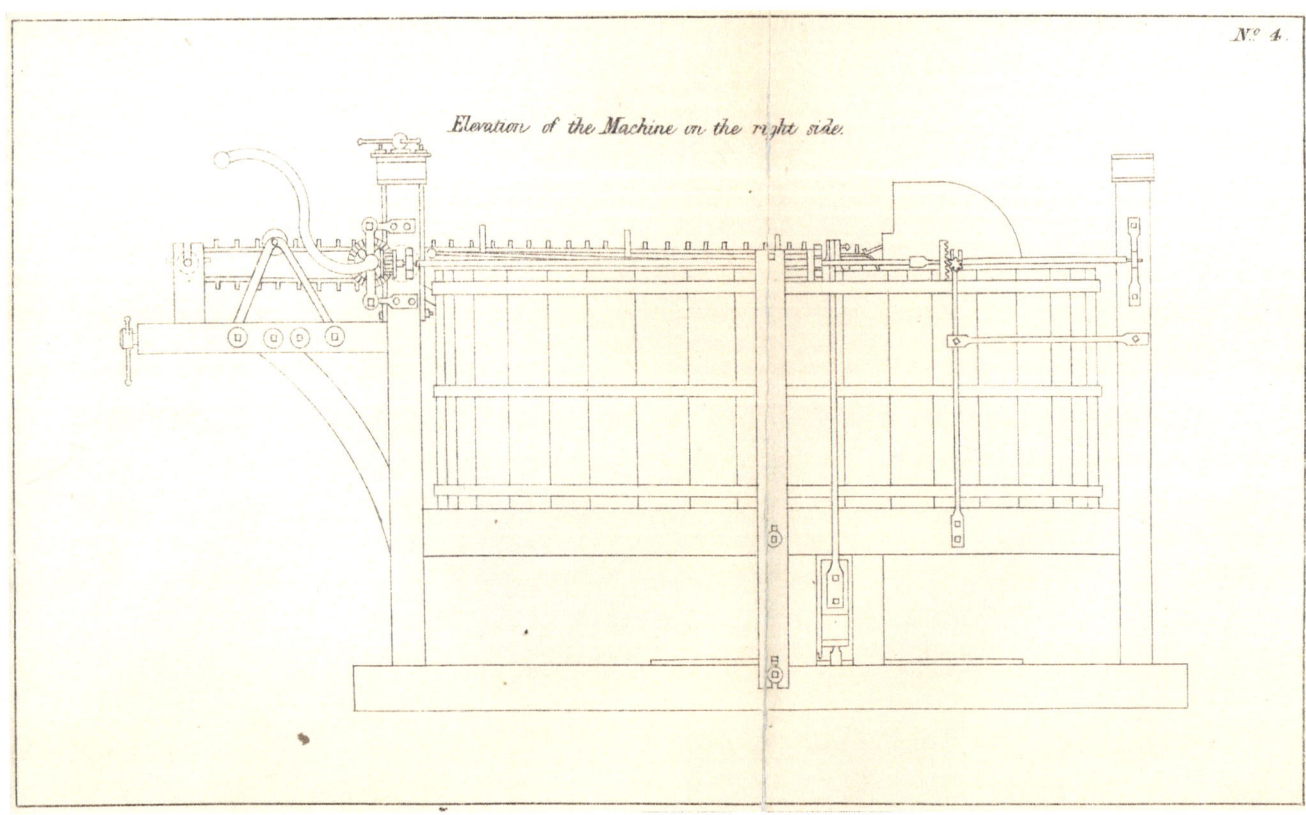

Douglas M. Parrish
Fourdrinier and the Papermaking Machine (c. 1960)
Oil on canvas, 36 × 28 in. (91 × 71 cm)
From Kimberly-Clark, *Graphic Communications Through the Ages*
Courtesy of RIT Cary Graphic Arts Collection

Notce the resemblance between the papermaking machine in the painting and that depicted in John Gamble's English patent of 1801. Even though the papermaking machine was invented by Robert, after it was developed in England by the Fourdrinier brothers, the machine became known as the Fourdrinier machine.

and 200 textile mills on the Arkwright model, while France had less than 1,000 jennies and only 8 textile mills on the Arkwright model.[30] In the paper industry, as late as 1812 only 21 percent of French papermakers employed Hollander beaters.[31] These were among the reasons why industrial processes in France, including the processes of book production, were mechanized somewhat later in France than in England.

To develop Robert's machine in England during the Napoleonic Wars, Didot enlisted the aid of his English brother-in-law, John Gamble, then employed in the office of the British Commissioner for exchanging prisoners of war in France. Gamble left Paris for London in March 1801 and, only a month later, received an English patent for the papermaking machine. Patent specification no. 2487 was granted for an "Invention of Making Paper in single Sheets, without Seam or Joining, from One to Twelve Feet [0.3 to 3.7 m] and upwards Wide, and from One to Forty-five Feet [0.3 to 13.7 m] and upwards in Length." Gamble's specification was essentially a translation and expansion of Robert's French *brevet*. The title of the specification, with its emphasis on the production of very large sheets, suggests that the original market for the product may have been expected to be wallpaper.

In London, Gamble was introduced to Henry and Sealy Fourdrinier, of the firm of Bloxam and Fourdrinier, the leading wholesale stationers in London, who took great interest in Robert's invention and set about trying to improve it. In 1802, the Fourdriniers hired a young English engineer, Bryan Donkin, to improve the machine and oversee its development at mills that they opened at St. Neots, Huntingdonshire, and at Frogmore and Two Waters, Hertfordshire. The Fourdriniers undoubtedly began manufacturing paper by machine shortly thereafter.

In 1806, Henry Fourdrinier obtained patent no. 2951 for a method of making a machine for manufacturing paper of an indefinite length, laid and wove, with separate moulds. However, Fourdrinier's patent for the endless chain-mould machine did not accompany his specification with drawings and did not describe the machine in much detail. For that reason, Didot and his associates, including Fourdrinier, thought it appropriate to patent a more detailed specification, which the Fourdriniers and John Gamble accomplished in 1807. Saint-Leger Didot (Leger Didot), then living in England, further improved the technology in 1812 when he was granted British patent no. 3568 for "Certain other Improvements upon the Said Machines for the Making of both Woven and Laid Paper." By this time, the papermaking machine originally invented by Robert became known as the Fourdrinier machine, the name it has retained to this day.

Because handmade papermaking was such a labor-intensive industry, numerous workers inevitably lost their jobs when paper mills mechanized. In some instances, they lost jobs that members of their family had held for generations. In protest, some workers broke machinery: Leonard Rosenband has usefully documented the participation of some English hand papermakers in the Swing Riots of 1830 and other Luddite-style protests.[32]

In addition to the obvious advantages of reducing labor problems by reducing the number of employees, the cost savings from machine papermaking were evident almost at once. In the fifth edition of his *Chronology of the Origin and Progress of Paper and Paper-Making*, American printer and printing and papermaking historian Joel Munsell wrote that, in 1806, the cost of making paper by machine was roughly 25 percent of the cost of making paper by hand.[33]

Nevertheless, like many new technologies, machine papermaking did not immediately catch on in the British paper industry. It also required numerous improvements over decades and generations before it became genuinely efficient. As had happened with certain other new technologies, some of the original inventors failed to benefit. The early pioneers who invented and developed the first papermaking machines—Louis-Nicolas Robert, Henry Fourdrinier, Saint-Leger Didot, and John Gamble—all died in comparative poverty, even though they witnessed the construction of many successful paper mills in which hundreds of papermaking machines were operating. As R. H. Clapperton describes:

> *The Bryan Donkin Company alone had built 197 papermaking machines before Henry Fourdrinier died, and by that time many other engineering firms were also building this type of machine. The Fourdrinier firm, of which Henry Fourdrinier was the head, lost at least £60,000 in the first ten years of the development of the machine, and became bankrupt in the process. Leger Didot lost his paper-mill and his business. Gamble lost his paper-mill at St. Neots to Matthew Towgood; and Robert was left completely out of it by everybody, and eventually got nothing but a statue and memorial many years after he died.*[34]

The early papermaking machine patents by Louis-Nicolas Robert, John Gamble, the Fourdriniers, and others make it clear that the objectives of the inventors of papermaking machinery were to produce paper in larger format than could be made by hand, to reduce costs, and to speed up production. Because of chronic labor disputes. another benefit to paper manufacturers was reduction of the number of employees in paper mills. Clues to the cost savings provided by mechanized printing can be found in *The Paper-Maker's Ready Reckoner, or Calculations to Shew the Prime Cost of Writing or Printing Paper*,[35] an unusual financial guide for the production of handmade paper by Jean Abbot Dusautoy, operator of the Lyng Paper Mill near East Dereham, Norfolk. Compiled from his own experience, and published just as the first papermaking machines began to be introduced, *The Paper-Maker's Ready Reckoner* provided valuable information for others in the hand papermaking industry. Its many tables show calculations not only for the prime cost of rags per ream, but also the expenses for the papermaker: rent, taxes, insurance, repairs, wages, interest, fuel, candles, stable expenses, traveling, utensils (felts, moulds, planes, boards, and levers), bad debts, oil, lard, soap, flannel, packing paper, various kinds of stationery, and, finally, customs duty. The last table is for calculating the wages of the journeymen papermakers on either a fixed weekly wage or by piecework. Published at the then-exorbitant cost of 5 guineas per copy, Dusautoy's *Ready Reckoner* was intended to be sold to other papermakers. My copy is signed by the author at the end of the explanation and numbered 112. Only a few copies would been sold, which also would explain why very few copies survived.36

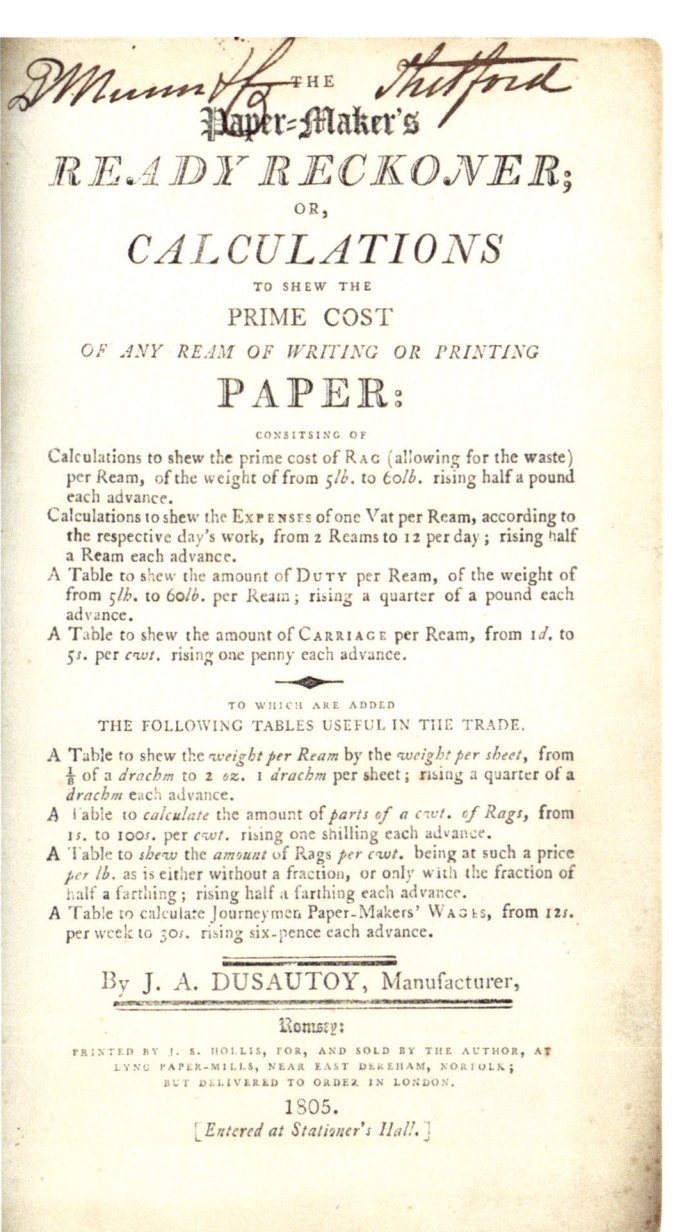

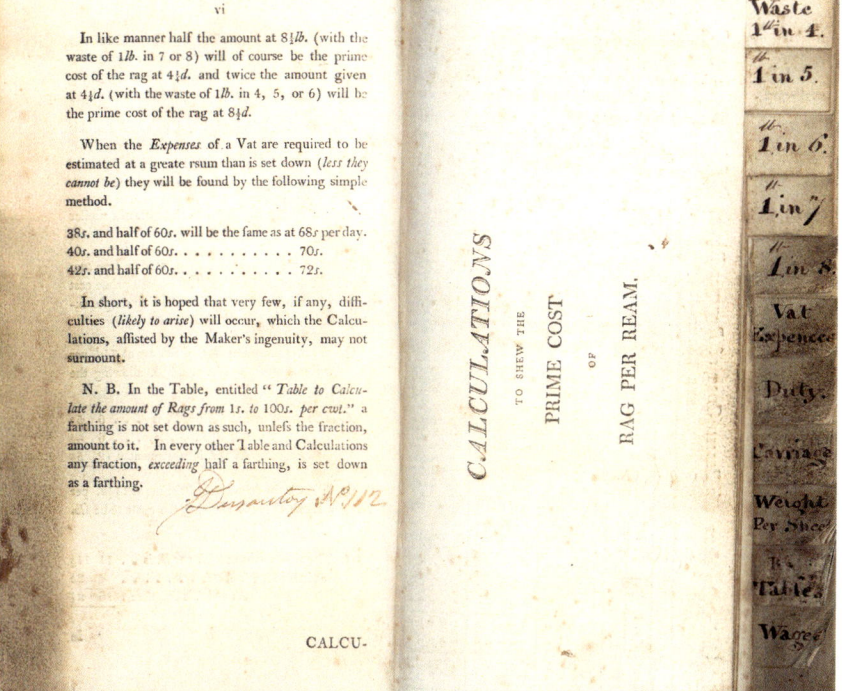

Jean Abbot Dusautoy, *The Paper-Maker's Ready Reckoner* (Romsey, 1805)

Published in a very small edition, this was a financial tool for the operators of hand papermills.

Further Developments in Papermaking Technology

Around 1820, the quantity of paper made by machine exceeded the quantity of paper made by hand. Besides its lower cost, machine-made paper had several advantages for printers using the new mechanized printing presses. Chief among those advantages was that machine-made paper could be made in much larger sizes than paper made by hand, allowing books to be printed in 16- or 32-page signatures. Another advantage was the speed at which machine-made paper could be made. The fact that mechanized paper mills could produce paper rapidly, and thus could meet whatever demand a printer faced, alleviated the need for printers to maintain several months' supply of paper on hand, thus reducing capital investment.

The first paper manufacturer to benefit financially from the papermaking machine was John Dickinson, an English paper dealer and inventor from Aspley, Hertfordshire. In July 1809, Dickinson patented the cylinder-mould papermaking machine.[37] Dickinson's concept, which he developed through his partnership with the publisher George Longman (who provided the initial working capital), was the first to allow for commercially viable machine production of paper. Of the early inventors in papermaking, Dickinson was the only machine papermaker to a develop a business that remained financially successful for generations.

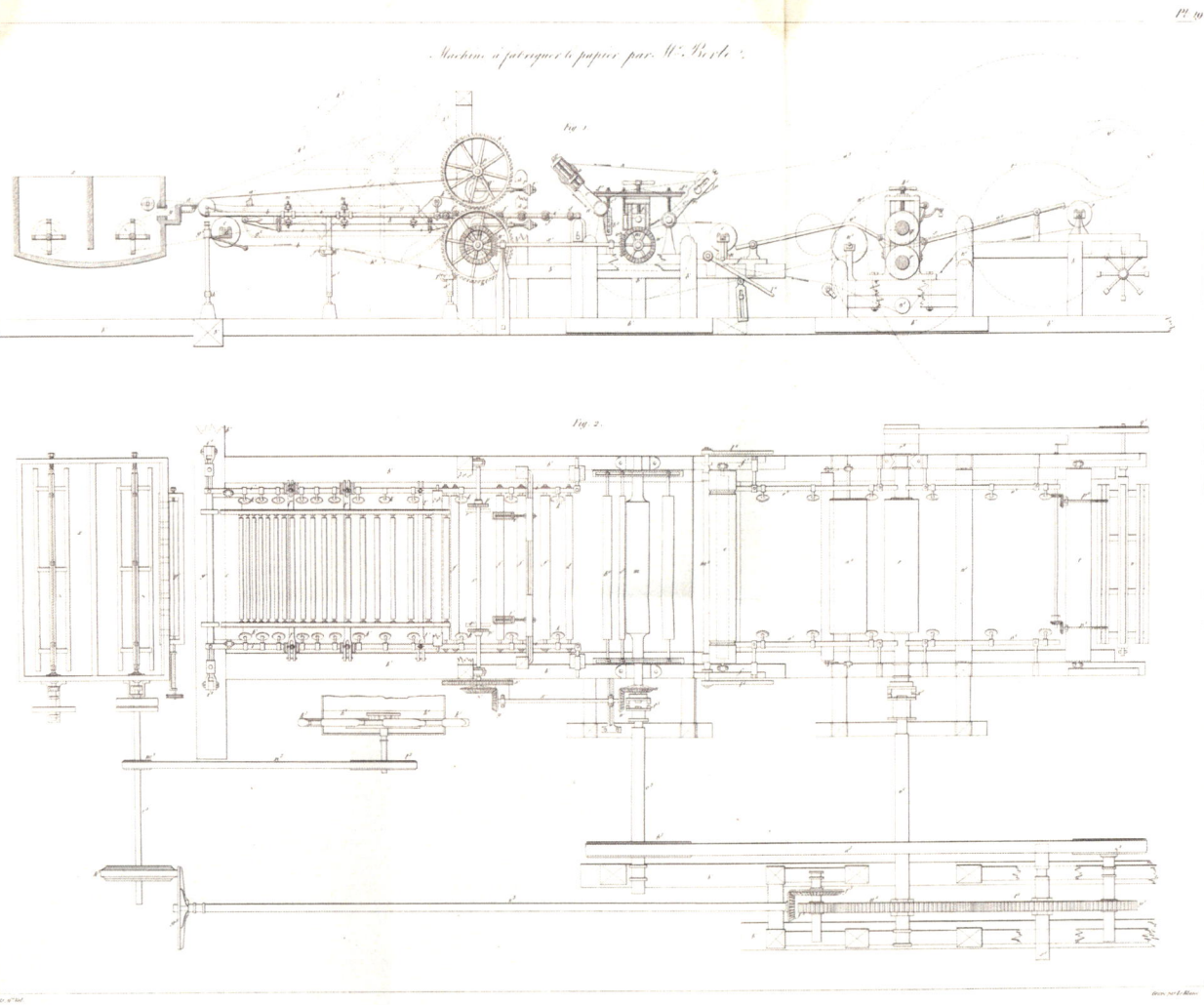

Antoine-François Berte, *brevet* 1455, granted in 1811, "Pour une machine à fabriquer le papier de différentes longuers" From *Description des machines et procedés spécifiés dans les brevets d'invention....* tome XVI (1828)

Berte's machine was the first operational papermaking machine invented in France after Robert's. In the twelve years after Robert's invention, papermaking technology advanced significantly.

Dickinson's machine consisted of a perforated cylinder of metal, with a closely fitting cover of finely woven wire, which revolved in a vat of pulp. The water from the vat was carried off through the axis of the cylinder, leaving the fibers of the pulp clinging to the surface of the wire. An endless web of felt, passed through what was known as a "couching roller" lying upon the cylinder, drew off the layer of pulp that, when dried, became paper.[38] The cylinder-mould machine took some time to perfect, and according to Coleman,[39] Dickinson initially installed machines of the Fourdrinier type at his mills.

It was not until 1811, twelve years after Robert patented his papermaking machine, that two other French inventors attempted to invent machines to mechanize the papermaking process. The first operational papermaking machine in France was developed by Antoine-François Berte, of the firm of Berte and Grenevich, who set up a papermaking machine at his establishment and obtained *brevet* no. 1455 for it on 11 October 1811. This was the first *brevet* after Robert's to be granted in France for a papermaking machine designed and built in that country. As one would expect after the elapse of twelve years and the advances that had occurred in England, Berte's patent was far more detailed and more advanced than the original Robert patent. Berte's initial machine was considerably inferior to the English Donkin machines, however, and it was Donkin machines that eventually established the French machine-papermaking industry, beginning with the first importation of a Donkin machine into France in 1822.[40] Nevertheless, Berte and Grevenich, and Saint-Leger Didot, were awarded gold medals in 1819 for developing machine-made paper in France.

The second French inventor after Robert to develop a papermaking machine was Ferdinand Leistenschneider of Poncey, who obtained *brevets* in 1811 for his machine that made paper of definite length, rather than a continuous sheet, such as those the Robert or Fourdrinier machines created. Leistenschneider's machine and its operation are described in a 32-page report to the Dijon Academy of Sciences by French engineers Philippe Leschevin and Pierre-Joseph Antoine, published in 1815; interestingly, the report appears to have been printed on handmade, rather than machine-made paper. It is probable that this pamphlet is the first separately issued publication on papermaking by machine in France, and one of the very earliest separate publications on papermaking by machine published anywhere. Louis Piette, the earliest historian of machine papermaking, discussed Leistenschneider's machine in detail, describing it as "quick" and "inexpensive," and stating that Leistenschneider was awarded four *brevets* between 1813 and 1827.

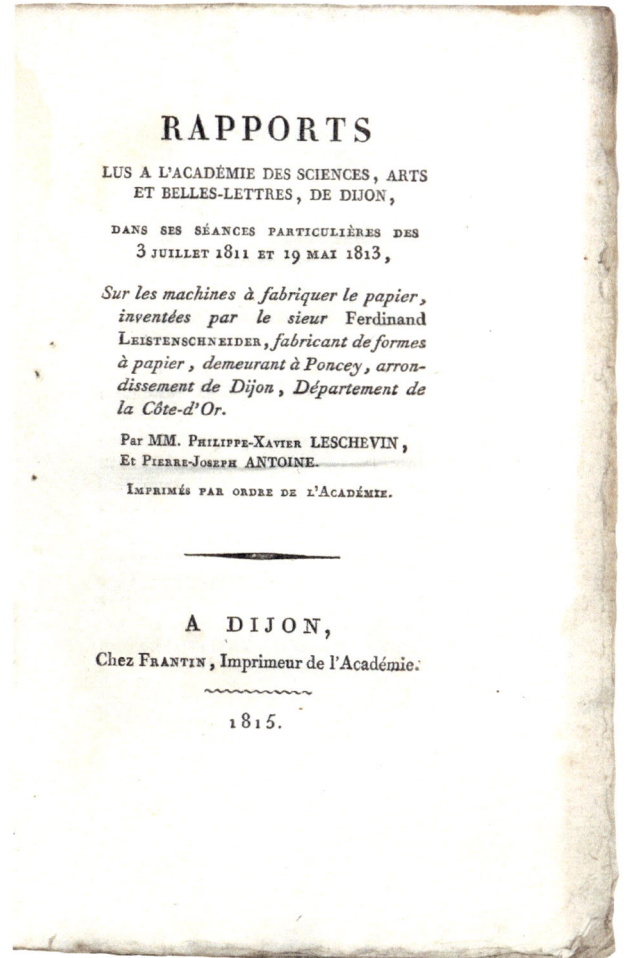

Philippe-Xavier Leschevin and Pierre-Joseph Antoine
Rapports lus à l'Académie des sciences . . . de Dijon . . . Sur les machines à fabriquer le papier inventées par le sieur Ferdinand Leistenschneider (Dijon, 1815)

This pamphlet on the papermaking machine invented by Ferdinand Leistenschneider of Poncey was probably the first separate publication ever published on a papermaking machine.

A primary source for the early years of the French machine papermaking industry is Piette's *Traité de la fabrication du papier. Contenant . . . descriptions détaillées des machines à faire le papier d'après les nouveaux procédés, etc.* (Paris: F. G. Levrault, 1831). In it,

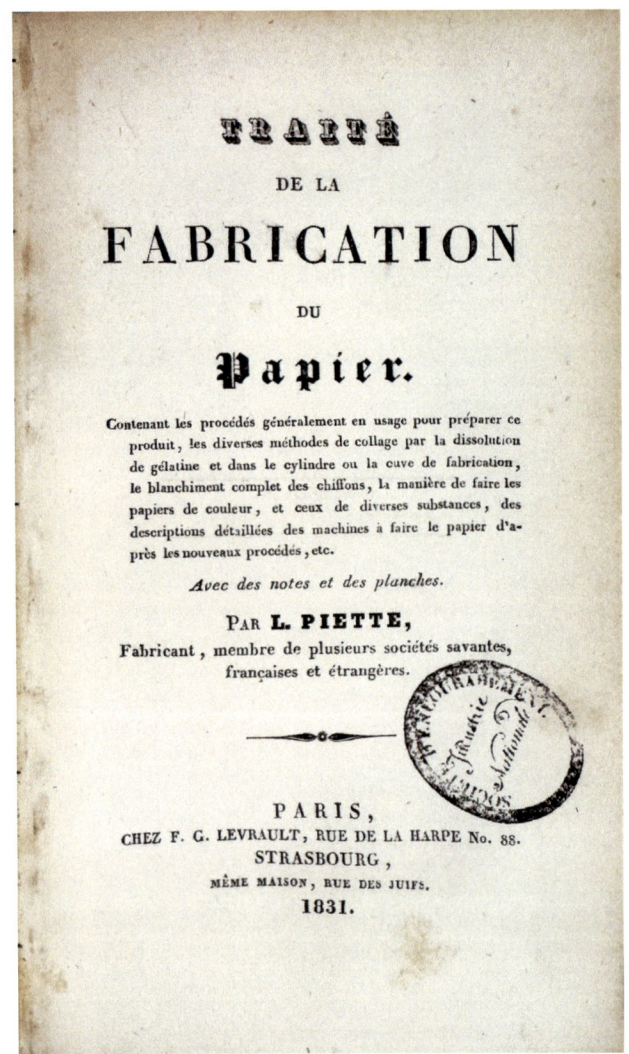

Louis Piette, *Traité de la fabrication du papier* (Paris, 1831)

Besides a technical treatise on papermaking, this work contains a detailed chronological record of the awards given to the earliest inventors of papermaking machines in England and France.

we learn that Berte and Grenevich, as well as Saint-Leger Didot, were awarded medals for paper made by machine in France as early as 1819. Piette, a French-Prussian paper manufacturer, inventor, publisher, and author, referred to a number of papermaking machines developed in France after Robert, as well as machines developed in England. He also included charts of significant patents awarded in England and France, and cited awards granted for contributions to papermaking.

In 2017, I acquired a volume of pamphlets once in the library of papermaker Paul Darblay, who, in 1860, bought the Essonnes paper mills where Louis-Nicolas Robert had invented the first papermaking machine. This volume contains four very scarce pamphlets recording nuances of a patent dispute between Didot and Berte over rights to the papermaking machine in France.

Précis et consultation pour le Sieur Léger Didot, contre le Sieur Berte [Paris, 1823]

Mémoire en réponse et consultation pour le Sieur A. F. Berte; contre le Sieur Leger Didot [Paris, 1823]

Réponse au factum calomnieux publié par M. Berte contre M. Didot-Saint-Léger, sous le titre: Sur quelques modèles faits à Londres en 1816, présentés en 1823 à l'exposition des produits de l'industrie française, et sur le colportage en Angleterre de la machine à fabriquer le papier continu, inventée en France [Paris, 1823]

Supplément a la réponse de M. Didot-Saint-Léger au mémoire de M. Berte. A Son Excellence le Secrétaire d'état, Ministre de l'intérieur [Paris, no date.]

Though these pamphlets provide some documentation regarding the earliest history of machine papermaking in France starting in 1810 and 1811, the titles of the first publications in France that were actually printed on machine-made paper appear to be unknown, except that historian Louis André mentions that, beginning in August 1816, Berte began supplying machine-made paper for the *Journal des débats*, a weekly French newspaper.[41]

Papermaking machinery crossed the Atlantic to the United States at about the same time that its development occurred in France. In 1817, papermaker Thomas Gilpin set up the first papermaking machine in America at his mill on Brandywine Creek, downstream from the DuPont Mills in Wilmington, Delaware. In 1816, Gilpin obtained a U.S. patent for the first continuous papermaking machine in the country, based on information secured by his brother in England; however, because of a fire in the U.S. patent office in December 1836, that patent no longer survives. Dard Hunter quoted a 29 November 1817 issue of Niles' *Weekly Register*, Baltimore, credited to the

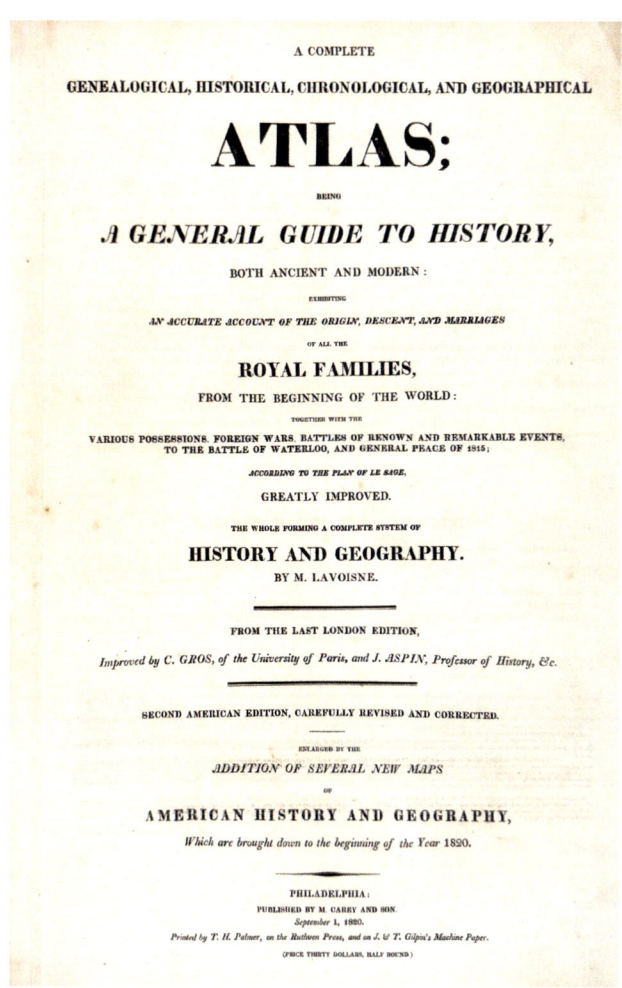

M. Lavoisne, *A Complete Genealogical, Historical, Chronological and Geographical Atlas* (Philadelphia, 1820)

Below the imprint date, the title page states that the book was "printed on J. & T. Gilpin's Machine Paper." This was the first American book to state that it was printed on machine-made paper. The Ruthven press on which the book was printed was a compact platen flatbed press invented and manufactured in Scotland by John Ruthven from 1819 to 1822. The design of the Ruthven press was unusual in that pressure on the inked type pressing against the paper was exerted from below the flatbed rather than from above.

Johann Carl Leuchs, *Darstellung der neuesten Verbesserungen in der Verfertigung des Papiers* (Nuremberg, 1821)

Leuchs's treatise on the latest developments in papermaking included, as the title page indicates, a thorough description of the Fourdrinier machine. This was the first book on machine papermaking in any language. The book was needed to stimulate the papermaking industry in Germany, since there were only a handful of papermaking machines in operation in Germany at the time.

Delaware Watchman that Gilpin's machine, which was based on the Dickinson cylinder-mould principle, did the work of ten vats in the handmade mills.[42]

Probably the first American book printed on Gilpin's machine-made paper was Mathew Carey's *General Atlas, Improved and Enlarged* (Philadelphia: M. Carey and Son, 1818). The first American book to advertise that it was printed on Gilpin's machine-made paper was Carey's edition of Lavoisne's *A Complete Genealogical, Historical, Chronological, and Geographical Atlas*

(1820). At the foot of its title page, the book's imprint states that it was "Published by M. Carey and Sons and printed by T. H. Palmer on the Ruthven Press, and on J. and T. Gilpin's Machine Paper" (the advertised price of $30 equates to about $814 today).

Regarding the earliest development of papermaking machinery in Germany, in 1821, German science writer Johann Carl Leuchs issued *Darstellung der neuesten Verbesserungen in der Verfertigung des Papiers, enthaltend insbesondere die Beschreibung und Abbildung der Maschine zur Verfertigung des Papiers ohne Ende* (Description of the latest improvements in paper manufacture, including in particular a description and illustration of the machine for making endless paper). This 90-page work, issued in Nuremberg, was the first published book on mechanized papermaking in any language; this is especially remarkable because there were only a handful of papermaking machines in Germany at the time. Leuchs provided specially detailed information on the Dickinson machines made in England, plus the few available German designs. He illustrated his book with three folding plates of papermaking machinery. His focus on substitutes for rags in the production of paper reflects the general scarcity of high-quality rags—scarcity that was aggravated as production of paper was increased by machinery.

Though tracing the earliest surviving examples of machine-made paper is difficult, the first book printed on machine-made paper, John Anastasius Freylinghausen's *An Abstract of the Doctrine of the Christian Religion*, was issued as early as 1804.[43] Though there is no statement in this edition that the paper on which it was printed was made by machine, in 1805 the antiquarian, collector, and scholar Richard Gough wrote an anonymous review of the book in *The Gentleman's Magazine* in which he stated, "The paper is the first specimen made by what is termed *the machine*, in which the sheets are extended to any dimensions at pleasure. Its texture is solid and even throughout, its colour good, and by a certain *roughness* on its surface, in which it resembles copper-plate paper, it is adapted to take the imprint advantageously."[44]

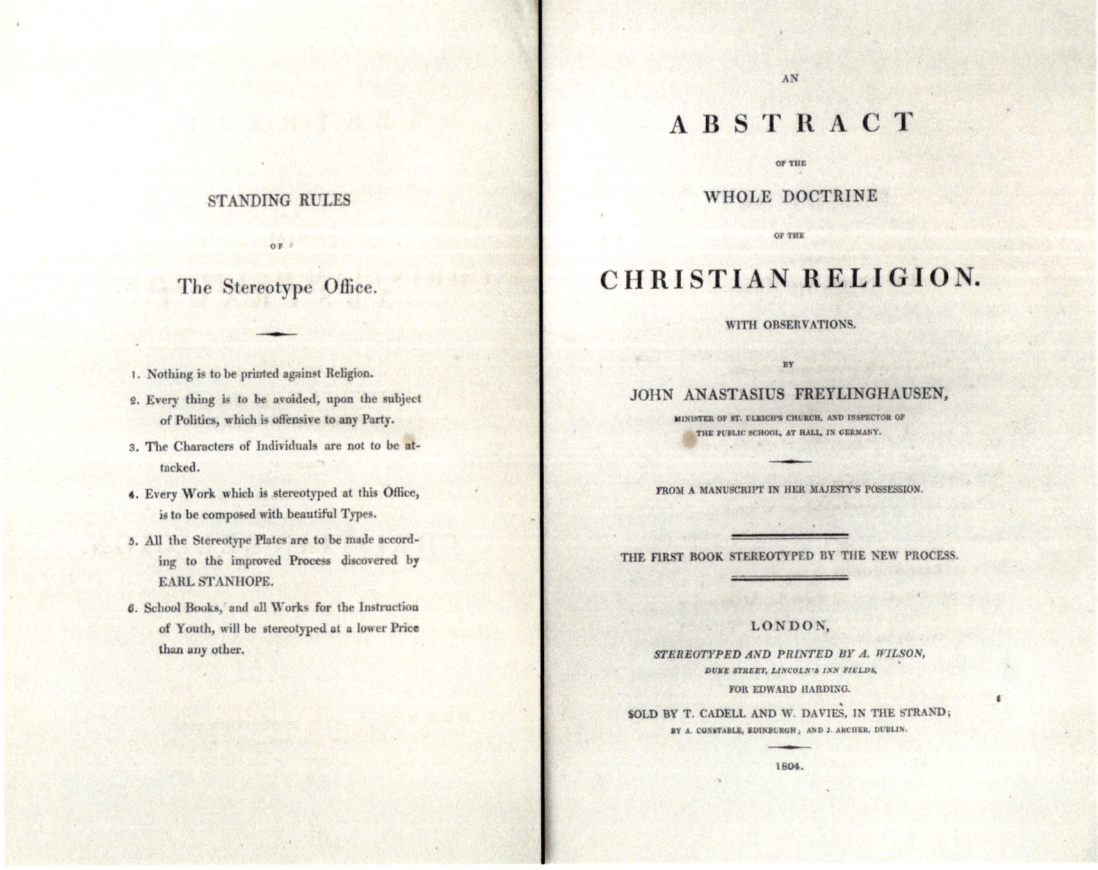

John Anastasius Freylinghausen, *An Abstract of the Whole Doctrine of the Christian Religion* (London, 1804)

The English translation of Freylinghausen, previously illustrated in the discussion of Lord Stanhope's improved method of stereotyping, was also the first book printed on machine-made paper.

Another clue concerning early machine-made paper used for printing is in a footnote to Joseph Wilson's three-volume *A History of Mountains*,[45] which, according to the book's subtitle, was "Accompanied by a Picturesque View of the Principal Mountains of the World, in Their Respective Proportions of Height above the Level of the Sea by Robert Andrew Ridell."[46] Ridell's plate was published on 1 January 1806, a year before the appearance of the first volume of Wilson's work. According to a footnote in the preface to the first volume, the plate

> *. . . is already much the largest that has ever been engraved on one plate of copper or printed on one sheet of paper, being 4 feet 6 inches by 3 feet [1.35 × 1 m], exclusive of margins; and it would be unjust in this place to omit stating that the possibility of executing it in its present dimensions is solely to be ascribed to the perfection to which the enterprize [sic] and spirit of two individuals [Messrs. Henry and Sealy Fourdrinier] have brought the important invention of papermaking by machinery; an invention by which manual labour is nearly rendered altogether unnecessary; the paper greatly improved in quality, while its size is unlimited, and which promises in its operation the utmost benefits to the artist, the manufacturer, the revenue, and the public.*[47]

Between 9 and 12 May 1837, various papermakers, printers, and publishers in England testified before the Select Committee of Parliament on Fourdrinier's Patent, a body tasked with determining whether the bankrupted Fourdrinier family should receive government compensation for inventing the Fourdrinier papermaking machine. From the testimonies, published in the parliamentary papers,[48] we see that there was a general consensus in the printing and publishing trades that papermaking machines produced paper that was larger, less expensive, and better suited for use with mechanized presses than handmade paper; and that the speed at which papermaking machinery produced paper (roughly 25 feet [7.6 m] per minute per machine) provided a much more reliable supply for printers and publishers, reducing their capital investment by eliminating the need for them to stockpile paper. For all these reasons, all the printers and publishers who testified preferred machine-made paper to handmade. As a result, they had great appreciation for the commercial value of the Fourdriniers' invention and believed that the family, which had invested so much money in its development but had never profited from it, was entitled to compensation. At the time this report was published, there were 240 Fourdrinier papermaking machines in operation in England.[49]

Charles Knight, publisher of *The Penny Magazine*, *The Penny Cyclopedia*, and many other publications, was among those called to testify; he noted the "positive cheapness" of machine-made printing paper "as compared with the former prices of printing paper made by hand," and pointed out "an additional cheapness arising from the improved quality of the paper; taking the paper, for example, made by hand, weighing 20 pounds, I should say all purposes of printing." John Davis, superintendent and publisher of the Religious Tract Society, told the Committee that "by means of this invention we obtain our supplies of paper with certainty, which we could not have done otherwise. We use nearly 25,000 reams annually; if we could not have the supply which we want, we should be compelled to keep a very large stock on hand, and that stock we do not keep." He also noted that "the price of paper during the last 20 years is greatly reduced, and I attribute that in a great degree to the facilities afforded by the [Fourdrinier] machine." George Clowes, of the large London printing firm of William Clowes and Sons, testified that Fourdrinier's invention "has been most beneficial to our trade, principally in reducing the price of paper, so enabling us to produce books at a much cheaper rate; and, in fact, the paper machine, in connexion with the printing machine, without which the printing machine would have been comparatively useless, has effected a complete revolution in our business; where we used to go to press with an edition of 500 copies, we now print 5,000." He also noted that, because of the Fourdrinier machine, "we are enabled to have paper of any size that we require, so that we can print a larger number of pages at one time; for instance, in printing a common demy octavo volume, instead of printing 16 pages in the sheet as we used to do, we can now obtain paper of twice the size, that is, double demy, and so print 32 pages in the same time that we used to print 16, so that we can print a book off, when it is ready for press, in half the time we used to do."

From *The Monthly Supplement of The Penny Magazine*, no. 96 (31 August to 30 September 1833)

An explanatory diagram of the workings of a Fourdrinier papermaking machine. Because of the very wide readership of *The Penny Magazine*, which claimed a circulation of about 200,000 subscribers at the time, this was probably the first time an image of a Fourdrinier machine was seen by a large number of people.

Drawn by Emile Bourdelin and engraved by H. Linton
Hand-colored woodcut, 10 × 14 in. (25.3 × 35.5 cm)
From Julien François Turgan, *Les grandes usines de France* (Paris, 1860)

A Fourdrinier papermaking machine at La Paperie d'Essone outside of Paris, created for Turgan's book. At this papermill, sixty-two years earlier, Louis-Nicolas Robert invented the first papermaking machine. The difference in scale and complexity between Robert's initial machine patented in 1799 and the machine shown in 1860 is significant.

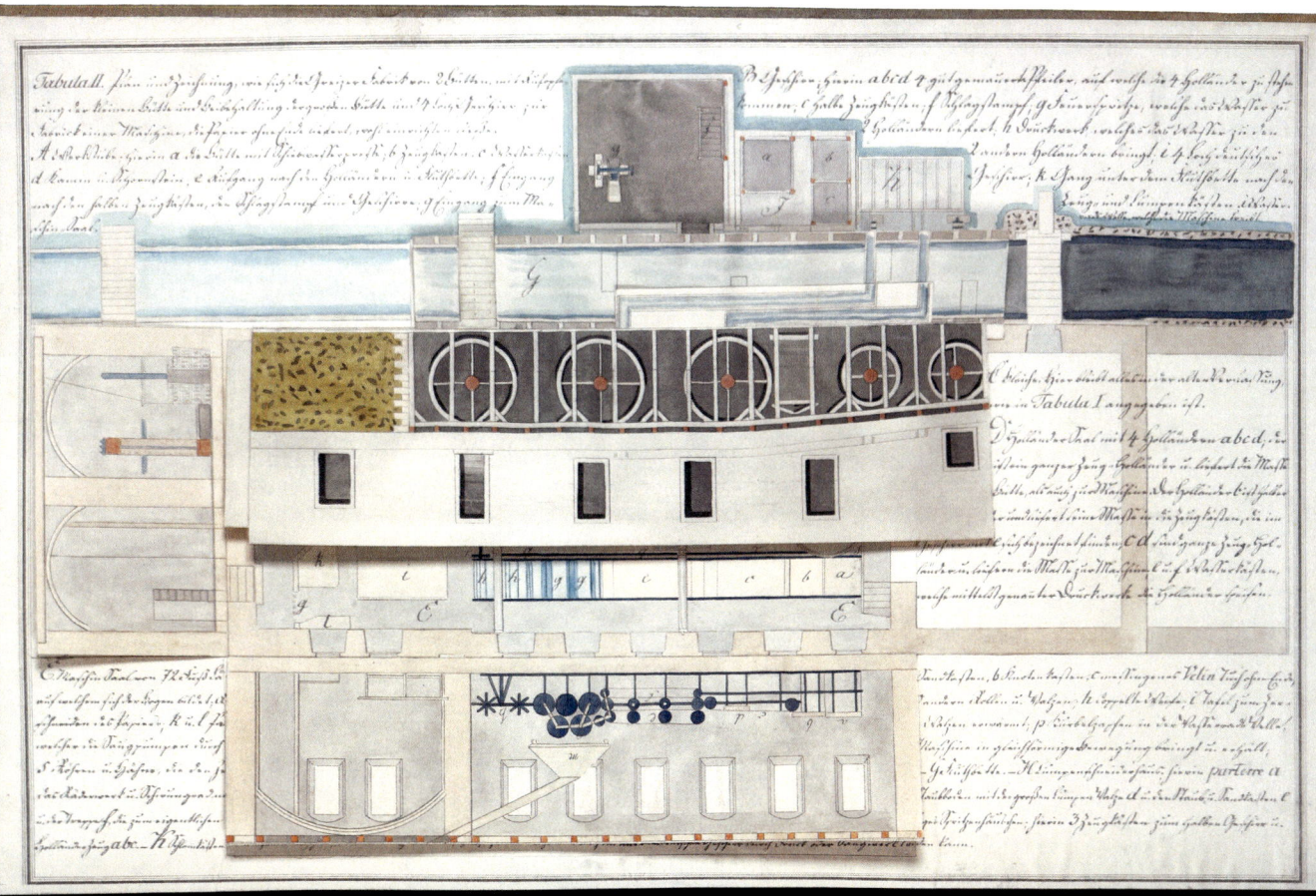
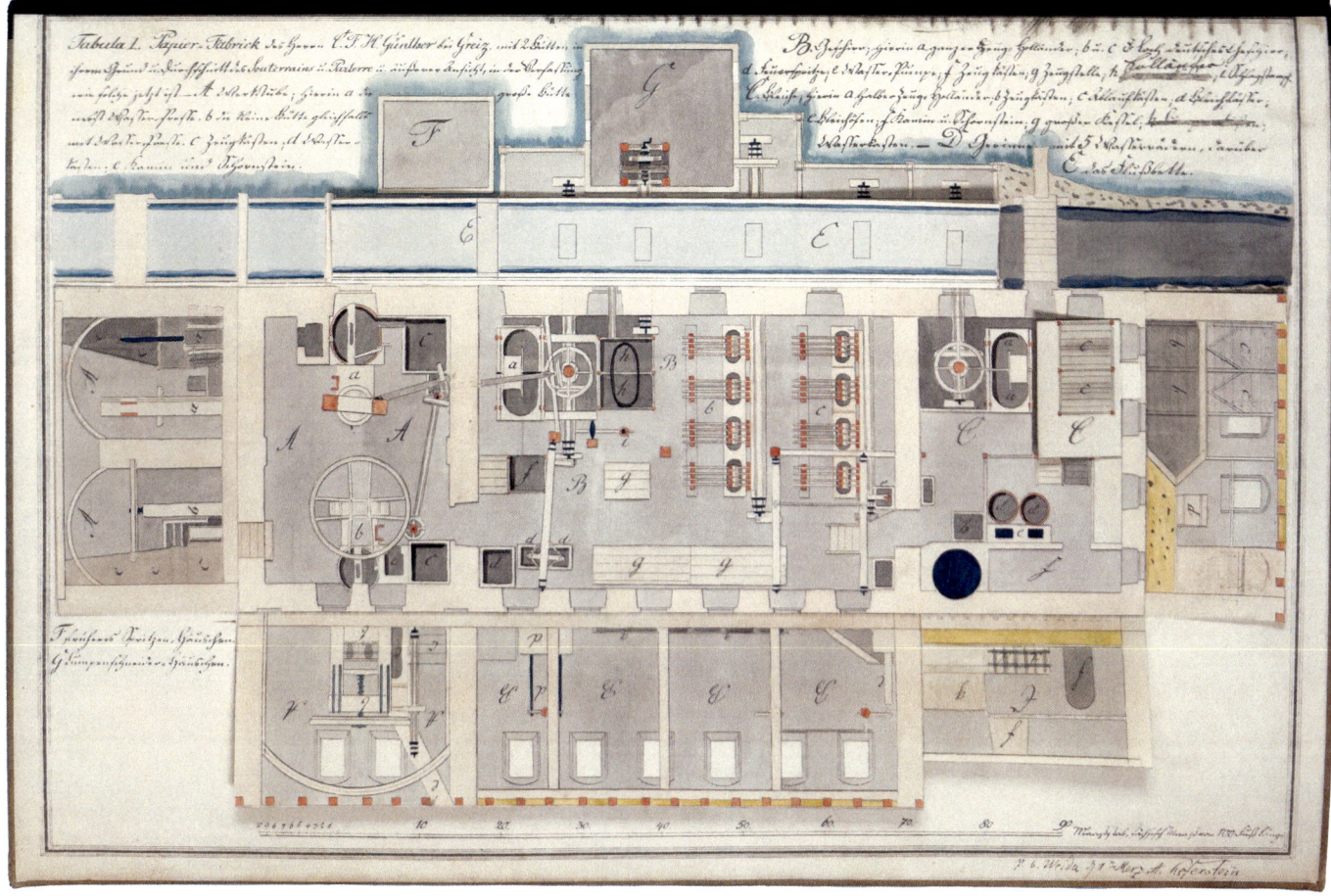

A. Heferstein (?)
Manuscript model
Watercolor on paper
(c. 1860?), 28⅓ × 21⅔ in.
(72 × 55 cm)

A three-dimensional manuscript model, with several fold-up flaps, of the papermaking machinery at the paper mill of C.F.H. Günther of Greiz, Germany, signed by architect A. Heferstein (?).

🖌 Anonymous British printmaker, after artist David Allan (1744–1796)
Representation of the New Shaving Machine, Whereby a Number of Persons May Be Done at the Same Time with Expedition, Ease, and Safety [no place, 1770]
11 × 8⅔ in. (27.5 × 22 cm)

This satire of mechanization and the factory system describes an imaginary machine offered for sale in Birmingham. A large model that could shave up to 20 people simultaneously was offered for 7 guineas, while the smaller model shown in the engraving cost 3 guineas. Notably, the shaving machine is powered by a man turning a hand crank because, at this stage in the Industrial Revolution, toward the end of the eighteenth scentury, no steam engine small enough for this task existed either in reality or in an artist's imagination.

🖌 Robert Seymour ("Shortshanks")
Shaving by Steam (undated, but c. 1825)
Hand-colored engraving, 11 × 16½ in. (28 × 42 cm)

This cartoon was one of a series of caricatures by Seymour, focusing on the growing spread of steam-powered mechanization to different aspects of life in England. Seymour was undoubtedly inspired to create this print by *Representation of the New Shaving Machine . . .* print that had been published in 1770, in which the machine was powered by a man turning a hand crank, before the availability of steam engines small enough to power such an apparatus.

Paul-Andre Basset (1759–1829)
Le premier et incomparable moulin à raser toutes les têtes à barbe et à cheveux coupés à la Cherubin et à la Titus &c. (Paris [1798])
15¾ × 26 in. (40 × 66 cm)

In this larger and much more elaborate satire on mechanizing the shaving process, published in Paris in 1798, before the socio-economic impact of mechanization or the factory system was felt in France, the prolific printmaker and print publisher Basset turned mechanized shaving into a social event with an orchestra playing and people socializing. To add to the absurdity or the exotic nature of the invention, which supposedly was a traveling sensation throughout Europe, the machine was powered by two camels.

Payall after a drawing by G. Morton, *A Steam Coach with Some of the Machinery Going Wrong* (detail). See p. 61.

CHAPTER 3

Friedrich Koenig Invents the Steam-Powered Printing Machine

The printing machine was invented in England, but not by an Englishman. In the second decade of the nineteenth century, Friedrich Koenig, an ingenious German inventor, was forced to travel to England to develop the first steam-powered press because Germany and the rest of Europe were lagging far behind England in the Industrial Revolution. Another factor controlling the timing of Koenig's invention was the availability of steam engines small enough to drive his machine. These were first available in England.

Even though we correctly associate the steam engine with powering machinery in the Industrial Revolution, for most of the Industrial Revolution small machines were generally driven by wind and waterpower, as well as by horse and manpower. This explains why several of the printing machines designed in the first half of the nineteenth century were designed to be driven by alternative power sources—horse or human power rather than steam. Others could be driven by human power if a steam engine was non-operational.

Courtesy of Koenig and Bauer AG, Würzburg

Paintings of Friedrich Koenig (left) and Andreas Bauer (right).

"Henry Maudslay Machine Making"
United Kingdom postage stamp (2009)
1½ × 1½ in. (3.5 × 3.7 cm)

Maudslay is shown with the table engine he invented and patented in 1807. Maudslay's relatively small steam engine became a practical power source for printing machines and other machines that were too small to be powered by traditional, very large steam engines.

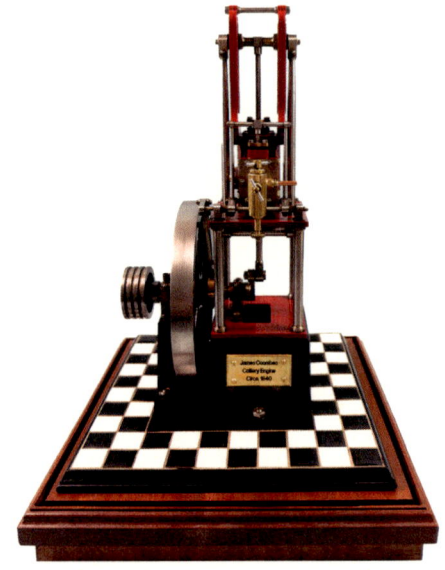

A small modern model of a James Combes colliery engine, built in 1840
14½ × 10⅔ in. (37 × 27 cm), excluding the base

This is a version of the original Henry Maudslay Table Engine, patented in 1807.

In the early years of the nineteenth century, engineers in England, after years of experimentation, obtained sufficient control over steam to develop small stationary steam engines, generally of 1 or 2 horsepower, that could power small industrial facilities such as printing presses. The first upright steam engine was developed in England by James Sadler in 1798. Sadler's engine was of the "house-built" type, meaning it was built in an engine house, like the large "house-built" Newcomen engines, which used the masonry of the engine house as an integral part of the support of the engine. In 1807, English engineer Henry Maudslay designed and patented the first self-supporting small steam engine that took its name "table engine" because it sat on an iron table and was independent of the structure on which it was standing. The table engine, which could be built small enough to operate in an ordinary room, became the engine of choice for driving steam-powered printing machines through most of the nineteenth century, typically in the range of 1.5 to 4 horsepower, though larger models could generate up to 40 horsepower. The boiler powering a table engine was typically located outside the workspace to reduce smoke and fire risk.[50]

Koening was born in Eisleben, Germany, on 17 April 1774. At the age of eight, he was forced to curtail his education due to financial difficulties following the death of his father. He became an apprentice as a printer and typesetter at Breitkopf and Hartel, music publishers in Leipzig, and qualified in 1794, nine months earlier than usual. After his apprenticeship, Koenig registered as student at the University of Leipzig to study mathematics, physics, and mechanics, while he earned his living as a printer and translator. He also made unsuccessful attempts at running a bookshop in Eisleben and a printing shop in Mainz.

As early as 1803, Koenig designed a power-driven printing device known as the Suhl press. Koenig's Suhl press mechanized the printing process except for the insertion of blank sheets and the extraction of printed sheets. However, the press, which was constructed out of a wooden frame, proved too complicated, and Koenig did not have the funds to make further improvements. Besides that, of the very few steam engines available in Germany at the time, none would have been small enough to power Koenig's machine even if it had been operational. What originally motivated Koenig to attempt to mechanize printing is unclear, since in Germany at the time there was no obvious need to speed up printing.

Since the Industrial Revolution had not yet reached Germany in the first decade of the nineteenth century,[51] Koenig found no backers in the country for the development of his printing machine. To solicit interest in his project, Koenig wrote a detailed technical description of his work in progress and sent copies to publishers and printers throughout Europe. He also traveled to Würzburg, Munich, Vienna, Hamburg, Lübeck, and even as far as St. Petersburg searching for sponsors. However, he met with nothing but rejection from publishers, university professors, the king of Bavaria, the emperor of Austria, and the tsar of Russia.[52] In this discouraging environment, Friedrich Koenig made the decision to travel to London in the hope of finding capital to develop his steam-powered printing machine. On 20 November 1806, one day before Napoleon began his Continental blockade of England, Koenig landed in London, where he initially earned a living as a printer and bookseller. Four months later, on 31 March 1807, he negotiated a contract to build a printing machine with the distinguished London printer Thomas Bensley, who was interested in innovative technology. To help finance the development of Koenig's machine, Bensley brought in two other partners, the printers George Woodfall and Richard Taylor.

While developing his steam-powered printing machine, Koenig partnered with Andreas Friedrich Bauer, a highly skilled engineer. Bauer was born in Stuttgart on 18 August 1783; after attending grammar school, he became an optician and instrument maker's apprentice, completing his apprenticeship in 1803. He then studied at the University of Tübingen, where he earned a master's degree in natural sciences and mathematics. Attracted to the more advanced industrial development in England, Bauer moved to London in 1805, where he later met Friedrich Koenig.

Thomas Bell
British patent no. 1378, for "Machinery for Printing Textile Fabrics," issued in 1782. Printed in 1856

A detail from Bell's patent drawing of the mechanism of his rotary press for printing textiles. Decades before rotary presses were practical for printing on sheets of handmade paper, there was a ready market in the textile industry for Bell's rotary printing machine for printing on textiles, since fabric came off the power looms on rolls, while handmade paper was made in relatively small single sheets. Powering Bell's printing machinery also presented no problems before small steam engines were available, since Bell's machines could be connected to the system powering the textile mills (waterpower or large steam engines).

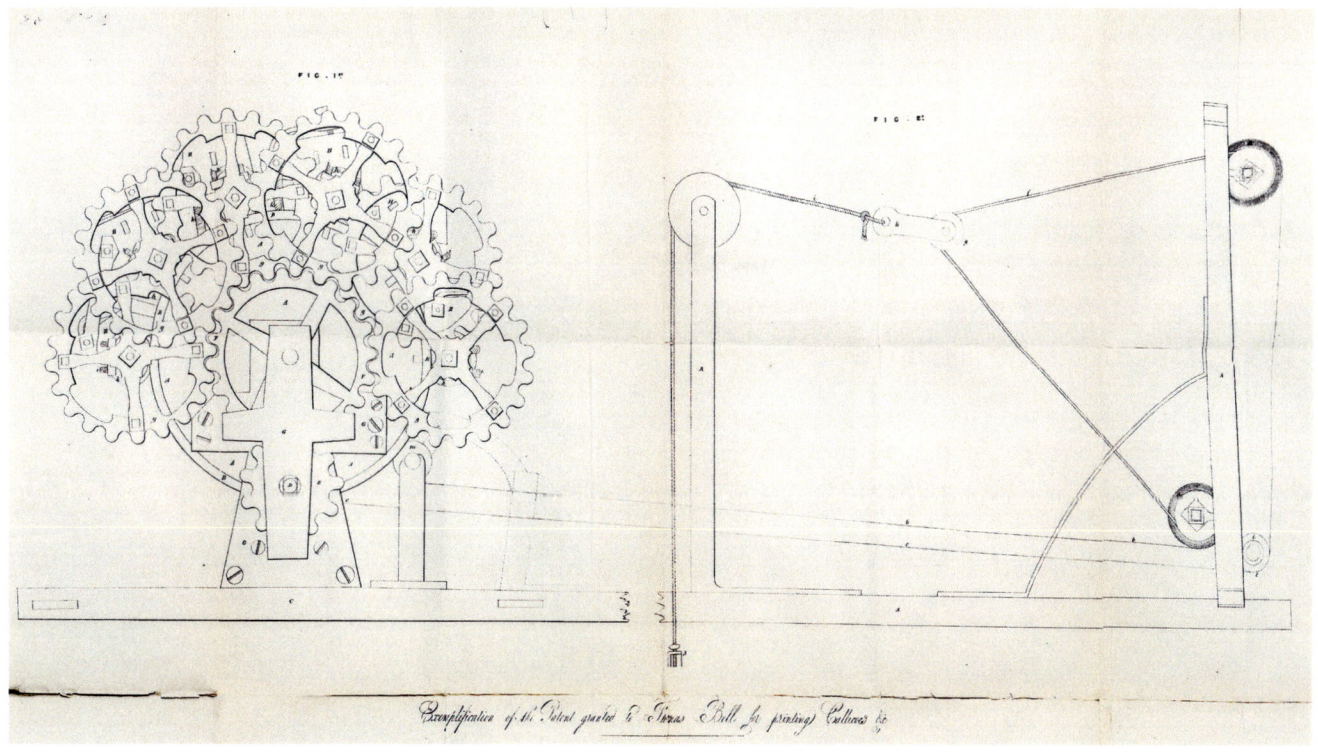

The two men started working together, signing a formal business partnership agreement in 1809. Bauer played an essential role in transforming Koenig's innovative ideas into functioning machines.

By the second decade of the nineteenth century, the disruptive social and economic impacts of the factory system were being felt increasingly by the working classes of England, causing widespread resentment against the replacement of traditional handcrafts by machinery. Most famously, while Koenig and Bauer were developing their machines in London, the Luddite machine-breaking riots against power looms in the textile industry occurred in the manufacturing districts of England in 1811 and 1812, with the trials and conviction of most of those accused taking place in 1812 and 1813. These trials reached a climax when, in 1817, Jeremiah Brandreth (tried for political issues), William Turner, and Isaac Ludlam, convicted for machine-breaking, were the last people sentenced to beheading by axe in England.

In 1783, several decades prior to Koenig's invention of the steam-powered rotary printing machine, Thomas Bell of Scotland had obtained patent no. 1378 for "Machinery for Printing Textile Fabrics." To print the fabric, Bell had invented a special-purpose cylinder press that could print up to five colors on "linnens." Regarding Bell's machine, Otto M. Lilien commented, "The patent drawing and description show a machine with all the essential parts of a present day industrial gravure machine, whether used for printing on paper, fabrics, or plastics, and explain in detail the ink duct, the cylindrical printing forme, the doctor blade for wiping the surplus ink from the cylinder and the impression roller."[53] Bell was undoubtedly inspired to develop a cylinder press for printing on textiles because fabric came off the power looms on rolls rather than in individual small sheets, the way that handmade paper was made, and was traditionally printed on a handpress. The demand for printed fabric from the rapidly growing mechanized textile industry was huge, and Bell's machine was rapidly adopted and improved.

In England, it was also understood that William Nicholson, a London chemist, translator, journalist, publisher, scientist, and inventor, had patented the general concept of the cylinder press for printing on paper and fabric in 1790; however, Nicholson never actually built such a press.[54] In his patent, Nicholson

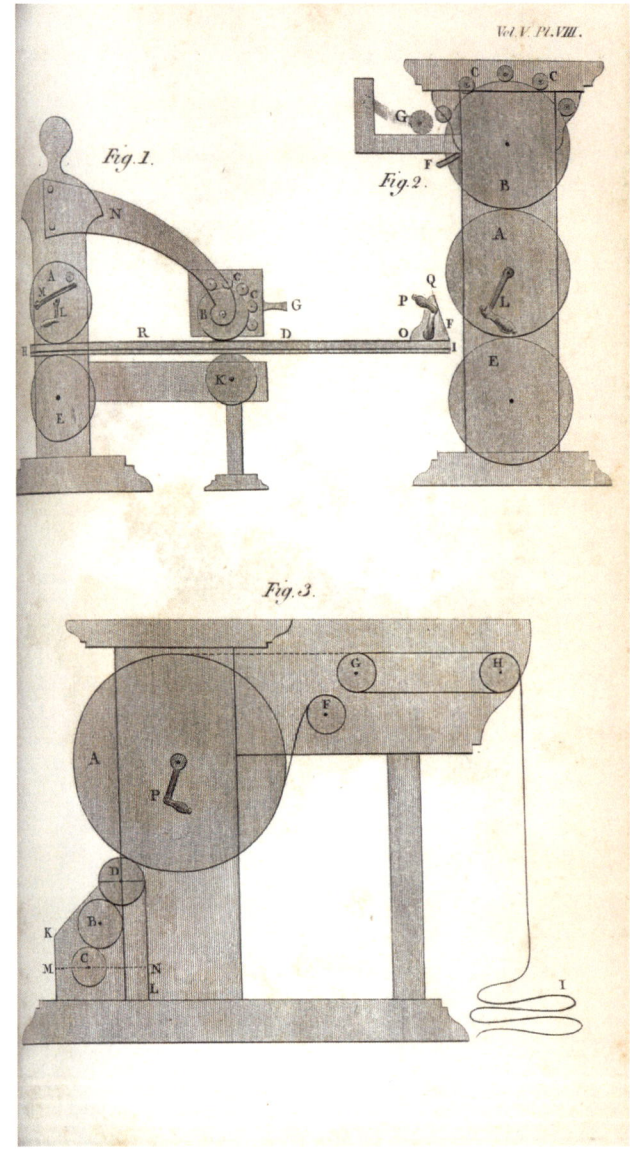

From *The Repertory of Arts and Manufactures*, vol. 5 (London, 1796)

An engraving illustrating William Nicholson's patent granted in 1790 for a rotary printing machine for printing on paper, linen, and other fabrics. Nicolson was the first to patent a design for a rotary printing machine for printing on paper, but it was understood that Nicholson never built an example of his machine.

made sketchy but prophetic proposals for printing with cylinders, and was also probably the first to refer to a printing device as a machine. "Nicholson's patent consisted of three parts. The first was for casting types in a multi-letter mould, so that 'two, three or more letters' could be cast at one pouring of the metal,

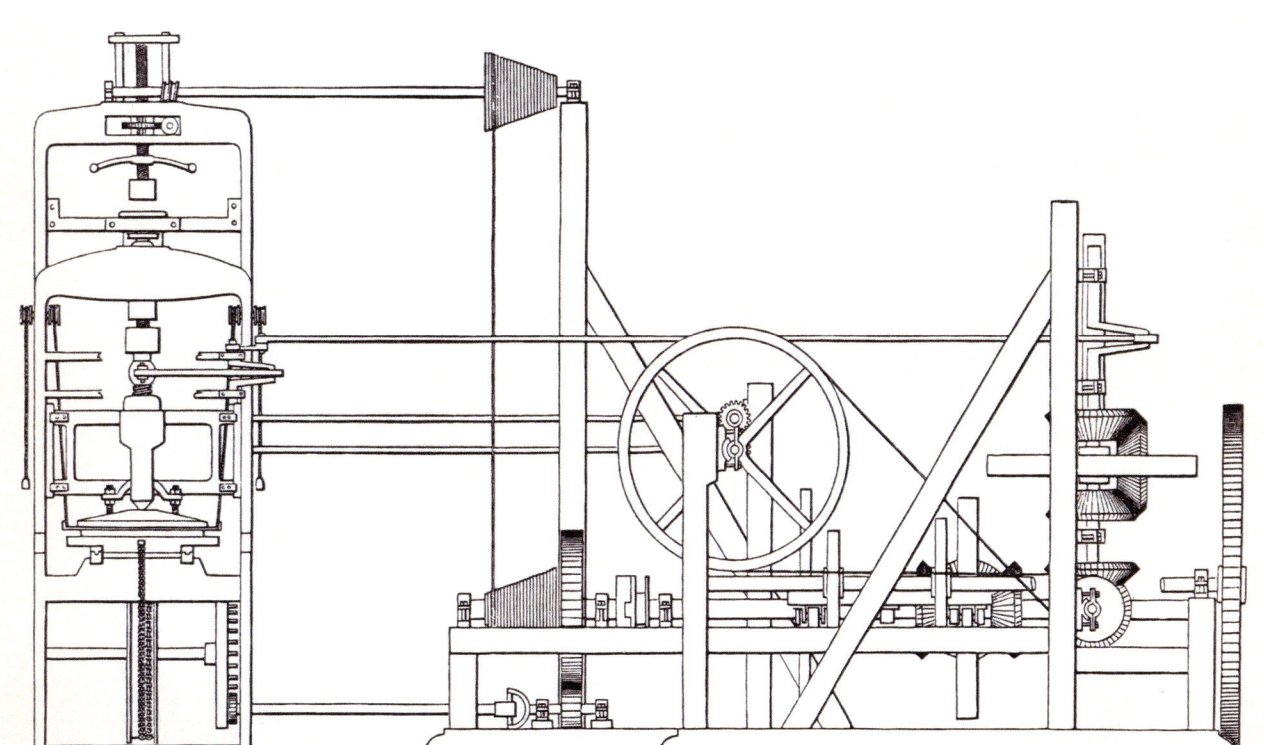

Patent 1810, Fig. II. Flachdruckmaschine. Machine with impression by a platen.

From *Die ersten Druckmaschinen. Erbaut in London bis zu dem Jahre 1818 von Friedrich Koenig und Andreas Friedrich Bauer* (The first printing machines. Constructed in London up to the year 1818 by Friedrich Koenig and Andreas Friedrich Bauer; Leipzig: Druck von F. M. Brockhaus auf einer Schnellpresse von Koenig und Bauer, 1851)

A diagram of Friedrich Koenig's first steam-powered printing machine, a powered platen press. In April 1811, 3,000 copies of sheet H (pp. 113-128) of *The New Annual Register . . . for 1810* were printed on this press—the first sheets ever printed on a printing machine not powered by hand.

but the resulting types were to be scraped into a shape so that they could be inserted around a cylinder. The second part called for cylinders covered with leather or cloth to distribute the ink. The third demanded that all printing was to be performed by passing paper or material to be printed between two cylinders, one of which 'has the block form, plate assemblance of types, or original, attached to or forming part of its surface'.[55]

Nicholson's patent was first published in 1796.[56]

At some point in his inventing process, Koenig might have benefited from reading Nicholson's patent, though after Koenig succeeded, Nicholson never asserted a prior claim to the invention. Apart from his printing machine patent, Nicholson published only a rather general article on the subject in volume 1 of his *Journal of Natural Philosophy, Chemistry, and the Arts* (1797),[57] titled "Observations on the Art of Printing Books and Piece Goods by the Action of Cylinders." This article discussed potential advantages of a cylinder press versus the centuries-old platen press and anticipated resistance from printers to such a radically new approach to printing.

With the support of Bensley, Woodfall, and Taylor, Koenig and Bauer were able to proceed with development of their first experimental printing machine. Koenig received his first British patent, no. 3321, on 29 March 1810, for "A method of printing by means of machinery," describing his powered platen press.[58]

In April 1811, at Richard Taylor and Co. Printers in

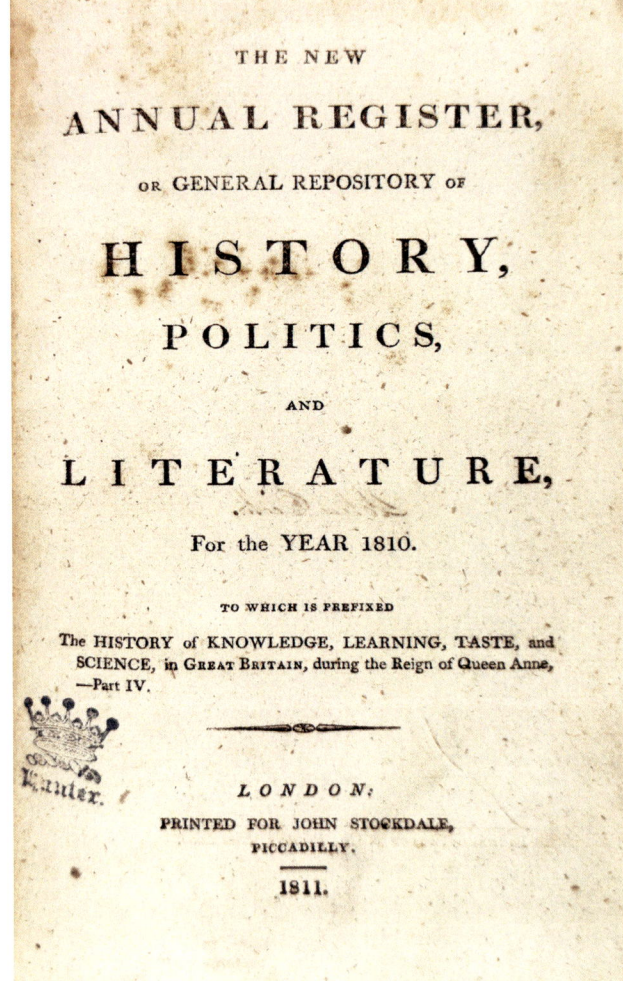

From *The New Annual Register, or General Repository of History, Politics, and Literature for the Year 1810* (London: John Stockdale, 1811)

Sheet H of this book was the first printed sheet printed by a printing press not powered by hand, as a test of Koenig's printing machine. There was more than one issue of this book. The correct issue, with sheet H printed by machine rather than a handpress, indicates on the verso of the title page that it was printed by Richard Taylor, Shoe Lane. Taylor, one of Friedrich Koenig's partners, financed development of the first printing machine.

Shoe Lane, London, Koenig conducted the first test of his steam-driven platen press, printing 3,000 copies of sheet H (pp. 113-128) of *The New Annual Register, or General Repository of History, Politics, and Literature for the Year 1810*. This was the first printing done by the first printing press not powered by hand, and, at the rate of 400 sheets per hour (according to some accounts),[59] it achieved more than double the speed possible with an iron handpress such as the Stanhope Press. Comparing the press work on sheet H with the sheets in the rest of the complete volume of the *Register for 1810* in my library, it is evident that the printing of this experimental sheet is different, and perhaps inferior. At any rate, Koenig and his partners did not consider the production gains from this steam-driven platen press sufficient, and Koening began developing a cylinder press.

Six months later, on 30 October 1811, Koenig received his second British patent, no. 3496, for "Further Improvements on My Method of Printing by Means of Machinery," describing the first cylinder flatbed press. This steam-driven machine, which was set to print 800 impressions per hour, still incorporated vestiges of the handpress.[60]

The first sheets printed by Koenig's cylinder flatbed press to be issued were sheets G (pp. 81–96) and X (pp. 305–320) of Thomas Clarkson's *Memoirs of the Private and Public Life of William Penn*, Vol. 1 (1813), printed by Koenig's sponsor, printer Richard Taylor, at his press in Shoe Lane, London. These were the first sheets ever printed on a totally cylindrical or rotary press, and "the results were in such an eminent degree satisfying, that the proprietor of the Times, Mr. J. Walter, when he saw it [the single cylinder machine] for the first time, and the [plan of the] double machine on the same [design] was explained to him, ordered after a few minutes two double machines for his own use."[61] A copy of this volume in the original boards, uncut, contains advertisements for Longman publications "Corrected to March 1813," suggesting that this book was issued in March or April of that year. Comparison of the inking between the gatherings printed on Koenig's machine and the remainder of the book printed on the handpress shows subtle differences.

Koenig's third British patent, no. 3725, for "Certain additional improvements in my method of printing by means of machinery," was issued on 23 July 1813; it

From *Die ersten Druckmaschinen. Erbaut in London bis zu dem Jahre 1818 von Friedrich Koenig und Andreas Friedrich Bauer* (The first printing machines. Constructed in London up to the year 1818 by Friedrich Koenig and Andreas Friedrich Bauer; Leipzig: Druck von F. M. Brockhaus auf einer Schnellpresse von Koenig und Bauer, 1851)

A diagram of Koenig's single-cylinder flatbed press, the technology of which was covered in his second and third patents of 1811 and 1813. This machine was first used to print sheets G and H of Clarkson, *Memoirs of the Private and Public Life of William Penn*, vol. 1. (London, 1813). Those were the first sheets ever printed on a cylinder press.

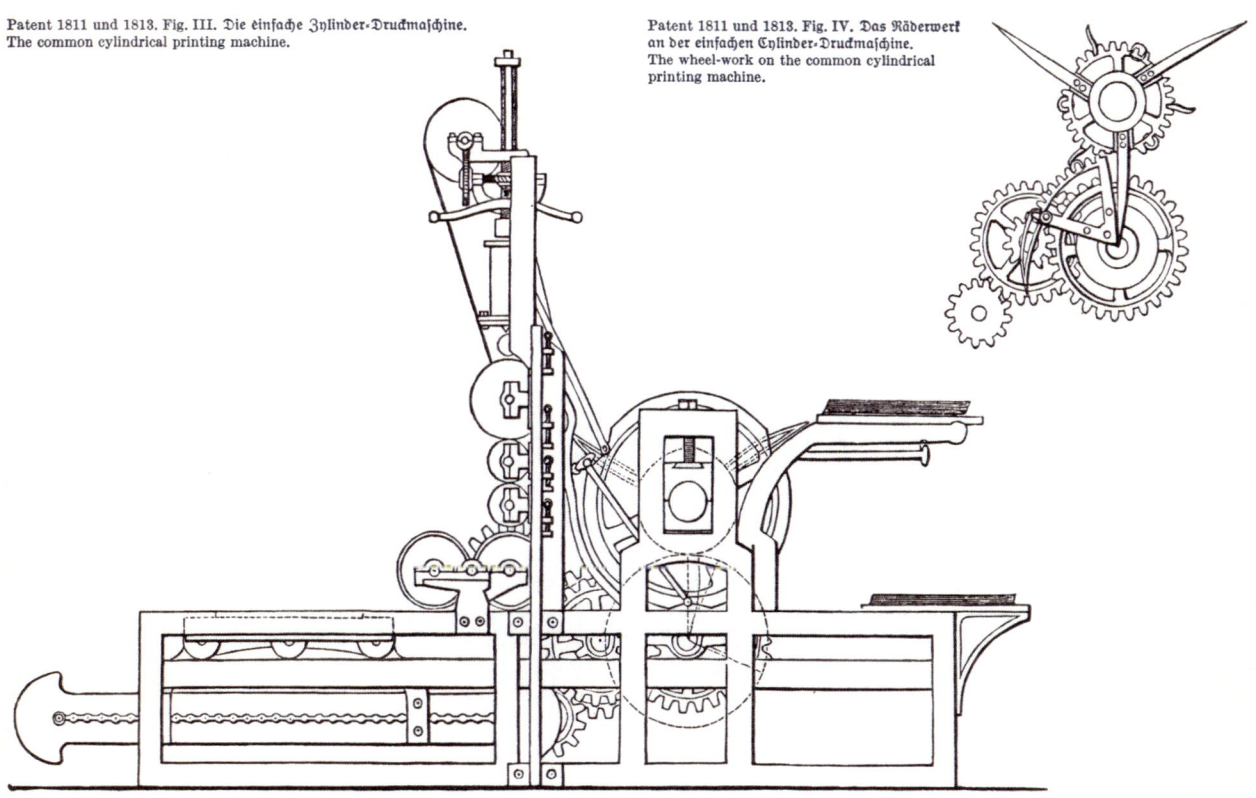

Douglas M. Parrish
Friedrich Koenig and the Cylindrical Press (c. 1960)
Oil on canvas, 36 × 27 in. (91 × 69 cm)
From Kimberly-Clark, *Graphic Communications Through the Ages*
Courtesy of RIT Cary Graphic Arts Collection

The machine partially shown in the painting is the double-cylinder machine used by John Walter II for printing *The Times* newspaper.

contained improvements on that of 1811 and served as the basis of the double machine. For this a second cylinder was added by which the return movement of the bed was made productive. While the printer cylinders were divided into three parts as before, each being covered with cloth with points attached, the "friskets" were abolished in favour of endless tapes conducted over rolls. The inking system underwent modification to meet the demands of double printing. The inking rollers were set transversely across the forme with their axles meeting on one side. In the patent the inking rollers were still described as covered in skin, but Koenig learned of the superiority of composition rollers during the year, otherwise The Times *machine could not have worked as effectively as it did.*[62]

Koenig's last English patent, no. 3868, for "Certain further improvements on my method of printing by means of machinery," granted on 24 December 1814, was the basis of an improved cylinder machine and of a perfecting machine—one that would print on both sides of a sheet of paper.[63]

Presumably, from the beginning of the development of the first printing machine, Koenig and his backers were aware that John Walter II, publisher of

From *Die ersten Druckmaschinen. Erbaut in London bis zu dem Jahre 1818 von Friedrich Koenig und Andreas Friedrich Bauer* (The first printing machines. Constructed in London up to the year 1818 by Friedrich Koenig and Andreas Friedrich Bauer; Leipzig, 1851)

A diagram of Koenig's double-cylinder flatbed press first used for printing *The Times* newspaper.

From the Office of *The Times*, *The History of the Times: "The Thunderer" in the Making 1785–1841* (London, 1935)

Portrait of John Walter II, publisher of the first newspaper printed by a printing machine.

The Times of London newspaper, was a potential customer for the new steam-powered press. According to the autobiography of writer and publisher Charles Knight, "For ten years Mr. Walter, the proprietor of 'The Times,' had been vainly endeavouring, at a heavy cost, to perfect some machinery by which he could send forth a greater number than the four thousand copies of his journal which he was able to produce by the utmost exertion of manual labour."[64]

John Walter II was faced with the limitations of having the newspaper printed on handpresses that could not be operated faster than 250 impressions per hour. At this time, a newspaper typically contained four pages: advertisements appeared on the first and last pages, which were printed together and consti-

Koenig's double-cylinder steam-powered printing machine, first used to print *The Times* on 29 November 1814, shown with that issue of the newspaper

Courtesy of Koenig and Bauer AG

tuted the "outer forme," while the news was printed on the inside spread, or "inner forme." Because the outer forme did not have the same time or "news" factor, it was printed ahead of the news pages, but there was always the problem of printing the 4,000 or more inside spreads within the limited time available before the newspaper had to be published. This problem was only partially solved if the inner forme was set up in type in duplicate so that it could be printed on two presses, especially since duplicate typesetting was an extra cost, limiting profitability of what was then only a marginally profitable business.

To surpass this production limit and print more than 4,000 copies of *The Times* newspaper in time for its distribution deadline, Koenig's double-cylinder, steam-powered printing machine was first put to practical use printing the 29 November 1814 issue of *The Times*. The output of the new machine was initially 1,100 sheets an hour—more than four times faster than the handpresses previously used by the newspaper.

Most published accounts of this very significant event in the history of printing and publishing describe it simply as a fait accompli. However, when steam power was first applied to printing at *The Times*, there were very strong feelings in the minds of working people against the introduction of machinery. Memories of the Luddite machine-breaking in the textile industry during 1811 and 1812, and the resulting convictions and executions in 1812 and 1813, were very fresh. It was felt, with much justification, that machinery mainly benefited the capitalist and deprived the working man of his right to labor. Along with other handcraftsmen, pressmen naturally shared this fear and hatred of mechanical innovations. Recognizing that threat, John Walter II secretly had Friedrich Koenig's printing machine set up in a separate building adjoining the office of *The Times* for fear that his pressmen, like Luddites, might smash the offensive machinery.

To maintain secrecy during the installation and first operation of the printing machine, Andreas Bauer bound his men with a £100 bond to divulge nothing. On the evening of 29 November, while the issue was being printed on Koenig and Bauer's machine, Walter told his staff that the presses had to be held for important news expected from the Continent. Walter subsequently avoided a confrontation with his workforce by threatening to meet objections to the new technology with the full weight of the law against combinations (unions) and also by offering displaced staff full wages until they found work elsewhere. Not that Walter was always diplomatic with his workers. In 1810, Walter had ruthlessly broken a union in *The Times* chapel (i.e., a compositors' union) when twenty-eight of his compositors walked out. One of the men Walter had put in prison died there.[65]

Extremely proud of printing the first issue of a newspaper by machine, John Walter II included this statement in the 29 November issue:

Our Journal of this day presents to the public the practical result of the greatest improvement connected with printing, since the discovery of the art itself. The reader of this paragraph now holds in his hand, one of the many thousand impressions of The Times *newspaper, which were taken off last night by a mechanical apparatus. A system of machinery almost organic has been devised and arranged, which, while it relieves the human frame of its most laborious efforts in printing, far exceeds all human powers in rapidity and dispatch. That the magnitude of the invention may be justly appreciated by its effects, we shall inform the public, that after the letters are placed by the compositors, and enclosed in what is called the form, little more remains for man to do, than to attend upon, and watch this unconscious agent in its operations. The machine is then merely supplied with paper; itself places the form, inks it, adjusts the paper to the form newly inked, stamps the sheet, gives it forth to the hands of the attendant, at the same time withdrawing the form for a fresh coat of ink, which itself again distributes, to meet the ensuing sheet now advancing for impression; and the whole of these complicated acts is performed with such a velocity and simultaneousness of movement, that no less than eleven hundred sheets are impressed in one hour.*

On 3 December 1814, five days after this historic achievement, *The Times* published a defense of their use of the new technology against the inevitable criticism that they faced for its introduction. Among the speed advantages from printing with the new machine that John Walter II mentioned was *The Times*' ability to report of the debates in Parliament on the same evening that they occurred. He also warned pressmen that he employed against interfering with the introduction of the new technology and pointed

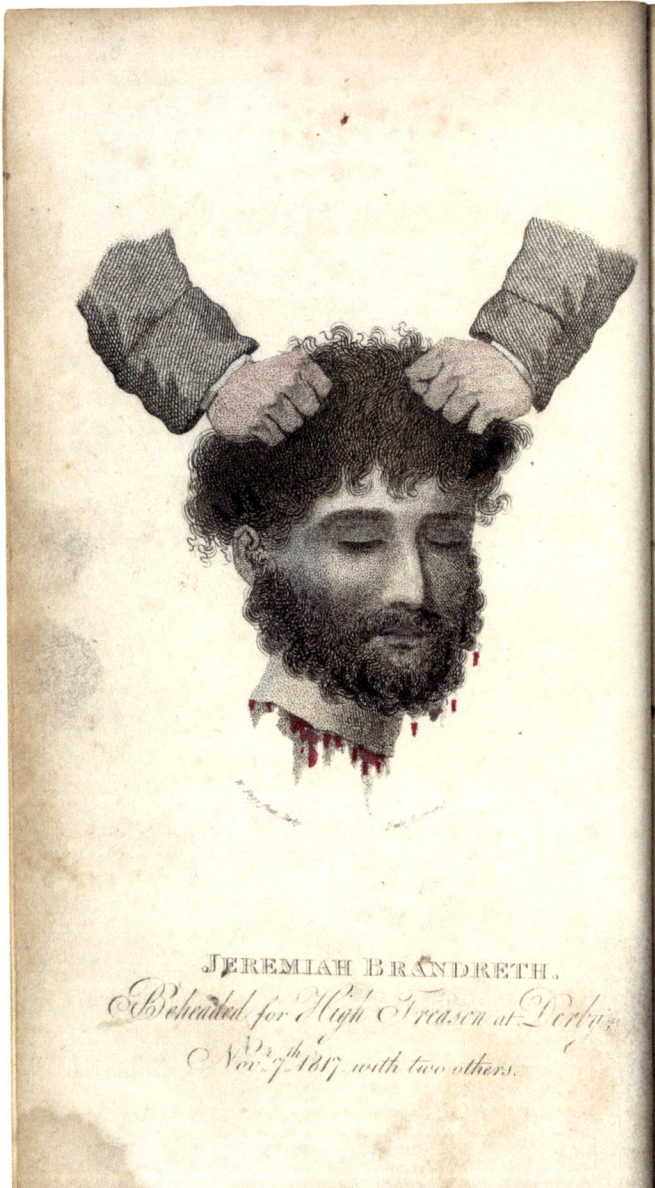

JEREMIAH BRANDRETH.
Beheaded for High Treason at Derby, Nov. 7th 1817, with two others.

Report
OF THE WHOLE OF THE
PROCEEDINGS,
UNDER THE
Special Commission,
HELD IN THE COUNTY HALL, AT DERBY,
In the Month of October, 1817,
INCLUDING
THE TRIALS
OF
JEREMIAH BRANDRETH,
ALIAS JOHN COKE,
Alias the Nottingham Captain;

WILLIAM TURNER,
ISAAC LUDLAM, THE ELDER,
AND
GEORGE WEIGHTMAN,
FOR
HIGH TREASON;
WITH THE
Speeches of the Counsel,
AND OTHER
INTERESTING PARTICULARS.

NOTTINGHAM:
PRINTED BY SUTTON AND SON,
Bridlesmith Gate.

---

100 Guineas Reward.

WHEREAS between the Hours of *eight* and *ten o'Clock* on Saturday Night last, the 31st of March, the

Machine Engine,

at the Street Pit, near the Long Bank, in the Parish of Lamesley, and County of Durham, the Property of the Executors of the late William Burdon, Esq. was maliciously set on Fire, by some evil disposed Person or Persons, when the House and a great Part of the Machinery was consumed and the other Part much injured.

Whoever will give such Information of the Offender or Offenders, so as he, she or they may be brought to Justice, shall, on Conviction, receive a Reward of TEN POUNDS from the Treasurer of the Lamesley Association, and a further Reward of NINETY FIVE POUNDS from the Executors of the late William Burdon, Esq.

Lamesley, April 2d, 1821.

Akenhead, Printers, Newcastle.

👆 *Report of the Whole of the Proceedings . . . in the Month of October 1817, Including the Trials of Jeremiah Brandreth, Alias John Coke, Alias the Nottingham Captain . . . for High Treason . . .* (Nottingham, 1817)

Throughout the eighteenth and early-nineteenth centuries, Luddite behavior was prosecuted very aggressively and punished severely. The hand-colored frontispiece of the proceedings of the trial for political activities of the Luddite Jeremiah Brandreth shows Brandreth's severed head held up by the executioner. This was the last punishment by beheading in England. Along with two other Luddites, Brandish was beheaded for high treason on 7 November 1817.

👈 "100 Guineas Reward" (broadside)
7 × 8¼ in. (18 × 21 cm)

The posting of a large reward in April 1821 for the capture of the Luddite or Luddites who damaged the steam-powered, pumping "machine engine" at the "street pit" of a mine owned by English academic, mine owner, and writer William Burdon, who died in 1819.

In the detail from an interior page of the issue, the publisher John Walter II pointed out the significance of this historic achievement.

out that, as a result of the new technology, he was employing only eight fewer men than previously.

In the *Times* issue for 9 December 1814, the inventor Friedrich Koenig published an article in which he personally recounted the early history of the

From the Office of *The Times*, *The History of the Times: "The Thunderer" in the Making 1785–1841* (London, 1935)

Receipt, dated 20 November 1816, acknowledging full payment of 2,800 pounds from John Walter II for the printing machines used to print *The Times*, signed by the partners printer Thomas Bensley, inventor Friedrich Koenig, and printer Richard Taylor.

steam-powered press's development, stating that he had the idea for the machine in Germany eleven years previously, explaining the impossibility of developing the machine in Germany, and its eventual successful development in England sponsored by Thomas Bensley, George Woodfall, and Richard Taylor. Among the reasons Koenig gave for the impossibility of developing the machine in Germany at the time was the lack in Germany of an effective system of patents.

Two years later, in 1816, Koenig added a perfector to *The Times* of London's steam-powered press, allowing the press to print almost as many copies on both sides of the sheet on one pass through the press as had been previously printed on one side only. By 1818, Koenig's steam-powered press had achieved an output of 2,400 impressions per hour.

Besides inventing the printing machine, Koenig also revolutionized the inking process. Prior to the development of the printing machine, type had been inked by hand, using leather balls that needed frequent cleaning and renewal. In the printing machine, the ink was held in a central reservoir, and after preparation between a system of rollers, it was passed over the surface of the type metal by a final roller. This eliminated the laborious process of hand inking.

Because of financial disputes with their partner, Thomas Bensley, Koenig, and Bauer left England in 1817 and returned to Germany, where they established the company known today as Koenig and Bauer at a secularized monastery at Oberzell, near Würzburg, Bavaria. Their working conditions in pre-industrial Germany were extremely difficult: tools, manufacturing machinery, pig iron, and coal all had to be imported from England and took months to arrive. Their greatest problem was finding suitable skilled workers. For the first few years, Koenig and Bauer had to train iron- and steelworkers to perform printing-machine factory tasks. Koenig understood what he was up against in this start-up: in a letter to John Walter II, Koenig wrote, "That's what happens when one attempts something beyond the cultural level of a country."[66] In Germany, adoption of printing machines occurred more slowly than in England, and even than in France. As late as the 1860s, there were attempts in Germany at imitating the industrialization that had taken place elsewhere in Europe. This imitation was only moderately successful. Only in 1871, when the modern German nation was created by Otto von Bismarck, were major industries founded that led to the full-fledged industrialization of Germany.

The first German newspaper to mechanize its printing was the *Haude und Spenersche Zeitung* in its eleventh issue, published on 25 January 1823.[67] This was the first publication printed in Germany by a printing machine made in Germany. The second was Johann

From *Die ersten Druckmaschinen. Erbaut in London bis zu dem Jahre 1818 von Friedrich Koenig und Andreas Friedrich Bauer* (The first printing machines. Constructed in London up to the year 1818 by Friedrich Koenig and Andreas Friedrich Bauer; Leipzig, 1851)

A diagram of the perfecting machine that Koenig and Bauer completed for Thomas Bensley in February 1816, before they departed London for Oberzell, Germany, in 1817. This was the first printing machine specifically designed to print books rather than newspapers. Koenig's machine was highly complex, with sixty wheels; in 1817, English printing engineer Edward Cowper accomplished the same result with an improved and greatly simplified machine that had sixteen wheels. Cowper's machine superseded the use of Koenig's machine in England and on the Continent.

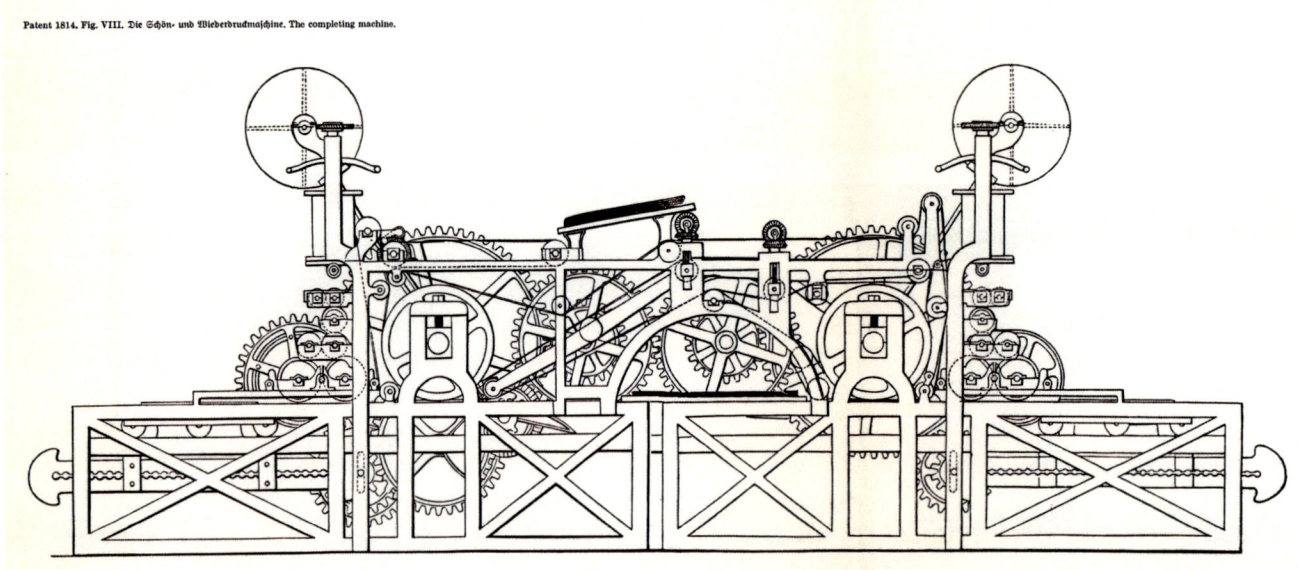

Friedrich Cotta's *Allgemeine Zeitung* in Augsburg. On 3 July 1824, Koenig did a test printing of the *Allgemeine Zeitung*. Nine days later, for the 12 July 1824 edition, Koenig was able to print 3,000 copies of the newspaper using his *Schnellpresse* (high-speed press); however, the steam engine driving the machine failed, and the press had to be driven by hand. With this awkward beginning, Koenig introduced high-speed printing to Germany, roughly ten years after he introduced it in England.

Anticipating that the young company would rapidly saturate the very limited market in Germany for high-speed printing machines, Koenig set up a demonstration machine in Paris. In 1825, his company delivered its first export order to the director of the Danish post office in Copenhagen; orders followed to Switzerland, the Netherlands, Spain, and France. An article describing the history and development of Koenig's *Schnellpresse* appeared in the *Allgemeine Zeitung* for 21 August 1825; this was probably the first detailed article on mechanized printing published in Germany. An additional article appeared in the 3 February 1827 issue. In July 1827, Koenig and Bauer published a list of the *Druckmaschinen oder Schnellpressen* (printing machines or fast presses) they had designed and manufactured, from the first machine for Mr. Bensley in London in 1813 to the most recent for Mr. Hostrup in Hamburg. Reflective of the slow development and limited availability of steam engines in Germany, most of the presses they sold between 1825 and 1827 were driven by hand cranks rather than steam engines.

In 1828, King Ludwig I of Bavaria granted Koenig and Bauer a special privilege guaranteeing Friedrich Koenig the industrial processing rights to his inventions for a period of ten years. The German government also granted tax, customs, and toll concessions, together with other benefits. In the same year, Koenig and Bauer established a paper mill as a side business in another former monastery in Bavaria. Two years later, when business was growing, the 1830 July Revolution in Paris brought export business to France, as well as domestic orders, to a sudden halt. The paper mill became the young company's sole source of income, and its workforce was cut back from 120 to 14. Three years later, on 17 January 1833, Friedrich Koe-

From *The Literary Gazette and Journal of the Belles Lettres, Arts, Politics, Etc.*, no. 50 (3 January 1818)
Image from HathiTrust

Below their masthead, the publishers confirmed that this was the first weekly magazine to be printed by steam power.

nig suffered a stroke and died before his fifty-fourth birthday. At Koenig's death, his partner Andreas Bauer became the company's managing director.[68]

Before Koenig and Bauer left London, they completed a perfecting machine for Thomas Bensley. This was the first printing machine specifically designed for printing books rather than newspapers. Bensley's machine was used to print the 3 January 1818 issue of *The Literary Gazette and Journal of the Belles Lettres,*

Johan Friedrich Blumenbach, *The Institutions of Physiology.* Translated . . . by John Elliotson, 2nd ed. (London, 1817) Images from Google Books

Printed on the perfecting machine that Koenig built for Thomas Bensley, this was the first complete book printed on a printing machine rather than a handpress. The printers confirmed the achievement in the postscript on p. iii.

Arts, Politics, Etc.—the first weekly magazine to be produced with the new technology. The publishers featured this achievement in the top center column of the first page of the issue, appearing to credit Bensley, rather than Koenig and Bauer, as the press's inventor:

It may also be interesting to our readers to know, that, commencing with the present Number, this Journal will be printed by Messrs. Bensleys' Patent Machine, an inventive improvement in the art of Printing which reflects honour on the present age, and exhibits a proof of the progress of the art of ingenious mechanism in this country. In this respect our Journal will enjoy an advantage over all other weekly papers, being the first ever printed by a steam-engine, and we shall thus be enabled to published at a very early hour on the Saturday morning.[69]

In the 10 January 1818 issue, the editors of the journal included an article on "The Patent Completing Machine":

In our last number, we mentioned, that the Literary Gazette *was the only Journal in the world printed by this most admirable machine; and as a matter of extraordinary mechanical interest we subjoin a brief account of the process by which about a thousand of these large sheets are per hour produced by this magical invention. The beauty of the movements, their rapidity, their precision are enhanced to the imagination by the nature of the operation they perform: it looks as if* mind *and not* matter *were at work.*[70]

Bensley first used the new machine to print a complete book in 1817. This was Johann Friedrich Blumenbach's *The Institutions of Physiology*, translated by John Elliotson and printed in London by Bensley and Son. At the end of Elliotson's preface, there is the following postscript, which shows that the printers regarded the book as a "typographical curiosity" rather than a significant innovation in book production:

P.S. The volume may be considered a typographical curiosity, being the first book ever printed by machinery. It is executed by Messrs. Bensley and Son's patent machine, which prints both sides of the sheet by one operation, at the rate of 900 an hour, and is the only one of the kind ever constructed.[71]

Virgil, Opera ... (London, 1822)

This edition of works of Virgil, printed in 1822 by Thomas Bensley's son, Benjamin, on Bensley's perfecting machine, was the first book to state specifically that it was printed by steam power, though the statement was made in Latin on the verso of the title page: *Vaporante machinâ excudebat*.

According to publisher Charles Knight, in 1824 there were only twenty steam engines in London set up to power printing machines.[72] In 1822, Thomas Bensley's son Benjamin Bensley used the completing machine built by Koenig and Bauer to print the first book to state specifically that it was printed by steam power. This was *P. Virgilii Maronis Opera*, an edition of Virgil's works edited by Charles de la Rue, which was published in London in 1822.[73] On the verso of the title page is the Latin phrase "*Vaporante machinâ excudebat* [printed by steam power], B. Bensley, Bolt Court, Fleet Street." Though Thomas Bensley or his son Benjamin may have used the steam press to print other books between the 1817 edition of Blumenbach and the 1822 Virgil's *Opera*, the latter work, printed entirely in Latin, was the first book to say so in print (it is possible that Benjamin Bensley added the notation just for the fun of it). The 1822 *Opera* may have been a widely adopted textbook at the time, since its imprint indicates that it was distributed by more than ten London publishers, implying a relatively large edition. Other books may have been printed on rotary presses in England by Bensley or others around this time, but they have not been identified.

Across the Atlantic, also in 1822, an edition of the New Testament was printed by Daniel Treadwell's Power Press, the first power press invented in North America. Since steam engines were scarce in America at this date, the power to print this edition on Treadwell's press was provided by horses. Treadwell had the habit of identifying books printed on his press within the books themselves, so several are known to have been issued between 1822 and 1826. Treadwell and his power press will be discussed in chapter 9.

Henry Heath, *The Pleasures of the Rail-Road—Showing the Inconvenience of a Blow Up* (London, 1831)
Hand-colored engraving, 9⅔ × 13¼ in. (24.5 × 33.5 cm)

Early steam engines powering railroads posed a real risk of exploding.

👆 Goldsworthy Gurney
The New Steam Carriage (London, 1828)
11¾ × 16¾ in. (30 × 42.5 cm)

In 1828, Gurney invented a steam-powered passenger carriage. The unusually long caption to this print explained how it operated.

👆 Payall after a drawing by G. Morton
A Steam Coach with Some of the Machinery Going Wrong
Hand-colored engraving, 8 × 11¾ in. (20 × 30 cm)

The risk of sitting next to a potentially explosive steam engine on a steam carriage like the one Goldsworthy Gurney invented must have been apparent to potential passengers.

The Effects of the Rail Road on the Brute Creation (c. 1830)
Printed textile, 23¼ × 19½ in. (59 × 49.5 cm)

By around 1830, "The Iron Horse" (the railroad) was starting to make actual horses obsolete as a mode of transportation.

CHAPTER 4

Railroads, Power Looms, and Bibles: Innovation Versus Tradition

In early-nineteenth-century England, the development of railroads and the expansion of the factory system contributed to an environment of rapid technological and socio-economic change. These changes, which were not always welcomed by everyone, were often depicted in prints and caricatures of the time, some of which are shown in this book. While society was undergoing many visible and social changes, the largest-selling book in England, the Bible, continued to be produced in very large editions by traditional handpress methods. What we might conclude from this dichotomy is that innovation in the second printing revolution tended to be applied where it was necessary and profitable. Initially, the demand for printing machines came from the printers of newspapers and large circulation magazines. In England, the efficiencies of steam-powered presses were not initially applied to printing Bibles, because the presses at Oxford and Cambridge Universities held a monopoly over most Bible printing in the United Kingdom, and while both presses were relatively early adopters of stereotyping for Bible production, until about 1840 both were able to fill their Bible orders by handpress.

During the first decade of the nineteenth century, the first steam-powered railroads were built in England to haul coal or textiles. In the 1830s, railroads began carrying people as well as freight, necessitating the creation of new publications, such as railroad guides and train schedules. Railroads were also used to distribute the new mass media of the second printing revolution, such as the Society for the Diffusion of Useful Knowledge's *The Penny Magazine*. Later, in 1848, W. H. Smith opened his first newsstand in a train station to sell books, newspapers, magazines, and other printed materials to railroad passengers.

On 21 February 1804, the first full-scale working railroad steam locomotive, built by British engineer Richard Trevithick, was used to haul a train along the tramway of the Penydarren Ironworks in South Wales. In 1822, George Stephenson's Hetton Colliery Railway, an 8-mile (13-km) private railway, began operation for the Hetton Coal Company at Hetton-le-Hole, County Durham, England. The Hetton railway was the first railway designed to be operated without animal power—that is, without carriages pulled by horses along rails. It ran between the Hetton colliery and a wharf along the River Wear, from which the coal was conveyed further by boat. On 27 September 1825, British engineer George Stephenson's Locomotion No. 1 pulled its first train on the 8-mile Stockton and Darlington Railway (S&DR). This was the first publicly subscribed passenger railroad, though it did not technically carry passengers under steam until 1833. The first line of the Stockton and Darlington Railway connected coal mines at Darlington to Stockton on the River Tees. From the beginning of its operation, the S&DR used steam locomotives to haul coal wagons, but until 1833, it carried passengers in coaches drawn by horses.

The success of the S&DR supported plans for con-

James Scott Walker
An Accurate Description of the Liverpool and Manchester Rail-Way (Liverpool, 1831)

The frontispiece shows the Liverpool and Manchester Rail-Way pulling textiles and other freight from the mills in Manchester to the port at Liverpool. This early railroad only carried freight before it added passenger service in 1830. This book represented an entirely new form of literature: the railroad guide.

James Scott Walker
An Accurate Description of the Liverpool and Manchester Rail-Way (Liverpool, 1831)

The L & M Rail-Way pulling passenger carriages through the Olive Mount excavation, a 2-mile (3.2-km) sandstone railway cutting. Early passenger railroad cars were carriages, similar to horse-drawn carriages, that were designed to roll on railroad tracks.

struction of a railroad between the port of Liverpool and the booming textile factories in Manchester. The Liverpool and Manchester Railway (L&MR), which opened for business on 15 September 1830 after seven years of construction, was the first railroad to rely exclusively on steam-powered locomotives, with no horse-drawn traffic permitted. It was also the first entirely double-track, inter-urban railroad, the first railroad with a signaling system, the first to follow timetables, and the first to carry mail. It was also one of first railroads to carry passengers on a regularly scheduled basis, and the first to provide tickets for all train travel. The average speed on these railroads was 12 miles per hour, but trains could reach a top speed of 30 miles per hour—a speed at which people had never traveled before.

The popular interest resulting from this exciting innovation was reflected in many publications of all kinds about the railroads, including a whole new genre of publishing—railroad travel guides. Between 1830 and 1831, three editions of James Scott Walker's *An Accurate Description of the Liverpool and Manchester Rail-Way* were issued in Liverpool by J. F. Cannell. This pamphlet contains two folding engravings of the train in operation, the train schedule and list of fares, and details of sights along the way. My copy of the third edition, published in 1831, has an inserted slip updating the schedule and list of fares. Walker's may be the earliest guide to the first passenger railroad to include images of the railroad in operation.

Along with railroad travel guides, the success of the Liverpool and Manchester Railroad led to another entirely new class of publications: railroad timetables. On 19 October 1839, English cartographer, printer, and publisher George Bradshaw of Manchester issued a very small clothbound book titled *Bradshaw's Railway Time Tables and Assistant to Railway Travelling*,

George Bradshaw, *Bradshaw's Railway Time Tables . . .*
(London, 19 October 1839)
4¾ × 3 in. (12 × 8 cm)

This very small volume was the first railroad timetable or railroad schedule. It was the first of a new class of publications. In England, these guides became known simply as "Bradshaws" until they ceased publication in 1961.

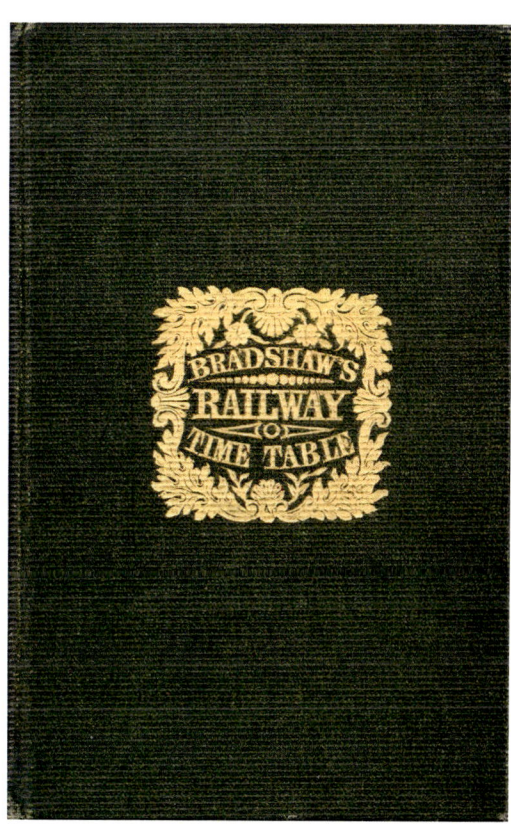

which sold for sixpence. This completely new class of publication contained route maps and train schedules for the new Liverpool and Manchester Railway. One week later, on 25 October 1839, Bradshaw issued a similar timetable, priced at 1 shilling, for the London and Birmingham Railway—the first intercity line to connect to London. The first railway timetables were necessary tools for travelers in the new and confusing railroad networks. The original Bradshaw timetables were published before the limited introduction of standardized railway time in November 1840, and its subsequent development into standard time. Bradshaw thus needed to provide an explanation of the time then employed by British railroads:

> *NOTE. LONDON TIME is kept at all the stations the Railway, which is 4 minutes earlier than Reading time; 7½ minutes before Cirencester time; 11 minutes before Bath and Bristol time; and 18 minutes before Exeter time.*

Beginning in the 1830s, when the first public railroads were established in England, it became possible to distribute books, pamphlets, newspapers, and mail over wider distances in less time. Railroad transportation and railroad stations, through which large numbers of people passed, provided a whole new market for printing, publishing, and bookselling. Inexpensive novels, or yellowbacks, sometimes also called railroad novels, were published at low cost to supply a wider range of society. As railroads developed, magazines such as *The Penny Magazine* relied on railroads to distribute their copies to subscribers over long distances throughout England. However, about twenty years would elapse before bookseller W. H. Smith recognized the strategic value of train stations as places to sell books and other printed matter. On 1 November 1848, the first W. H. Smith railway bookstall opened in London's Euston Station.

As the century advanced, publishers in the United States also exploited the developing railroad market in that country with series such as *Putnam's Semi-Monthly Library for Travellers and the Fireside* and *Appleton's Railway and Steam Navigation Guide*.

At the same time that railroads were developing in England, dramatic improvements and expansions were occurring in the textile industry. In 1822, inventor Richard Roberts, from the village of Llanymynech

Three yellowbacks of the type sold in train stations.

in Wales, received British patent no. 4726 for "machinery applicable to the process of weaving plain or figured cloths or fabrics, and which may be applied to looms now in common use; construction of looms for such purpose; working looms either by hand, steam, or other power." This loom, which became known as the Roberts loom, was made of cast iron, and could operate at significantly higher speed than the traditional wooden power loom invented by Edmund Cartwright. The Roberts loom was eventually installed in most British textile factories during the Industrial Revolution. However, the transition from handlooms to power looms was by no means sudden; in 1836, a quarter of a million handlooms were still in operation for cotton weaving in England.[74] In 1825 and 1830, Roberts also patented the self-acting or automatic spinning mule, thus eventually revolutionizing production in both the spinning and weaving industries. By 1850, there were 260,000 power looms, primarily of the Roberts type, in operation in England.

The Roberts power looms, as installed in Thomas Robinson's factory in Stockport, were illustrated in a folding plate in Andrew Ure's *The Philosophy of Manufactures*, published by Charles Knight in 1835. Ure's book was one of the most informed surveys of the developing factory system in England. Regarding the scale of employment in textile factories in the United Kingdom of Great Britain and Ireland at that time, Ure published a "Statistical Table of the Textile Factories of the United Kingdom, subject to the Factories Regulation Act, in which the machinery is worked by Mechanical Power." At this date, Ure apparently felt it was worthwhile to distinguish between the larger factories driven by steam engines or waterpower and some smaller factories that might have continued to be driven by horses or humans turning cranks.

One of the early commentators on textile factory life was Harriet Martineau, who published what has been called the first factory novel, *A Manchester Strike*.[75] Martineau did not advocate factory workers striking against factory owners. She regarded this novel as an exploration of the economic challenges of working in factories, and intended the novel, which was printed in a small, inexpensive format, to educate working-class people.

Contemporaneous with Martineau, the Manchester surgeon Peter Gaskell published two critiques of

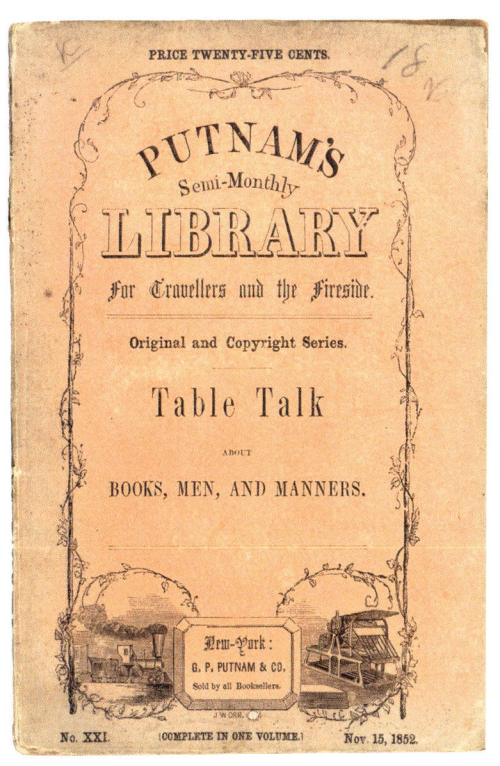

Putnam's Semi-Monthly Library for Travelers and the Fireside (New York, 1852) and *Appleton's Railway and Steam Navigation Guide* (New York, 1869)

Examples of publications directed toward railroad and steamship passengers in mid-nineteenth-century America. Note the printing machine illustrated at the foot of the cover of *Putnam's Semi-Monthly Library*.

❞❞ From Henry Sampson, *A History of Advertising* (London, 1875)

By 1874, printed posters plastered the walls of train stations. A W. H. Smith bookstall is shown in the lower right corner of the image.

MODERN ADVERTISING: A

RAILWAY STATION IN 1874.

the effects of the factory system on the working people involved, *The Manufacturing Population of England, its Moral, Social and Physical Conditions Which Have Arisen from the Use of Steam Machinery . . .* (1833) and *Artisans and Machinery: The Moral and Physical Condition of the Manufacturing Population Considered with Reference to Mechanical Substitutes for Human Labor* (1836). In the second book, Gaskell referred to the Roberts power looms as automata, or Iron Men, observing that the power looms enabled girls and boys to accomplish what used to be the work of men, causing men who traditionally worked in the textile mills to become unemployed. He explains:

The time, indeed, appears rapidly approaching, when the people, emphatically so called, and which have hitherto been considered the sinews of a nation's strength, will be even worse than useless; when the manufactories will be filled with machinery, impelled by steam, so admirably constructed as to perform nearly all the processes required in them; and when land will be tilled by the same means. Neither are those visionary anticipations; and these include but a fraction of the mighty alterations to which the next century will give birth.[76]

At around the same time as the development of power looms, the British and Foreign Bible Society was formed, in March 1804, with the intention of selling Bibles at affordable prices in many languages throughout the world. Within a very a few years, the society became one of the largest book publishers in England, and the largest customer of the Cambridge and Oxford University Presses—the two presses, which along with the King's Printer, held the "privilege" for printing the authorized edition of the Bible in England and Wales. One of the early goals of the society was to distribute affordable Bibles in Welsh to Welsh-speaking Christians, but the first book that the society published, in 1804, was a translation of the Gospel of John into Mohawk. The society was a success from its beginning: *The First Report of the British*

From Andrew Ure, *The Philosophy of Manufactures* (London, 1835)

The frontispiece shows hundreds of Roberts power looms in operation at Thomas Robinson's "loom factory" in Stockport, a town in Greater Manchester, England. The looms and the elaborate flat belt system driving them dwarfed the humans operating the machinery.

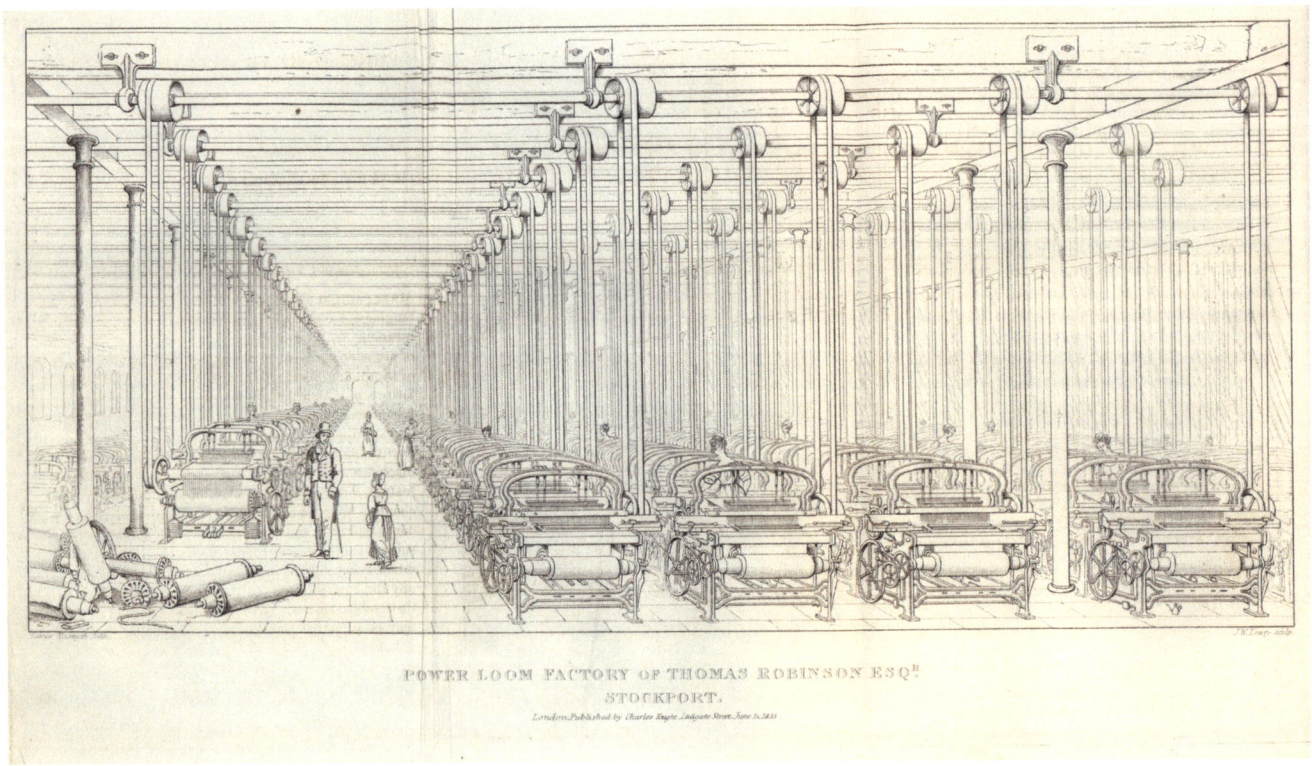

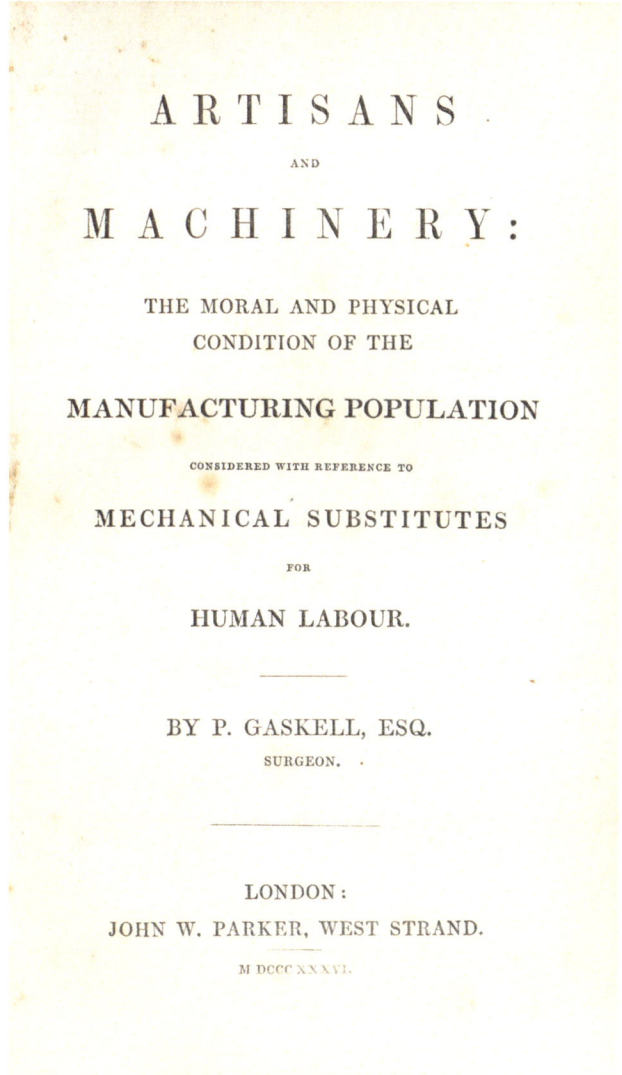

Peter Gaskell, *Artisans and Machinery* ... (London, 1838)

As often happened during the Industrial Revolution, innovation caused social and economic disruption. In his book, Manchester surgeon Peter Gaskell referred to the Roberts power looms, such as those in Robinson's factory, as automata, or "Iron Men," observing that the power looms enabled girls and boys to accomplish what was formerly the work of men. Mechanization of the textile mills undoubtedly caused some men who traditionally worked in the industry to become unemployed.

and Foreign Bible Society for 1805 required 47 pages to list the names of its many hundreds of donors.

In 1805, Cambridge University Press issued the Cambridge Stereotype Edition of the Bible for the British and Foreign Bible Society, using the improved Stanhope stereotyping process promoted by Stanhope's printer, Andrew Wilson. This was the first edition of the Bible ever printed from stereotype plates. The copy illustrated is of the 1806 printing; its calf binding bears the circular blind stamp of the British and Foreign Bible Society on the upper cover. Characteristically the British and Foreign Bible Society employed blind stamps rather than the more conspicuous and possibly more expensive stamping of their name in gilt on the upper cover of bindings.

In 1807, Andrew Wilson produced for the society 7,000 copies of a pocket-sized edition of the New Testament in French for distribution to some of the 100,000 French prisoners of war imprisoned in England during the Napoleonic Wars. This is confirmed in the society's Fourth Report for 1808. Why they decided to publish and donate 7,000 copies, rather than more or fewer, is unknown, but presumably they took into account the limited percentage of prisoners who would have been literate, and the expectation that copies would be shared.

The paper on which my copy was printed is slightly browned, indicating that the society paid for inexpensive paper; otherwise, it shows little signs of use, suggesting that it escaped the notoriously awful prison conditions of the time. The binding is a kind of cat's-paw calf, and on the upper cover is the circular blind stamp of the British and Foreign Bible Society. There is no spine label; it looks like an economical-edition binding. This was probably the first book printed in French using the Stanhope process of stereotyping.

Though the British and Foreign Bible Society began with funding from the upper class, by 1816 or earlier it was also appealing to the working classes for funding. The broadside illustrated in this chapter is *An Appeal to Mechanics, Labourers, and Others, respecting Bible Associations*. The society was very direct in its approach, stating that "A labouring man, who can just support his family, may well afford a penny a week to a Bible Association, which will enable him, at the year's end, to be a benefactor of a man poorer than himself."

Though the Cambridge and Oxford University Presses were relatively early adopters of stereotyping, they were late among larger book printers in England in adopting mechanized printing for book production. The Cambridge University Press did not operate steam-powered printing machines until 1839.[77] Oxford University Press adopted steam-powered printing machines a few years earlier, in 1834, but may not

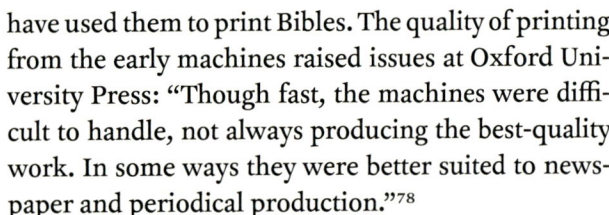

The First Report of the British and Foreign Bible Society (London, 1805)

The British and Foreign Bible Society was a success from its beginning in 1804. This first report listed many subscribers and benefactors.

The Holy Bible . . . Cambridge Stereotype Edition (Cambridge, 1806)

Cambridge University Press was an early adopter of the Stanhope innovative method of stereotyping. This is the 1806 printing of the first edition of the Bible printed from stereotype plates, following the first printing in 1805. The original leather cover bears the circular blind stamp of the British Bible Society on the upper cover.

have used them to print Bibles. The quality of printing from the early machines raised issues at Oxford University Press: "Though fast, the machines were difficult to handle, not always producing the best-quality work. In some ways they were better suited to newspaper and periodical production."[78]

By 1841, the British and Foreign Bible Society had donated or sold 22,288,706 Bibles, most of them probably printed on handpresses, often from stereotype plates. That was a remarkable number of copies sold between 1804 and 1841. These statistics come from James Wyld's specially published world map titled *The World, Designed to show the Languages and Dialects into which The British & Foreign Bible Society has translated the Scriptures or aided in their distribution The Position of the Places Where Societies Have Been*

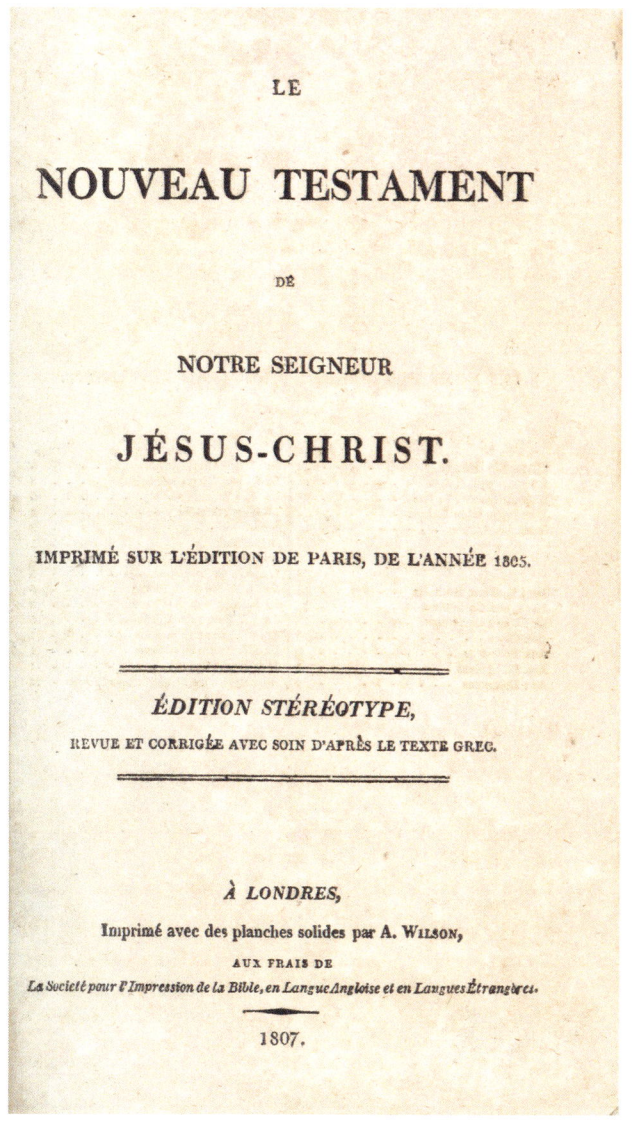

Le Nouveau Testament . . . (London, 1807)

This French-language edition of the New Testament was stereotyped in London and issued by the British and Foreign Bible Society in 1807 for distribution to French prisoners of war imprisoned in England during the Napoleonic Wars.

Formed The Population of those Countries For which Versions Have Been Prepared & the Relative Proportions of Christianity, Mahomedanism and Paganism.

A printed key at the lower right of the map explains the society's work "Distributing, Printing or Translating of the Scriptures" directly into 67 languages and dialects and indirectly into 68 languages and dialects, including 105 "translations never before printed." A total of about twenty-two million bibles are cited as having been distributed by the society as of March 1841, at a cost of nearly 2,800,000 pounds sterling.

A second key, to the left of the first, breaks down the Bible societies in the United Kingdom and "Colonies and Other Dependencies," with a total of 2,636 societies counted. At bottom center, a color key identifies the colors used on the map:

Christianity / Roman Catholic: light orange
Christianity / Protestant: pink
Christianity / Greek Church: green
Mahomedanism [Islam]: blue
Paganism: blue/grey

A table across the lower margin delineates the distribution of Bibles around the world by regions and subregions, including Russia, Persia, India, Central India, Indo Chinese, Chinese Empire, Hither Polynesia, Further Polynesia, Africa, North America, South America, Western Europe, Eastern Europe, Northern Europe, and Central Europe.

James Wyld's very specialized world map was probably printed by traditional handpress methods. It was definitely colored by traditional hand coloring rather than color printing. For that matter, most, if not all the new books published about railroads mentioned in this chapter were also probably printed on handpresses. We have no way to confirm that for sure, but during the first half of the nineteenth century in England, unless an edition of a book was unusually large or production had to be done unusually rapidly, it would have been printed on an iron handpress or multiple handpresses. During the first half of the nineteenth century, while newspapers and magazines increasingly adopted high-speed, steam-powered printing machines to increase their circulation and meet their constant deadlines, most traditional book printers continued to rely upon handpresses for typical editions of books.

An Appeal to Mechanics, Labourers, and Others Respecting Bible Associations (Bradford, 1816)
Broadside, 12½ × 7⅔ in. (31.5 × 19.5 cm)

A broadside soliciting financial support for the British and Foreign Bible Society from members of the working classes.

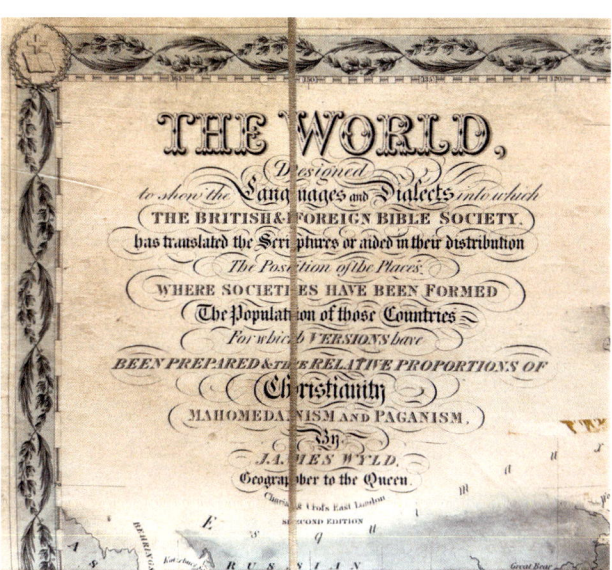

👆 James Wyld, *The World, Designed to show the Languages and Dialects into which The British & Foreign Bible Society has translated the Scriptures or aided in their distribution. The Position of the Places Where Societies Have Been Formed, The Population of those Countries For which Versions Have Been Prepared & the Relative Proportions of Christianity, Mahomedanism and Paganism* (London, 1841)
Hand-colored map, 36 × 51 in. (91 × 130 cm)

This very specialized world map shows the number of Bibles and their locations worldwide sold by the British and Foreign Bible Society up to 1841. The detailed captions below the map specify the number of copies sold in different countries. By 1841, the organization had distributed 22,288,706 Bibles at a cost of nearly 2,800,000 pounds sterling.

👈 Detail from the upper corner of James Wyld's map

Robert Seymour ("Shortshanks"), *The March of Intellect* (detail). See p. 95.

CHAPTER 5

Henry Brougham, Charles Knight, and the Society for the Diffusion of Useful Knowledge Use the Printing Machine to Reach Large Audiences

One of the first political leaders and educational reformers to see the cost savings and educational potential of the new technologies for the production of books, pamphlets, and periodicals, in addition to newspapers, was the English politician and educational reformer Henry Brougham, 1st Baron Brougham and Vaux. In 1825, in a pamphlet entitled *Practical Observations upon the Education of the People, Addressed to the Working Classes and Their Employers*, Brougham commented upon John Limbird's wide-selling popular magazine, *The Mirror of Literature*, as follows:

> *The Mirror, a weekly publication, containing much matter of harmless and even improving amusement, selected with very considerable taste, has besides, in almost every number, information of a most instructive kind. Its great circulation must prove highly beneficial to the bulk of the people. I understand, that of some parts upwards of 80,000 were printed, and there can be no doubt that the entertainment which is derived from reading the lighter essays, may be made the means of conveying knowledge of a more solid and useful description—a consideration which I trust the conductor will always bear in mind.*

Brougham's cheaply printed pamphlet, which underwent about twenty reprints in London in 1825 and was reprinted elsewhere in England that year, represented a kind of call to arms to harness the new mechanized printing and papermaking technologies to produce editions of informative works that were affordable for any working person. Brougham noted in the pamphlet that, even with all the manufacturing expertise in England up until then, England had never succeeded in printing books at a cost less than double the price of books printed on the Continent.

Industrialization of papermaking and printing were not the only forces behind the growth of lower-priced publications in the 1820s and 1830s. Economic forces also played a role. The Panic of 1825, a stock market crash that started in the Bank of England, was felt most acutely in England, where it forced the closing of six London banks and banks in sixty countries; it also impacted the markets of Europe, Latin America, and the United States. This financial crisis has been called the first modern economic crisis not attributable to an external event, such as a war, and is considered the beginning of modern economic cycles.[79]

With respect to the publishing industry in England, the crash did not reduce the number of publishers between 1825 and 1827, but it radically altered the nature of the industry. As a result of the economic downturn, there was declining demand for expensive luxury books and an increased market for less-expensive publications, leading to demand for cheap publications and encouraging serialization. One of the first British publishers to capitalize on this demand was Archibald Constable of Edinburgh, who launched *Constable's Miscellany of Original and Select Publications*, a series of inexpensive, clothbound, small-octavo vol-

The Mirror of Literature, Amusement, and Instruction, no. II (9 November 1822)

The preposterous illustration and article about a mermaid graced the first page of the second issue of *The Mirror* magazine. Sensationalism was undoubtedly a factor in the commercial success of this magazine.

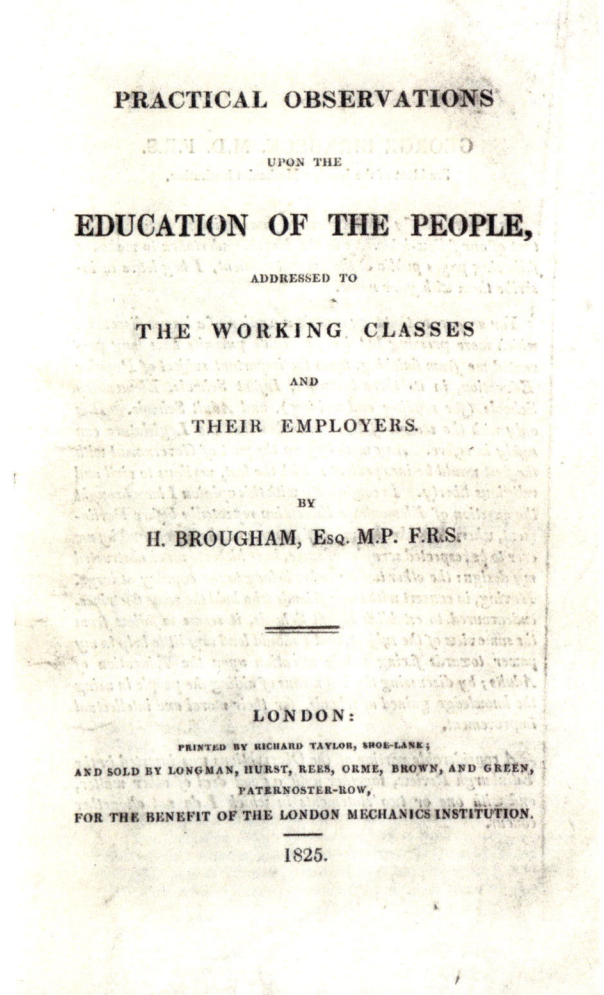

Henry Brougham, *Practical Observations upon the Education of the People, Addressed to the Working Classes and Their Employers* (London, 1825)

In this very widely reprinted pamphlet, Brougham laid out reformist ideas that were adopted two years later by the Society for the Diffusion of Useful Knowledge.

umes, in 1826. Another was the London publisher John Murray, who began issuing his series *The Family Library* in 1829; the series' small-octavo volumes were bound in printed cloth, with their retail price (5 shillings) printed on the upper cover (an innovation in book publishing; the technology for printing on the cloth had existed since 1783, when Thomas Bell invented machinery for printing on textiles, but Murray was the first to apply it to bookbindings).

In November 1826, Brougham and a large committee of fellow educational reformers, many of them Whig or radical MPs and lawyers, founded the Society for the Diffusion of Useful Knowledge (SDUK). The SDUK's purpose was to educate what they hoped would be the widest range of readers by using the new high-speed, steam-powered printing technology to publish inexpensive informative works, and by taking advantage of the new high-speed, steam-powered transportation technology—that is, railroads—to distribute them as cheaply and widely as possible.

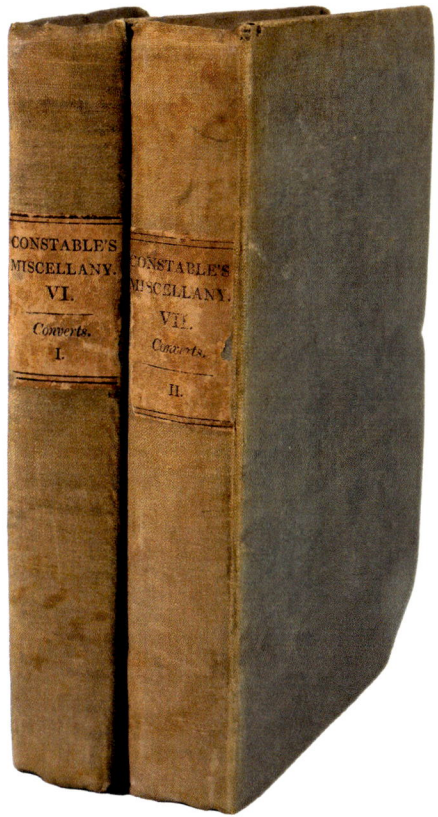

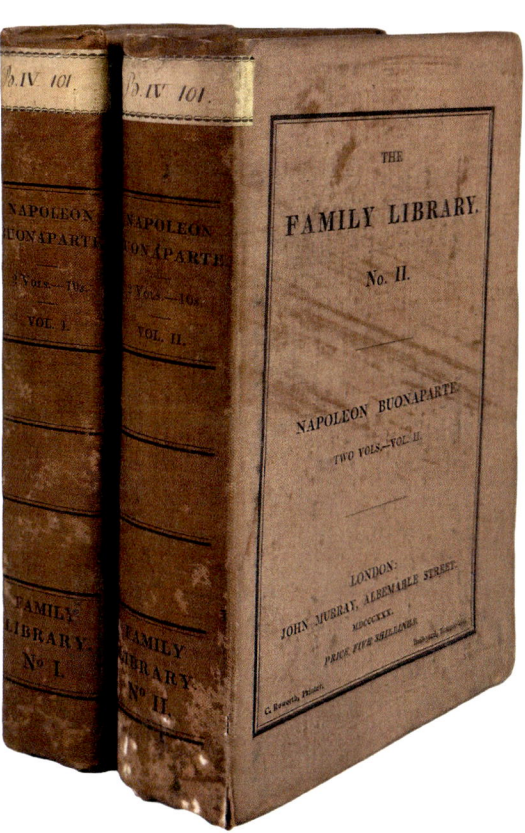

Converts from Infidelity: *Constable's Miscellany of Original and Select Publications*, vol. VI (London, 1827)
Covers measure 6½ × 4 in. (163 × 100 mm)

A set in *Constable's Miscellany*, a series of books issued in small-octavo format, in cloth bindings, designed to be sold at reasonable prices.

The History of Napoleon Buonaparte. The Family Library, nos. 1 and 2 (London, 1830)

Publisher John Murray issued *The Family Library* in competition with *Constable's Miscellany*. The format was identical, with covers measuring 6½ × 4 in. (163 × 100 mm). *The Family Library* was issued in some of the earliest printed cloth bindings.

Though the SDUK hoped to bring education to the lowest classes of society and the growing middle class, they eventually found that the primary market for their teaching and publications was the English middle class. Not everyone agreed that the SDUK was necessarily a positive force for change. In Thomas Love Peacock's 1831 comic novel *Crochet Castle*, one of the characters, Dr. Folliott, called the SDUK the "Steam Intellect Society" and linked the march of technological and efforts at progress in education to folly, rural protest, and the rise in crime.

In 1827, Brougham launched the SDUK's Library of Useful Knowledge with an introductory treatise that he wrote to promote science: *A Discourse of the Objects, Advantages and Pleasures of Science*. The pamphlet offered a brief survey of mathematics, natural philosophy, the solar system, electricity, and the workings of the steam engine; this was followed by a statement of the aims of the SDUK and its initial publishing program. Brougham's pamphlet found a ready audience, reaching a sale of 42,000 by 1833. In July 1829, the SDUK hired London publisher Charles Knight to superintend its publications, and in 1829, Knight became the SDUK's publisher and editor. Knight had suffered considerable financial losses as a result of the Panic of 1825, which put him out of business, but in the years afterward, Knight was able to rebuild his business through his association with the SDUK.

In 1829, as a complement to the SDUK's Library of Useful Knowledge, Knight and the SDUK launched a second publication series titled the Library of Entertaining Knowledge. Up to this date, printing on the steam-powered rotary presses used for SDUK pub-

Carte de visite photograph of Henry Brougham, 1st Baron Brougham and Vaux, presumably with a grandchild (c. 1850)

A politician and reformer, Brougham was frequently portrayed and often caricatured.

Henry Brougham, *A Discourse of the Objects, Advantages, and Pleasures of Science* (London, 1827)

Lord Brougham launched the Society for the Diffusion of Useful Knowledge (SDUK) with this widely reprinted pamphlet.

lications had been unillustrated, but in April 1829, the printing firm William Clowes Ltd. was able to figure out how to print high-quality, wood-engraved illustrations on their rotary presses. Knight and the SDUK were thus able to publish the first part of the first extensively illustrated book ever printed on a mechanized printing press—a three-volume work called *The Menageries*, written by Knight himself but issued anonymously.[80] In his autobiography Knight described this significant step in book production technology:

In looking back at some correspondence of September, 1828, I am enabled to form an accurate conception of the technical difficulties of producing a cheap book with excellent wood-cuts. I had arranged to have my 'Menageries' illustrated with representations of animals drawn from the life. I was fortunate in securing the assistance of several rising young men, who did not disdain what some painters might have deemed ignoble employment. Two of these are now Royal Academicians. There were not many wood-engravers then in London; and this art was almost invariably applied to the production of ex-

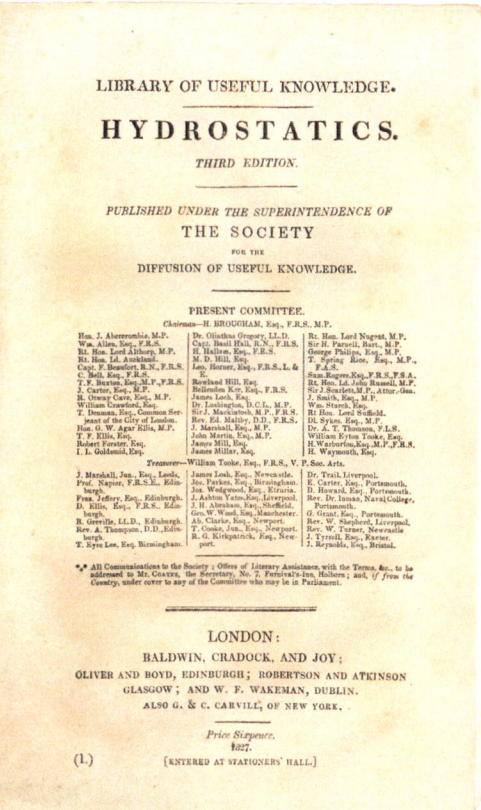
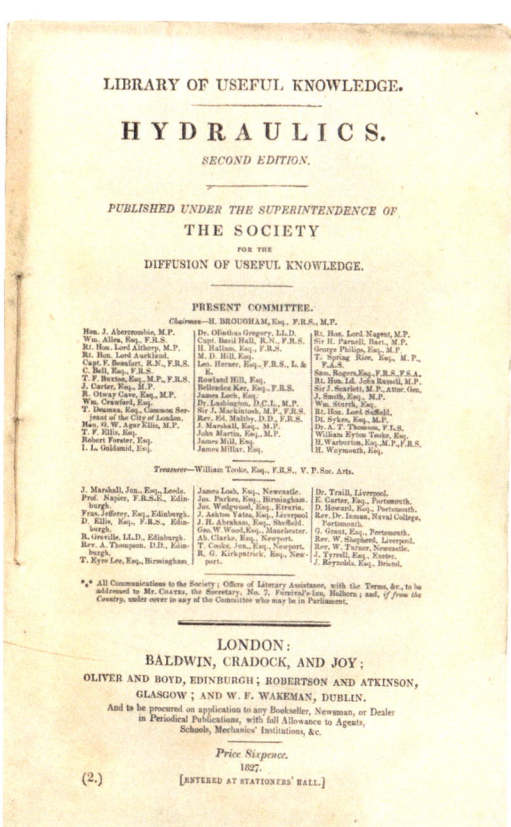
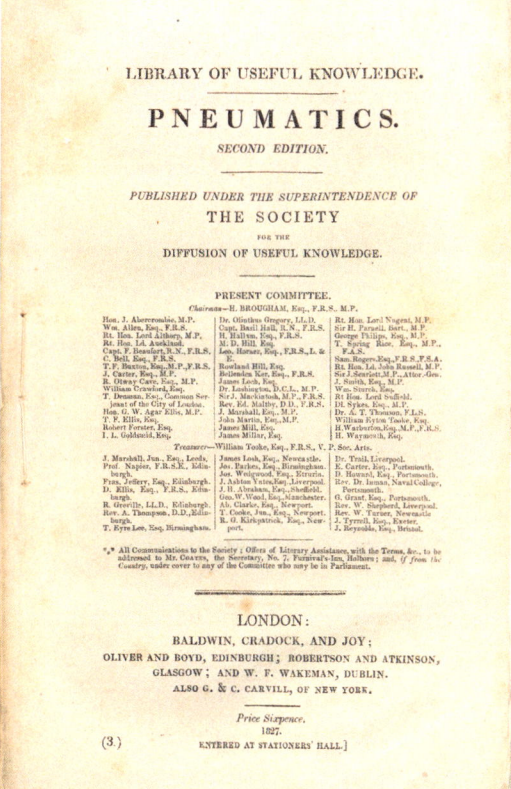

🕯 Library of Useful Knowledge: *Hydrostatics, Hydraulics, Pneumatics* (London, 1827)

The SDUK began its extensive publishing program with the Library of Useful Knowledge in 1827, of which these were the first of many pamphlets issued. These and later publications from the SDUK carried the names of all SDUK committee members on their upper cover.

👉 Engraved portrait of Charles Knight from his obituary in *The Illustrated London News* (22 March 1873)

A prolific publisher and writer, Knight was one of the greatest promoters of the use of printing machines to produce low-priced, high-circulation publications.

PUBLISHED UNDER THE SUPERINTENDENCE OF THE SOCIETY
FOR THE DIFFUSION OF USEFUL KNOWLEDGE.

THE LIBRARY OF ENTERTAINING KNOWLEDGE.

Vol. I.—Part I.

THE MENAGERIES:
QUADRUPEDS,
DESCRIBED AND DRAWN FROM LIVING SUBJECTS.

COMMITTEE.

Chairman—H. BROUGHAM, Esq., F.R.S., M.P.
Vice Chairman—LORD JOHN RUSSELL, M.P.
Treasurer—WILLIAM TOOKE, Esq., F.R.S., V.P. Soc. Arts.

Rt. Hon. J. Abercrombie, M.P.
W. Allen, Esq., F.R.S.
Lord Althorp, M.P.
Rt. Hon. Lord Auckland.
Capt. F. Beaufort, R.N., F.R.S.
C. Bell, Esq., F.R.S., L. & E.
T. F. Buxton, Esq., M.P., F.R.S.
R. Otway Cave, Esq., M.P.
John Conolly, M.D.
William Coulson, Esq.
Wm. Crawford, Esq.
Fred. Daniell, Esq., F.R.S.
T. Denman, Esq.
Hon. G. Agar Ellis, M.A., M.P.
T. F. Ellis, Esq., M.A.
I. L. Goldsmid, Esq., F.R.S.
B. Gompertz, Esq., F.R.S.
H. Hallam, Esq., F.R.S., M.A.
M. D. Hill, Esq.
Rowland Hill, Esq.
Edwin Hill, Esq.
Leonard Horner, Esq., F.R.S.
David Jardine, Esq.
Henry B. Ker, Esq., F.R.S.
J. G. S. Lefevre, Esq., F.R.S.
Edward Lloyd, Esq., M.A.
James Loch, Esq., M.P., F.G.S.
J. W. Lubbock, Esq.
Dr. Lushington, D.C.L., M.P.
Sir J. Mackintosh, M.P.
R. H. Malkin, Esq., M.A.
Rev. Ed. Maltby, D.D., F.R.S.
James Manning, Esq.
F. O. Martin, Esq.
J. Marshall, Esq., M.P.
John Herman Merivale, Esq.
James Mill, Esq.
James Morrison, Esq.
Rt. Hon. Lord Nugent, M.P.
Sir H. Parnell, Bart., M.P.
Professor Pattison.
T. Spring Rice, Esq., M.P., F.A.S.
Dr. Roget, Sec. R.S.
C. E. Rumbold, Esq., M.P.
J. Smith, Esq., M.P.
Rt. Hon. Lord Suffield.
Dr. A. T. Thomson, F.L.S.
William Tooke, Esq.
A. N. Vigors, Esq., F.R.S.
H. Warburton, Esq., M.P., F.R.S.
H. Waymouth, Esq.
J. Whishaw, Esq., M.A., F.R.S.
Mr. Serjeant Wilde.
John Wrottesley, Esq., M.A.

THOMAS COATES, *Secretary.*

LONDON:
CHARLES KNIGHT, PALL-MALL EAST;
LONGMAN, REES, ORME, BROWN, AND GREEN, PATERNOSTER-ROW;
OLIVER AND BOYD, EDINBURGH; ROBERTSON AND ATKINSON, GLASGOW;
W. F. WAKEMAN, DUBLIN; E. WILLMER, LIVERPOOL; BAINES AND CO.,
LEEDS; AND G. AND C. CARVILL, NEW YORK.

Part I.   (*Price Two Shillings.*)   April 1, 1829.

PUBLISHED UNDER THE SUPERINTENDENCE OF THE SOCIETY
FOR THE DIFFUSION OF USEFUL KNOWLEDGE.

THE LIBRARY OF ENTERTAINING KNOWLEDGE.

Vol. I.—Part II.

THE MENAGERIES:
QUADRUPEDS,
DESCRIBED AND DRAWN FROM LIVING SUBJECTS.

COMMITTEE.

Chairman—H. BROUGHAM, Esq., F.R.S., M.P.
Vice Chairman—LORD JOHN RUSSELL, M.P.
Treasurer—WILLIAM TOOKE, Esq., F.R.S., V.P. Soc. Arts.

Rt. Hon. J. Abercrombie, M.P.
W. Allen, Esq., F.R.S.
Lord Althorp, M.P.
Rt. Hon. Lord Auckland.
W. B. Baring, Esq., M.P.
Capt. F. Beaufort, R.N., F.R.S.
C. Bell, Esq., F.R.S., L. & E.
T. F. Buxton, Esq., M.P., F.R.S.
R. Otway Cave, Esq., M.P.
John Conolly, M.D.
William Coulson, Esq.
Wm. Crawford, Esq.
Fred. Daniell, Esq., F.R.S.
T. Denman, Esq.
Hon. G. Agar Ellis, M.A., M.P.
T. F. Ellis, Esq., M.A.
I. L. Goldsmid, Esq., F.R.S.
B. Gompertz, Esq., F.R.S.
H. Hallam, Esq., F.R.S., M.A.
M. D. Hill, Esq.
Rowland Hill, Esq.
Edwin Hill, Esq.
Leonard Horner, Esq., F.R.S.
David Jardine, Esq.
Henry B. Ker, Esq., F.R.S.
J. G. S. Lefevre, Esq., F.R.S.
Edward Lloyd, Esq., M.A.
James Loch, Esq., M.P., F.G.S.
J. W. Lubbock, Esq.
Dr. Lushington, D.C.L., M.P.
Sir James Mackintosh, M.P.
R. H. Malkin, Esq., M.A.
Rev. Ed. Maltby, D.D., F.R.S.
James Manning, Esq.
F. O. Martin, Esq.
J. Marshall, Esq., M.P.
John Herman Merivale, Esq.
James Mill, Esq.
James Morrison, Esq.
Rt. Hon. Lord Nugent, M.P.
Sir H. Parnell, Bart., M.P.
Professor Pattison.
T. Spring Rice, Esq., M.P., F.A.S.
Dr. Roget, Sec. R.S.
C. E. Rumbold, Esq., M.P.
J. Smith, Esq., M.P.
Rt. Hon. Lord Suffield.
Dr. A. T. Thomson, F.L.S.
William Tooke, Esq., F.R.S.
A. N. Vigors, Esq., F.R.S.
H. Warburton, Esq., M.P., F.R.S.
H. Waymouth, Esq.
J. Whishaw, Esq., M.A., F.R.S.
Mr. Serjeant Wilde.
J. Wood, Esq., M.P.
John Wrottesley, Esq., M.A.

THOMAS COATES, *Secretary.*

LONDON:
CHARLES KNIGHT, PALL-MALL EAST;
LONGMAN, REES, ORME, BROWN, AND GREEN, PATERNOSTER-ROW;
OLIVER AND BOYD, EDINBURGH; ROBERTSON AND ATKINSON, GLASGOW;
W. F. WAKEMAN, DUBLIN; E. WILLMER, LIVERPOOL; BAINES AND CO.,
LEEDS; AND G. AND C. CARVILL, NEW YORK.

Part 3.   *Price Two Shillings.*   July 1, 1829.

PUBLISHED UNDER THE SUPERINTENDENCE OF THE SOCIETY
FOR THE DIFFUSION OF USEFUL KNOWLEDGE.

THE LIBRARY OF ENTERTAINING KNOWLEDGE.

Vol. VII.—Part I.

THE MENAGERIES:
QUADRUPEDS.
Vol. 2.

COMMITTEE.

Chairman—H. BROUGHAM, Esq., F.R.S., M.P.
Vice Chairman—LORD JOHN RUSSELL.
Treasurer—WILLIAM TOOKE, Esq., F.R.S., V.P. Soc. Arts.

W. Allen, Esq., F.R.S.
Viscount Althorp, M.P.
Rt. Hon. Visct. Ashley, M.P.
Rt. Hon. Lord Auckland.
W. B. Baring, Esq., M.P.
Capt. F. Beaufort, R.N., F.R.S.
C. Bell, Esq., F.R.S., L. & E.
T. F. Buxton, Esq., F.R.S., M.P.
R. Otway Cave, Esq., M.P.
John Conolly, M.D.
William Coulson, Esq.
Wm. Crawford, Esq.
Fred. Daniell, Esq., F.R.S.
John F. Davis, Esq., F.R.S.
T. Denman, Esq., M.P.
Hon. G. Agar Ellis, M.A., M.P.
T. F. Ellis, Esq., M.A.
Thomas Falconer, Esq.
I. L. Goldsmid, Esq., F.R.S.
B. Gompertz, Esq., F.R.S.
H. Hallam, Esq., F.R.S., M.A.
M. D. Hill, Esq.
Rowland Hill, Esq.
Edwin Hill, Esq.
John C. Hobhouse, Esq., M.P.
Leonard Horner, Esq., F.R.S.
David Jardine, Esq.
Henry B. Ker, Esq., F.R.S.
J. G. S. Lefevre, Esq., F.R.S.
Edward Lloyd, Esq., M.A.
James Loch, Esq., M.P., F.G.S.
George Long, Esq., A.M.
J. W. Lubbock, Esq., F.R.S.
Dr. Lushington, D.C.L., M.P.
Zachary Macaulay, Esq., A.M.
R. H. Malkin, Esq., M.A.
Rev. Ed. Maltby, D.D., F.R.S.
James Manning, Esq.
F. O. Martin, Esq.
J. Marshall, Esq.
John Herman Merivale, Esq.
James Mill, Esq.
James Morrison, Esq., F.G.S.
Sir H. Parnell, Bart. M.P.
Professor Pattison.
T. S. Rice, Esq., M.P., F.A.S.
Dr. Roget, Sec. R.S.
C. E. Rumbold, Esq., M.P.
J. Smith, Esq., M.P.
Rt. Hon. Lord Suffield.
C. T. Thomson, Esq., M.P.
Dr. A. T. Thomson, F.L.S.
N. A. Vigors, Esq., F.R.S.
H. Warburton, Esq., M.P., F.R.S.
H. Waymouth, Esq.
J. Whishaw, Esq., M.A., F.R.S.
Mr. Serjeant Wilde.
J. Wood, Esq., M.P.
John Wrottesley, Esq., M.A.

THOMAS COATES, *Secretary*, 4, South Square, Gray's Inn.

LONDON:
CHARLES KNIGHT, PALL-MALL EAST;
LONGMAN, REES, ORME, BROWN, AND GREEN, PATERNOSTER-ROW;
OLIVER AND BOYD, EDINBURGH; THOMAS ATKINSON AND CO., GLASGOW;
W. F. WAKEMAN, DUBLIN; E. WILLMER, LIVERPOOL;
BAINES AND CO., LEEDS; AND G. AND C. CARVILL, NEW YORK.

PART 13.   (*Price Two Shillings.*)   August, 1830.

PUBLISHED UNDER THE SUPERINTENDENCE OF THE SOCIETY
FOR THE DIFFUSION OF USEFUL KNOWLEDGE.

THE LIBRARY OF ENTERTAINING KNOWLEDGE.

Vol. VII.—Part II.

THE MENAGERIES.
Vol. II.

COMMITTEE.

Chairman—The Right Hon. LORD CHANCELLOR.
Vice Chairman—The Right Hon. LORD JOHN RUSSELL, M.P.
Treasurer—WILLIAM TOOKE, Esq., F.R.S.

W. Allen, Esq., F.R. & R.A.S.
Rt. Hon. Visc. Althorp, M.P.
Rt. Hon. Visct. Ashley, M.P., F.R.A.S.
Rt. Hon. Lord Auckland.
W. B. Baring, Esq., M.P.
Capt. F. Beaufort, R.N., F.R., and R.A.S.
C. Bell, Esq., F.R.S., L. & E.
John Conolly, M.D.
William Coulson, Esq.
Wm. Crawford, Esq.
J. Fred. Daniell, F.R.S.
Lt. Drummond, R.E., F.R.A.S.
Viscount Ebrington, M.P.
T. F. Ellis, Esq., M.A., F.R.S.
John Elliotson, M.D., F.R.S.
Hon. and Elphinstone, Esq., M.A.
Thomas Falconer, Esq.
I. L. Goldsmid, Esq., F.R. and R.A.S.
B. Gompertz, Esq., F.R. and R.A.S.
G. B. Greenough, Esq., F.R. & L.S.
H. Hallam, Esq., F.R.S., M.A.
M. D. Hill, Esq.
Rowland Hill, Esq., F.R.A.S.
Edwin Hill, Esq.
John C. Hobhouse, Esq., M.P.
Leonard Horner, Esq., F.R.S.
David Jardine, Esq., A.M.
Henry B. Ker, Esq., F.R.S.
J. G. S. Lefevre, Esq., F.R.S.
Edward Lloyd, Esq., M.A., F.G.S.
George Long, Esq., A.M.
J. W. Lubbock, Esq., F.R., R.A., and L.S.
Dr. Lushington, D.C.L.
Zachary Macaulay, Esq.
B. H. Malkin, Esq., M.A.
A. T. Malkin, Esq., M.A.
Rev. Ed. Maltby, D.D., F.R.S.
James Manning, Esq.
F. O. Martin, Esq.
James Mill, Esq.
James Morrison, Esq., M.P.
Rt. Hn. Sir H. Parnell, Bt., M.P.
Professor Pattison.
T. Spring Rice, Esq., F.A.S., M.P.
Dr. Roget, Sec. R.S., F.R.A.S.
Sir M.A. Shee, P.R.A., F.R.S.
J. Smith, Esq., M.P.
Wm. Sturch, Esq.
Dr. A. T. Thomson, F.L.S.
N. A. Vigors, Esq., F.R.S.
H. Warburton, Esq., F.R.S., M.P.
H. Waymouth, Esq.
J. Whishaw, Esq., M.A., F.R.S.
Mr. Serjeant Wilde, M.P.
J. Wood, Esq., M.P.
John Wrottesley, Esq., M.A., Sec. R.A.S.

THOMAS COATES, *Secretary*, 59, Lincoln's Inn Fields.

LONDON:
CHARLES KNIGHT, PALL-MALL EAST;
LONGMAN, REES, ORME, BROWN, AND GREEN, PATERNOSTER-ROW;
OLIVER AND BOYD, EDINBURGH; THOMAS ATKINSON AND CO., GLASGOW;
W. F. WAKEMAN, DUBLIN; E. WILLMER, LIVERPOOL;
BAINES AND CO., LEEDS; AND G. C. AND H. CARVILL, NEW YORK.

PART 22.   (*Price Two Shillings.*)   August, 1831.

Group of Animals of opposite natures living in the same cage.

The Library of Entertaining Knowledge: *The Menageries*, vols. 1 and 2 in the original 4 parts (London, 1829–1831)

The SDUK's Library of Entertaining Knowledge was meant to be less "serious" and more "entertaining" than the SDUK's Library of Useful Knowledge. The first volume in the series, the first part of *The Menageries*, was written by its publisher, Charles Knight, but was published anonymously on 1 April 1829. It was the first extensively illustrated book printed on a printing machine. As in their Library of Useful Knowledge, the SDUK published the names of all their committee members on the upper cover of each part. Whether that was for purposes of self-congratulation, or whether they believed that publishing all their names would stimulate sales, is unclear.

pensive books, printed in the finest style. The legitimate purpose of wood-engraving was not then attained. It is essentially that branch of the art of design which is associated with cheap and rapid printing. In the costly books of period a single woodcut introduced into a sheet to be worked off with the types, enhanced the cost of manual labour in a proportion which would now seem incredible. In engraving the wood-cuts for the 'Menageries,' some attention of the artist was necessary to give his shadows the requisite force, and his lights the desired clearness, so as to meet the uniform application of the ink, and the cylindrical pressure, in the printing-machine process. It was long before this excellence could be practically attained. Without this explanation it would appear ludicrous that Mr. Hill should write to me from Mr. Brougham's house,—'Everybody here is in raptures with the proofs of the wood-cuts; but we have misgivings about the machine.' A sheet of my book was to be set up with the engravings in their due place, and a hundred or two were to be printed off by the rapid operation. 'Mr. Loch is here,' writes Mr. Hill. 'We have held a committee. He will be in London in a fortnight, quite at leisure, and anxious to attend to our affairs. He has promised to assist at Clowes's [the printers]. I hope you will succeed in assembling everybody.' 'Everybody' not only meant the patentee of the machine, the wood-engraver, the stationer, the ink-maker, and the ingenious overseer of the printing office, but as many of the committee as I could get together. Imagine a learned society thus employed! Imagine a hard-worked editor thus exhorted to interference with a printer's proper duty! Yet such was part of my editorial duty at a time when the great revolution in the production of books to be accomplished by the printing machine, was almost as imperfectly realised as when Caxton first astonished England by the miracles of the printing press. We succeeded in partially overcoming the difficulties of making an illustrated volume not despicable as a work of art, and yet cheap—something very different from the lesson books with blotches called pictures, that puzzled the school-boy mind half a century ago.[81]

The achievement that Charles Knight marked by the SDUK's edition of *The Menageries* relates to the quality and number of woodcut illustrations reproduced by printing machine in a single publication. It was not necessarily the first illustrated book printed by printing machine, though that claim may be made for it considering how few pamphlets and books can be confirmed to have been printed by printing machine by this date. It was not even the first illustrated publication issued by the SDUK, since some of their earliest pamphlets in their Library of Useful Knowledge were illustrated with crude woodcuts. With the exception of newspapers, we generally don't know which publications, illustrated or not, might have been printed by machine up to this date. It is also difficult to know whether issues of periodicals like Limbird's magazine, *The Mirror*, which typically contained a woodcut image on the first page and were said to have sold up to 80,000 copies, were printed by machine or by handpress.

Charles Knight was not only a publisher but a prolific writer on many subjects, including economics. In October 1830, under the auspices of the SDUK, he published an anonymous eight-page pamphlet titled *An Address to the Labourers, on the Subject of Destroying Machinery*, a response to the destruction of threshing machines and other agricultural machinery that had taken place during the Swing Riots of that year. The pamphlet, one of the scarcest and least-known of Knight's early writings, was the first of Knight's public-service publications explaining the advantages of technology to workers threatened by the increasing mechanization of agriculture and the textile industry,

particularly those workers who destroyed machinery in the Luddite tradition. Knight returned to this topic many times in his writings, both in respect to mechanization in general and to mechanization within the book-production industries.

On 31 March 1832, Charles Knight and the SDUK launched *The Penny Magazine*, the first extensively illustrated, low-cost, mass-circulation weekly magazine. This magazine, which was published until 31 October 1845, became the most successful publishing venture for both Knight and the SDUK. Each weekly issue of the magazine, which was marketed to the English working classes and the developing middle class, consisted of eight pages—one printed sheet, on which four pages were printed on each side, and then folded twice—filled with wholesome educational content and liberally illustrated with wood engravings. The images allowed even those with limited reading skills to derive enjoyment from its pages; this was particularly important as, at that time, a significant portion of the British population was illiterate.[82] Using illustrations, Knight and the Society for the Diffusion of Useful Knowledge were able to appeal to an audience with limited reading skills, while also enabling self-education. The illustrations became highly popular with Knight's target audience; during the years of its existence, the magazine published 1,887 illustrated articles. However, the cost of the wood engravings forced Knight eventually to raise the price of the magazine to 4 pence per issue, reducing its circulation and commercial success.

So positive, and so free of controversy, scandals, or sensationalism was *The Penny Magazine* that some critics considered it insipid and condescending to the working-class readers that it served. The reviewer "F. H.," in the 15 September 1832 issue of *Mechanics' Magazine*, stated:

The contents, too, were not so very amazing; if they were to be considered as the product of the long array of Lord Chancellors, Cabinet Ministers, Noble Lords, Right Honourable Gentlemen, distinguished philosophers, dignified clergymen, celebrated authors, and liberal characters of every shade, whose names invariably grace the covers of the Society's works. After a pompous and inflated address. . . . Here was novelty—here was a proof of the advantages of a learned Society undertaking the task of enlightening the lower orders—here was a sample of the heretofore-inaccessible information 'now set within the reach of the humblest among them.' Can they be sufficiently grateful for such condescension, and ought they not to pass a unanimous vote of thanks to their benefactors?[83]

An anonymous satirical print entitled "Frontispiece for the Penny Magazine" encapsulated popular criticism of the magazine. The margins of my copy are trimmed. A copy in the Wellcome Collection states that it was drawn by C. J. Grant and lithographed by G. Davies. In the lower right corner, there is a message, also cut off on my copy, stating, "Take Notice; All other Frontispieces are Illegal."

The idea of publishing a large-circulation magazine to promote "useful knowledge" among the working and middle classes did not originate with Knight or the SDUK. Six months earlier, in October 1831, the French journalist, publisher, and politician Emile de Girardin launched the monthly *Journal des connaissances utiles*; however, that magazine had a more modest format than *The Penny Magazine* and included only very few basic woodcuts. The title page of the first trimester (October, November, and December 1831) of Girardin's *Journal* boldly stated that 100,000 copies of the magazine were printed—a circulation for a periodical that had not previously been achieved in France, or anywhere else, for that matter. Printing 100,000 copies with the very limited number of printing machines that were available in France in 1831 required "9 printing machines operated by 5 Parisian printers. Production of each number took a month: composition took 8 days, corrections 3 days, stereotyping 4 days, printing 8 days, and 7 days for folding and distribution."[84] That is why Girardin was forced to issue *Journal des connaissances utiles* as a monthly.

In 1833, Girardin issued his *Almanach de France*, which announced on its front cover that that it had been published in 1,300,000 copies at the price of only "dix sous." Girardin was very aggressive in marketing this almanac as a bargain, taking the highly unusual step of also stating on the cover that it contained 224 pages containing 600,000 letters of type, and that a volume of this size would ordinarily cost 6 francs. He also claimed that the *Almanach* would be on sale in 37,200 communes in France. Whether or not the *Almanach de France* actually achieved a circulation of 1,300,000 copies during 1833, it was probably

The Penny Magazine, no. 1 (31 March 1832)

The first issue of the wildly successful *Penny Magazine* had only three small woodcuts, but as more weekly issues were published, the number and size of woodcuts per issue increased.

Frontispiece for *The Penny Magazine of the Society for the Diffusion of Useful Knowledge*, vol. 1 [no date]
11 × 8½ (28 × 21.5 cm)

This print, probably published in the mid-1830s, satirized the wide and socially diverse impact *The Penny Magazine* was having on society. A caricature of its instigator, Lord Brougham, appears in the center of the print.

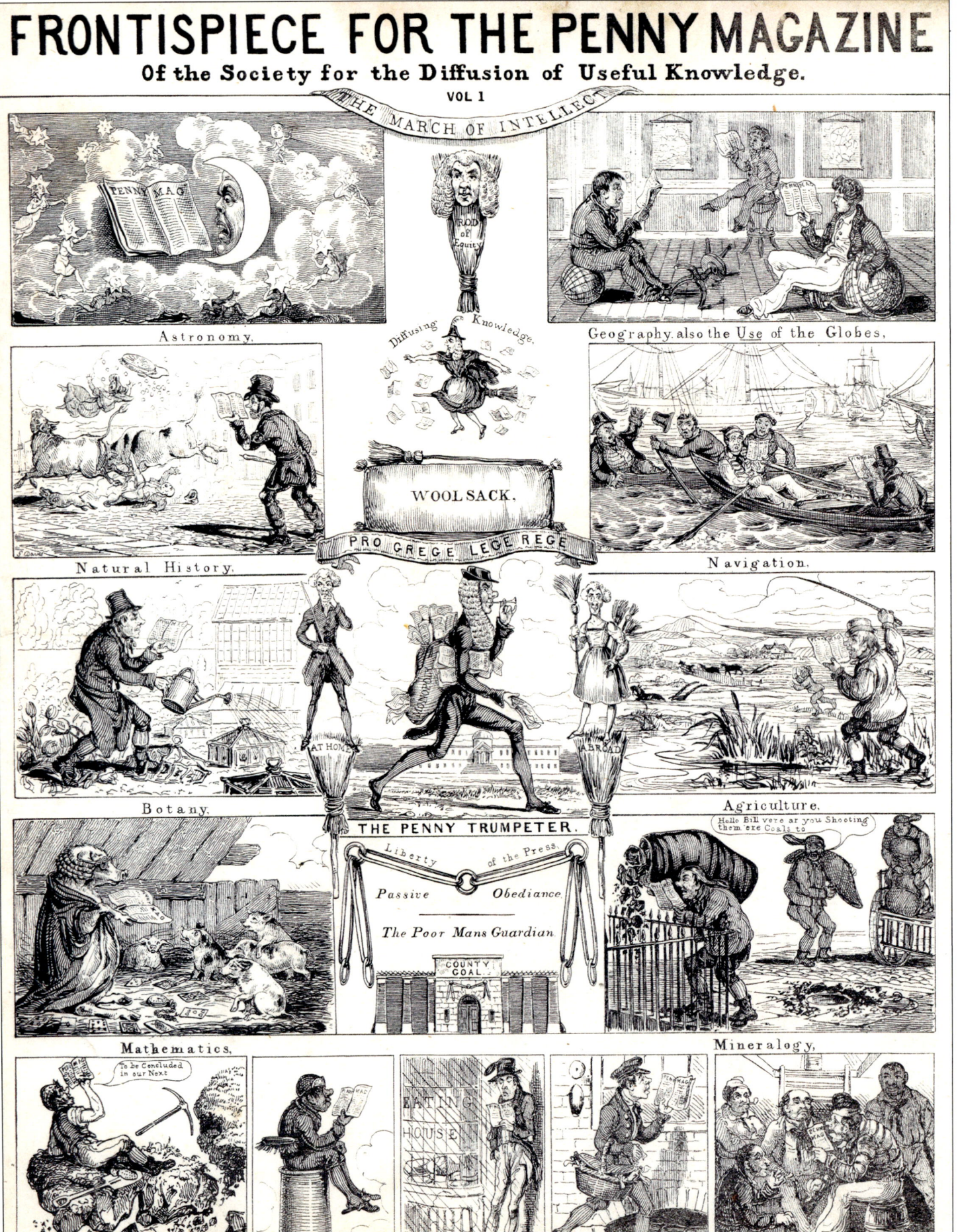

Portrait of Émile de Girardin (c. 1835)
9½ × 6½ in. (24 × 16.5 cm)

Almost simultaneously with the launch of the SDUK's *The Penny Magazine*, innovative Parisian publisher Émile de Girardin began publishing inexpensive periodicals for a mass audience in France.

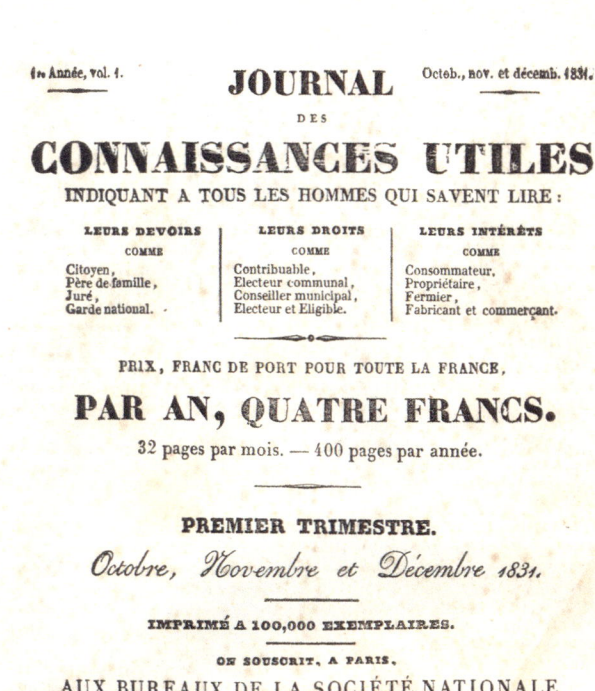

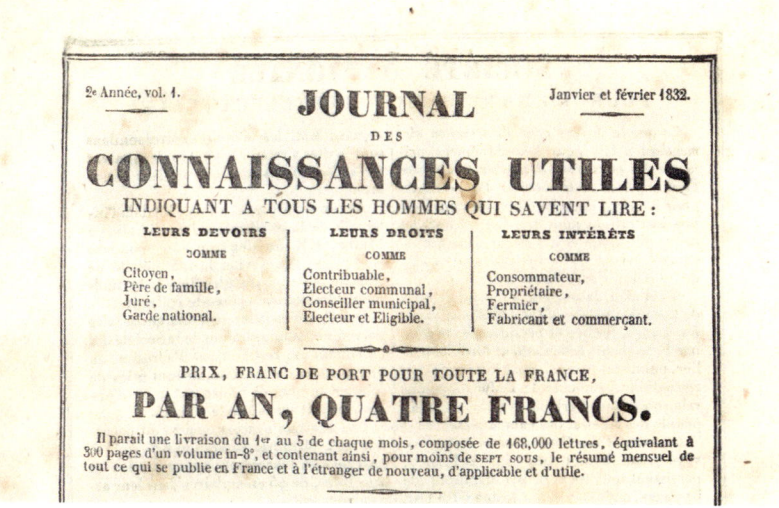

Journal des connaissances utiles, vol. 1 (October 1831) and vol. 2 (January 1832)

On the title pages of the first issues of his monthly *Journal des connaissances utiles*, a journal of "useful knowledge" "for all men who know how to read," Émile de Girardin claimed that 100,000 copies were printed. In the second year of publication, he also made a point on the cover of each issue that each issue contained 168,000 individual letters, as if that actual number of letters (rather than words) would impress his customers.

the first publication in the world to claim a circulation that large on its cover.

One month before the first issue of *The Penny Magazine*, Scottish publishers William and Robert Chambers launched *Chamber's Edinburgh Journal* at the price of one penny per issue. Its circulation eventually reached 84,000. The Chambers brothers did not state that they were printing the magazine on a steam-powered press, but in December 1833, they indicated in an issue of *Chambers' Historical Newspaper* that it was "From the Steam-Press of W. and R. Chambers." For distribution of *Chambers's Edinburgh Journal* in England, when demand justified it, the brothers arranged for the journal to be printed from stereotype plates in London by Bradbury and

Evans. This would have happened at least as early as 1835, based on a reprint of the first volume of the journal printed by Bradbury and Evans, which is in my library. In the thirty-fifth installment of their encyclopedia, *Chambers's Information for the People*, the Chambers brothers illustrated their double-cylinder printing machine, which appears to be an adaptation of an Applegath and Cowper double-cylinder perfecting machine. The machine was "constantly employed in printing 'Chambers's Edinburgh Journal,' 'Historical Newspaper,' and the present work, 'Information for the People,' and which it executes, as we have said, in a style not surpassed by any other species of press."[85]

Charles Knight sometimes communicated with his readers by personally writing articles for *The Penny Magazine*. At the end of the first year of publication, he wrote a preface to the first volume of the magazine, stating that its circulation had reached 160,000 by the end of its first month and 200,000 at the end of its first year. From this he assumed that the magazine, delivered all over England by coach and the new railroads, was being read or at least looked at each week by a million people. Knight noted that, only forty years earlier, Edmund Burke had written that there were only 80,000 readers in all of England.

The success of *The Penny Magazine* motivated numerous publishers in Europe to emulate the magazine and its appearance. Virtually all the emulators

L'Almanach de France (1833)

Girardin took the unusual step of claiming on the cover of the first edition of his pocket *Almanach* for 1833 that 1,300,000 copies were printed, and perhaps even more unusually, he stated that the text was composed of 600,000 individual letters. The *Almanach* was published by the Société pour l'Émancipation Intellectuelle (the society for intellectual emancipation). Girardin claimed that it would be sold in 37,200 towns in France.

From William and Robert Chambers, *Information for the People*, no. 35 (Edinburgh, 1835)

The Chambers brothers devoted number 35 of their *Information for the People* to the history of printing and printing and publishing technology. In it, they illustrated their new printing machine and explained its operation in detail. Their printing machine was a variant of the Applegath and Cowper double-cylinder perfecting machine.

in France, Germany, the Netherlands, Italy, and other countries installed printing machines for more efficient and faster production, if they were available. Opposition by publishers to the introduction of printing machinery does not appear to have been frequently recorded in print—mainly, perhaps, because publishers, who controlled the content of their newspapers and magazines, were motivated to introduce the machinery to increase productivity and circulation. The only emulator of *The Penny Magazine* who refused to install printing machines was J. S. Folds, publisher of *The Dublin Penny Journal*, which began publication on 30

June 1833, just three months after *The Penny Magazine* was launched. In Ireland, Folds emulated Knight's *The Penny Magazine* in concept and format, but not in printing technology, and it is evident from his reference to Knight's book, in the quotation below, that Folds was well aware of the arguments supporting introduction of the new technologies. Nevertheless, Folds refused to employ steam-powered printing machines in the production of *The Dublin Penny Magazine*, which had a relatively small circulation of only a few thousand copies. The topic must have continued to concern Folds, since in the twenty-eighth issue of the magazine, published on 5 January 1833, Folds explained his refusal and urged caution regarding technical innovation:

Yet wherever machines have been introduced, for the first time, the working people have in general combined to put them down, or endeavoured by various means to arrest their progress, or impede their exertions. Are we to attribute this to ignorance? ... The opposition to machinery amongst workmen does not proceed from simple ignorance. No workman would say that it would be better for us that the old times were revived, when sailing vessels used to take a week and ten days in summer, and two or three weeks in winter, to cross from Dublin to Liverpool, instead of going over now in fourteen hours in a steam boat.

Every printing press is, of course, a machine. Each press employs two individuals; and every thousand copies of the DUBLIN PENNY JOURNAL takes up four hours in printing, there being two presses and four men constantly employed upon it, frequently working day and night. When any thing delays the sending it to press ten days before the date of publication, severe extra labour is required to bring up the lost time, and produce the number of copies requisite; and frequently the delay has very seriously interfered with the interests and circulation of the periodical. Here, then, is a case in point. The printer would be justified in the eye of common sense in procuring a machine, by which the Journal could be all printed off in a few days, and thus not only the annoyance, the loss of time, the extra expense be avoided, but the Journal enabled in every respect to compete with the English and Scotch periodicals. Yet from the fear of setting an example of reckless indifference to the interests of men, from the fear of awakening that spirit of avaricious emulation, which would indiscriminately introduce machinery, which would supersede manual labour, he has hitherto abstained. Doubtless, the case still stands, that it is but exchanging one machine for another. Yet as numbers of men depend for subsistence on their labour at the old machines, the printing presses, the new machines should not be recklessly introduced; and as Ireland is yet comparatively guiltless of machinery, let it be introduced cautiously and deliberately, least [sic] in breaking up the soil for her future improvement, we hastily and wantonly plough through the hopes, the prospects, and the interests of her working classes.[86]

As bold as Folds was in holding the line against the new technology, he gave up shortly thereafter and sold *The Dublin Penny Journal* in 1833 to poet, printer, and publisher Philip Dixon Hardy. Very soon after the purchase, Hardy installed the first steam-powered printing machines in Ireland to print the magazine, writing about them with pride in the issue of 10 May 1834.

In the second year of publication of the SDUK's *The Penny Magazine*, Charles Knight wrote and published a memorable series of articles in four "Monthly Supplements" to the regular issues under the general title "The Commercial History of a Penny Magazine."[87] These articles represent the best illustrated introductions in English to the history and technology of printing, woodcut illustration, and binding as practiced during the first third of the nineteenth century. They also appear to be the earliest widely circulated general descriptions of the new processes of machine papermaking and high-speed printing technology. The final supplement, "Printing Presses and Machinery—Bookbinding," featured a full-page woodcut of an Applegath and Cowper steam-powered press as used in the machine room at William Clowes Ltd.,[88] with detailed explanatory captions. This was undoubtedly the first widely seen image of a high-speed double-cylinder perfecting press.

The long quotation below is probably the best description of a new machine printing facility. In it, Charles Knight took his readers on a virtual tour of the Clowes industrial printing establishment, where there were, at the time, more printing machines in operation than anyplace else in the world—sixteen steam-powered, double-cylinder perfecting machines of the type shown in the illustration, powered by two steam engines. In his description of the operation of one of the printing machines, Knight described the special process of making ready, the particular arrangement of the stereotype plates before they could be correctly printed on a rotary press. He explains:

One man, and sometimes two men, are engaged in what is technically called making ready; *and this with stereotype plates is a tedious and delicate operation. The plates are secured upon wooden blocks by which they are raised to the height of moveable types; but then, with every care in casting, and in the subsequent turning operation, these plates, unlike moveable types, do not present a perfectly plane surface. There are hollow parts which must be brought up by careful adjustment; and this is effected by placing pieces of this paper under any point where the impression is faint. This process often occupies six or seven hours, particularly where there are casts from wood-cuts. Let us suppose it completed. Upon the solid steel table at each end of the machine lie the eight pages which print one side of the sheet. At the top of the machine, where the laying on boy stands, is a heap of wet paper. The visitor will have seen the process of wetting previously to entering the machine-room. Each quire of paper is dipped two or three times, according to its thickness, in a trough of water; and being opened is subjected, first to moderate pressure, and afterwards to the action of a powerful press, till the moisture is equally diffused through the whole heap. If the paper were not wetted, the ink, which is a composition of oil and lamp-black, would lie upon the surface and smear. To return to the machine. The signal being given by the director of the work, the laying-on boy turns a small handle, and the moving power of the strap connected with the engine is immediately communicated. Some ten or twenty spoiled sheets are first passed over the types to remove any dirt or moisture. If the director is satisfied, the boy begins to lay on the white paper. He places the sheet upon a flat table before him, with its edge ready to be seized by the apparatus for conveying it upon the drum. At the first movement of the great wheels the inking apparatus at each end has been set in motion. The steel cylinder attached to the reservoir of ink has begun slowly to move,—the 'doctor' has risen to touch that cylinder for an instant, and thus receive its supply of ink,—the inking-table has passed under the 'doctor' and carried off that supply—and the distributing-rollers have spread it equally over the surface of the table. This surface having passed under the inking-rollers, communicates the supply to them; and they in turn impart it to the form which is to be printed. All these beautiful operations are accomplished in the fifteenth part of a minute, by the travelling backward and forward of the carriage or table upon which the form rests. Each roller revolves upon an axis which is fixed. At the moment when the form at the back of the machine is passing under the inking-roller, the sheet, which the boy has carefully laid upon the table before him, is caught in the web-roller and conveyed to the endless bands of tapes which pass it over the first impression cylinder. It is here seized tightly by the bands, which fall between the pages and on the outer margins. The moment after the sheet is seized upon the first cylinder, the* form *passes under that cylinder, and the paper being brought in contact with it receives an impression on one side. To give the impression on the other side the sheet is to be turned over; and this is effected by the two drums in the centre of the machine. The endless tapes never lose their grasp of the sheet, although they allow it to be reversed. When the impression has been given by the first cylinder, the second form of tapes at the other end of the table has been inked. The drums have conveyed the sheet during this inking upon the second cylinder; it is brought into contact with the types; and the operation is complete.*[89]

Enthralled by the new technologies, and motivated by the dramatic success of *The Penny Magazine*, Knight launched a new periodical in February 1834 titled *The Printing Machine: A Review for the Many*, with an image of the double-cylinder Applegath and Cowper printing machine on the masthead. Under the masthead, Knight printed the following quotation: "What the PRINTING-PRESS did for the instruction of the masses in the fifteenth century, the PRINTING-MACHINE is doing in the nineteenth. Each represents an era in the diffusion of knowledge; and each may be taken as a symbol of the intellectual character of the age of its employment." Knight repeated this masthead in the March issue, but in April he renamed the periodical *The Printing Machine: Companion to the Library* and removed the image of the Applegath and Cowper machine. In August 1834, he terminated what must have been an unprofitable venture.

In 1835, as a byproduct of the successful *Penny Magazine*, Knight issued a volume of *One Hundred and Fifty Wood Cuts, Selected from The Penny Magazine; Worked, by the Printing-Machine from the Original Blocks*, whose

THE PRINTING MACHINE:
A REVIEW FOR THE MANY.

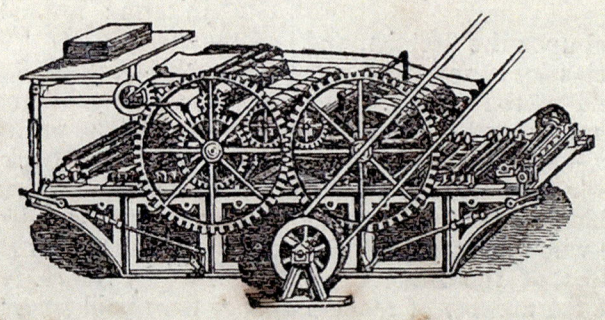

No. 1.
SATURDAY,
FEBRUARY 15, 1834.
Price 4d.

TO BE
CONTINUED
MONTHLY.

"What the PRINTING-PRESS did for the instruction of the masses in the fifteenth century, the PRINTING-MACHINE is doing in the nineteenth. Each represents an æra in the diffusion of knowledge; and each may be taken as a *symbol* of the intellectual character of the age of its employment."—*Penny Magazine.*

The Printing Machine, vol. 1, no. 1 (15 February 1834)

During the peak of success of *The Penny Magazine*, Charles Knight launched this review, lauding the achievements possible with printing machines. On its masthead, we see a much-reduced copy of the image of the Applegath and Cowper double-cylinder perfecting machine published in *The Penny Magazine*.

From the *Monthly Supplement of the Penny Magazine*, no. 112 (30 November to 31 December 31 1833)

The woodcut, with its detailed explanatory caption, illustrates an Applegath and Cowper double-cylinder perfecting machine of the type used by William Clowes to print *The Penny Magazine*.

VIRGIN AND CHILD.

Engraved by J. Jackson, from the Print by Raffaelle Morghen, after Raffaelle.

[Vol. ii. p. 417.]

ONE HUNDRED AND FIFTY

WOOD CUTS,

SELECTED FROM

The Penny Magazine;

WORKED, BY THE PRINTING-MACHINE, FROM THE ORIGINAL BLOCKS.

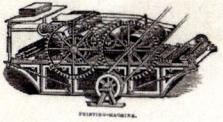

LONDON: CHARLES KNIGHT, 22, LUDGATE STREET.
MDCCCXXXV.
PRICE FOURTEEN SHILLINGS, BOUND IN CLOTH, WITH GILT EDGES.

[Charles Knight,] *One Hundred and Fifty Wood Cuts, Selected from The Penny Magazine; Worked, by the Printing-Machine from the Original Blocks* (London, 1835)

This was the first art book printed by a printing machine. In very small type at the end of the list of plates, Charles Knight explained what he meant by the expression "Worked by the Printing-Machine . . . " in the title, suggesting that the images printed from the original woodblocks might be more faithful to the originals than the prints in the magazine made from stereotype plates. He also indicated that woodcuts intended to be printed on rotary presses were engraved slightly differently than those intended to be printed on a handpress. He wrote, "The Cuts in the 'Penny Magazine' are worked from stereotype casts of the original blocks, which blocks remain as clear and sharp as when they came from the hands of the engraver. The impressions in this volume are from those original blocks; but they are worked, as is the 'Penny Magazine,' by the uniform pressure of the Printing Machine, and not, as wood-cuts generally are, by the Hand-Press. The cuts are purposely executed with reference to this mode of printing, the lights being produced by lowering the surface of the block itself." Notably, in this book, Knight credited the engraver of every image—details omitted from the issues of *The Penny Magazine*. Most of the images that Knight selected were engraved by John Jackson (1801–1848), a pupil of Thomas Bewick.

title page incorporated a small woodcut of the Applegath and Cowper double-cylinder perfecting press. This was the first "art book" printed on a printing machine rather than a handpress, and the title makes the point that the impressions of the woodcuts were made from the original wood blocks rather than from the stereotype plates used to print the same woodcuts in *The Penny Magazine*. Knight had this book printed by William Clowes on fine paper in a small-folio format, and it is likely that the edition was relatively small. The title page indicates that the book sold for "Fourteen shillings, bound in cloth, with gilt edges." Captions to the book's illustrations sourced them to their original appearance in *The Penny Magazine*.

For the rest of his life, Knight remained one of the leading proselytizers of the use of the new high-speed printing technology to produce inexpensive, high-quality literature. He was also a prolific writer of non-fiction and a historian of the history of books and printing, fond of comparing the limitations of handpress book production with the advances gained through machine-made paper and steam-powered printing. It was perhaps fitting that, in 1844, he initiated the series of monographs called Knight's Weekly Volume for All Readers with his *William Caxton, the First English Printer*, a biography of England's first printer and publisher.[90] Like other books published in this series (most of which were written by other authors), this was a quality work of over 200 pages with illustrations, but in small-octavo format and using relatively small type—the format selected, of course, to retail the book at very reasonable cost.[91]

Knight ended his biography of Caxton with a postscript extolling the advances in book production made possible by the new technology:

The Printing-Machine has done for the commerce of literature what the mule [spinning invention for textiles] and Jacquard have done for the commerce of silk. It has made literature accessible to all. It has given us the power of producing a Weekly Volume for One Shilling, for All Readers. We say For ALL Readers. The lowly and the exalted have long each stood upon a common ground—that of the Faith which knows no distinction of persons. We venture to believe that they should meet in the same manner upon the common ground of knowledge.[92]

Acknowledging Knight's argument that mechanization of book, newspaper, and magazine production benefited society by making print more affordable to more people, how did mechanization affect the working environment of the men and women working in the new printing plants? Was the working environment better or worse than in traditional handpress printing environments? In December 1839, writer and politician Sir Francis Bond Head, a former lieutenant governor of Upper Canada, published a long article in *The Quarterly Review* titled "The Printer's Devil." In contrast to Knight's constant emphasis on the socio-economic gains to readers from high-speed printing, Head focused more on the human side of the new technology, writing sympathetically about the workdays of the staff at William Clowes (including the many young boys and girls who were employed there) and describing the actual processes of composition and printing by machinery. He did not feature Clowes's handpress department, which was still used for short-run printing. Head was not critical of the working conditions at Clowes's printing plant; if anything, the working conditions were probably better than in many other factories at the time. Instead, Head wanted to describe his experience visiting this modern mechanized printing establishment. He left a vivid description of the Clowes press rooms in operation:

The visitor, on approaching the northern wing of Mr. Clowes's establishment, hears a deep rumbling sound, the meaning of which he is at a loss to understand, until the doors before him being opened, he is suddenly introduced to twenty-five enormous steam-presses, which, in three compartments, are all working at the same time. The simultaneous revolution of so much complicated machinery, crowded together in comparatively a small compass, coupled with a moment's reflection on the important purpose for which it is in motion, is astounding to the mind; and as broad leather straps are rapidly revolving in all directions, the stranger pauses for a moment to consider whether or not he may not get entangled in the process, and, against his inclination, as authors generally say in their prefaces, go "to press."[93]

Robert Seymour ("Shortshanks")
The March of Intellect (London, 1828)
Hand-colored print, 11 × 15¾ in. (28 × 40 cm)

Satirizing, and probably exaggerating the social impact of steam-powered printing, this print, signed with Robert Seymour's pseudonym "Shortshanks," was published in 1828, just as reformers, including Henry Brougham, Charles Knight, and the SDUK, were exploiting mechanized printing and low-priced publications as a tool for educational, governmental, and social reform. It shows a steam-powered automaton made of printing machine parts with a head of books topped by a university sweeping away quackery, delayed parliamentary bills, and court cases. Seymour's unusually long caption to the print, engraved with numerous misspellings, reads, "I saw a Vision, A Giant form appeard it's eyes where burning lights even of Gass [sic] and on its learned head it bore a Crown of many towers, Its Body was an Engine yea of steam its warms where wrn and the legs with which it strode like unto presses that men called printers use, from whence fell over and anon small Books that fed the little people of the earth, It rose and in its hand it took a Broom to sweep the rubish [sic] from the face of the land, the Special pleaders & their wigs also of the Quack Doctors also and the ghosts & those that whear [sic] Horns, & the Crowns of those that set themselves above the laws & the Delays in Chancery it utterly destroy'd, likewise it swept from the Clergy every Plurality, Nevertheless the lawyers & the Parsons & divers others kickt up a great Dust!!!"

From *Album de L'industrie française commerciale, manufacturière et agricole* (detail). See p. 102.

CHAPTER 6

Image-Reproduction Methods Appropriate for Rotary Printing Machines: Wood Engraving, Lithography, Steel Engraving, Electrotyping

Between 1814 and 1828 most printing done on the new printing machines was text because of the difficulty of printing illustrations on rotary presses. Once the technical problems of printing illustrations on rotary presses were solved, four methods of reproducing illustrations were successively adapted for use on printing machines, but not strictly in the order in which the processes were invented. The methodologies were wood engraving, steel engraving, electrotyping, and lithography; lithography, invented by Alois Senefelder toward the end of the eighteenth century, was the last of the processes to be adapted for mechanized printing. Ironically, it turned out that one of the oldest media for printing illustrations, woodcuts—used for reproduction of illustrations since the fifteenth century or earlier—turned out to be very well suited for reproduction of images on printing machines, after the invention of wood engraving by British engraver and natural history author Thomas Bewick at the end of the eighteenth century. It was by this method that the first illustrated publications printed by mechanized presses were illustrated.

At the end of the eighteenth century, Bewick published *A General History of Quadrupeds. The Figures Engraved on Wood by T. Bewick* (1790), which introduced the technique of wood engraving. Wood engraving, an innovation in the traditional method of creating woodcuts, became one of the most widely used methods of image reproduction in the nineteenth century,

"The Stag, or Red Deer"
From Thomas Bewick, *A General History of Quadrupeds* (London, 1790)

Bewick introduced his innovative method of engraving woodcuts in this book. With some technical modifications, forty years later, Bewick's method of wood engraving proved to be an excellent method of reproducing illustrations on rotary presses.

especially because wood blocks engraved using Bewick's techniques could withstand many thousands of impressions from machine presses with almost no deterioration. Bewick's wood engravings differed from earlier woodcuts in two ways. First, rather than using woodcarving tools such as knives, Bewick used an engraver's burin; this allowed him to create thin, delicate lines, often for the purpose of creating large dark areas in the composition. Second, while the older woodcut technique used wood's softer side grain, Bewick used the wood's harder and more durable end grain, allowing the creation of more detailed images. When applied in mechanized printing, Bewick's new method of wood engraving drove a rapid expansion of illustrations in nineteenth-century publications. And advances in stereotyping enabled the reproduction of wood engravings onto metal, where they could be mass-produced for sale to other printers.

In the 1790s, roughly contemporary with developments such as stereotyping and handpress printing on early iron handpresses, German actor and playwright Alois Senefelder invented lithography, or printing from stone, as a cheaper way of publishing his plays. The method involved using a greasy, acid-resistant ink as a resist on smooth, fine-grained limestone from Solnhofen in Bavaria, halfway between Nuremberg and Munich. Senefelder then discovered that he could print from the flat surface of the stone alone. This was the first planographic process of printing—that is, printing from a flat surface as distinct from relief printing (typography) or intaglio printing (engraving). It was also the first radically new method of printing since Gutenberg's invention of printing by movable type. For the process, Senefelder invented a special type of printing press.

Senefelder called his process stone printing or chemical printing, but the French name *lithographie* (lithography) became widely adopted. On 20 June 1801, Senefelder received British patent no. 2518 for "A New Method and Process of performing the Various Branches of the Art of Printing on Paper, Linen, Cotton, Woollen and other Articles." This patent, with eighteen pages of text and nine figures on a large folding plate, represents Senefelder's earliest technical description of the process of lithography. As the specification of the patent indicated, Senefelder foresaw the wide range of applications of his process beyond strictly printing on paper. By 1803, Senefelder had adapted zinc plates as substitutes for limestone in the process of lithography. Zinc plates eliminated the necessity of using smooth, fine-grained limestone, and also made it possible to lithograph larger images with plates that were much lighter in weight and thus more manageable in the press than limestones of equivalent dimensions.

After issuing various publications using lithography, in 1818 Senefelder published a manual of lithography in Munich entitled *Vollständiges Lehrbuch der Steindruckerey* (Complete textbook of lithography). This outstanding and comprehensive manual, which included many different examples of lithography, also introduced chromolithography with a two-color plus black lithographic reproduction of the first page of the 1457 *Mainz Psalter*, reproducing its large two-color initial letter. Senefelder's book was translated into French and published Paris in 1819. The same year, the book appeared in English, published in London by Ackermann as *A Complete Course of Lithography . . . Accompanied by Illustrative Specimens of Drawings. To which is prefixed a history of lithography*. Of the three editions of Senefelder's textbook, it has been argued that the English edition had the most impact in spreading the technique of lithography around the world.[94]

The first lithographed illustration published in a book printed by letterpress was in *Antiquities of Westminster*, a large quarto printed in London by Thomas Bensley for the author John Thomas Smith, with a title page dated 9 June 1807 (a few years later, Bensley would sponsor Friedrich Koenig's development of the steam-powered printing machine). According to its title page, Smith's work contained 246 engravings of topographical subjects, of which 122 were no longer in existence when the book was published. These engravings were published on 38 plates, nearly all either drawn or engraved by Smith, of which 2 were tinted and 12 were hand-colored (2 heightened with gold).

In *Antiquities of Westminster*, Smith experimented with various print media, including etching, steel engraving, mezzotinting, aquatint, and lithography. For his plate of the "Ceiling of the Star Chamber," Smith, a painter, engraver, antiquary, and sometime keeper of prints at the British Museum, used an old steel saw blade in its unsoftened state as a medium. He broke a number of burins in the process of trying to engrave the steel plate, and it took him two months to com-

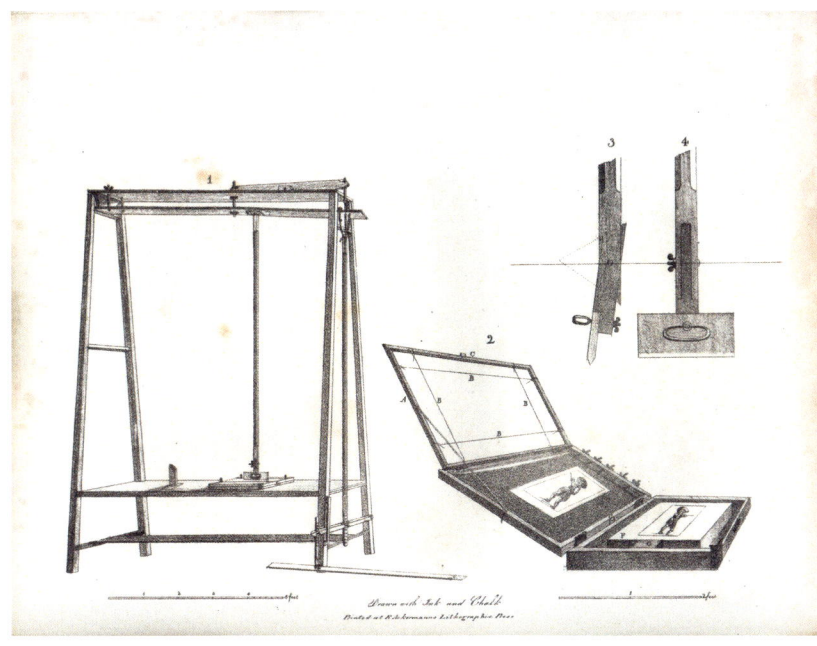

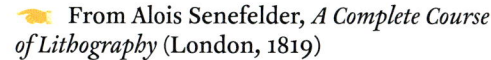 From Alois Senefelder, *A Complete Course of Lithography* (London, 1819)

The first planographic system of printing, lithography required entirely different kinds of presses than were used for relief printing. Senefelder invented two types of lithographic presses, illustrations of which he published in his *Complete Course of Lithography*.

From John Thomas Smith, *Antiquities of Westminster* (London, 1807)

This illustration is the first known use of lithography to illustrate a book printed by letterpress. Smith produced this experimental print ten years before Senefelder's manual of lithography was published.

plete the plate instead of the two days it would have taken to engrave the plate on copper. Steel engraving had been invented by an American, Jacob Perkins, in 1792, but Perkins's method used special tools and a special press that Smith would not have seen since Perkins did not emigrate to Britain until 1818. Because of the difficulty with his steel engraving, Smith did not produce any further prints by that method.

His image published in 1807 was, however, the first steel-engraved book illustration.

Smith's "Internal view of the painted chamber" is the first known instance of a lithograph being used to illustrate a book printed by letterpress. Smith's original intention was to illustrate the whole edition with one plate only produced by lithography; but after 300 prints were printed, the stone was ruined, and Smith

VISIT OF THE PRINCE OF WALES AND PRINCE ALFRED TO MESSRS. DAY AND SON'S LITHOGRAPHIC ESTABLISHMENT.—(SEE PAGE 486.)

From *The Illustrated London News* (3 May 1856)

A woodcut showing the Prince of Wales and Prince Alfred visiting the leading London lithographic establishment of Day and Son. In 1856, Day and Son had only flatbed lithographic handpresses in operation.

decided to revert to etching on copper for the remaining copies. As Smith explained in the text, only the first 300 copies of the edition have both the lithographed and etched versions.[95]

The first lithograph published in the United States was a rather dull-looking image of a home by the American artist Bass Otis (1784–1861), who is chiefly known for his portraits. The image appeared in 1819 in volume 14 of *The Analectic Magazine*.[96]

For about the first half of the nineteenth century, lithography remained a handpress process, with roughly half the output of handpress letterpress printing: 100 to 120 impressions per hour. On 3 May 1856, *The Illustrated London News* published a woodcut illustration of Prince Albert, the prince of Wales, and Prince Alfred (Albert's brother) visiting the lithographic establishment of Day and Son; the image shows only lithographic handpresses in operation.[97]

One of the first lithographic presses designed to be operated under steam power and intended for chromolithography was that invented by Jean Baptiste Huguet of Paris, which received British patent no. 1623 on 29 June 1864.[98] Another machine, also developed

🐦 From M. Knecht, *Nouveau manuel complet du dessinateur et de l'imprimeur lithographe* (Paris, 1867) Chromolithograph, 3 × 4¾ (8 × 12 cm)

A small format specimen of chromolithography printed on Théodore Dupuy's mechanized lithographic press.

🐦 From M. Knecht, *Nouveau manuel complet du dessinateur et de l'imprimeur lithographe* (Paris, 1867)

Dupuy's rotary press for printing from lithographic stones and his machine for grinding the stones in preparation for their use in lithography. Both machines could be powered by hand or by steam.

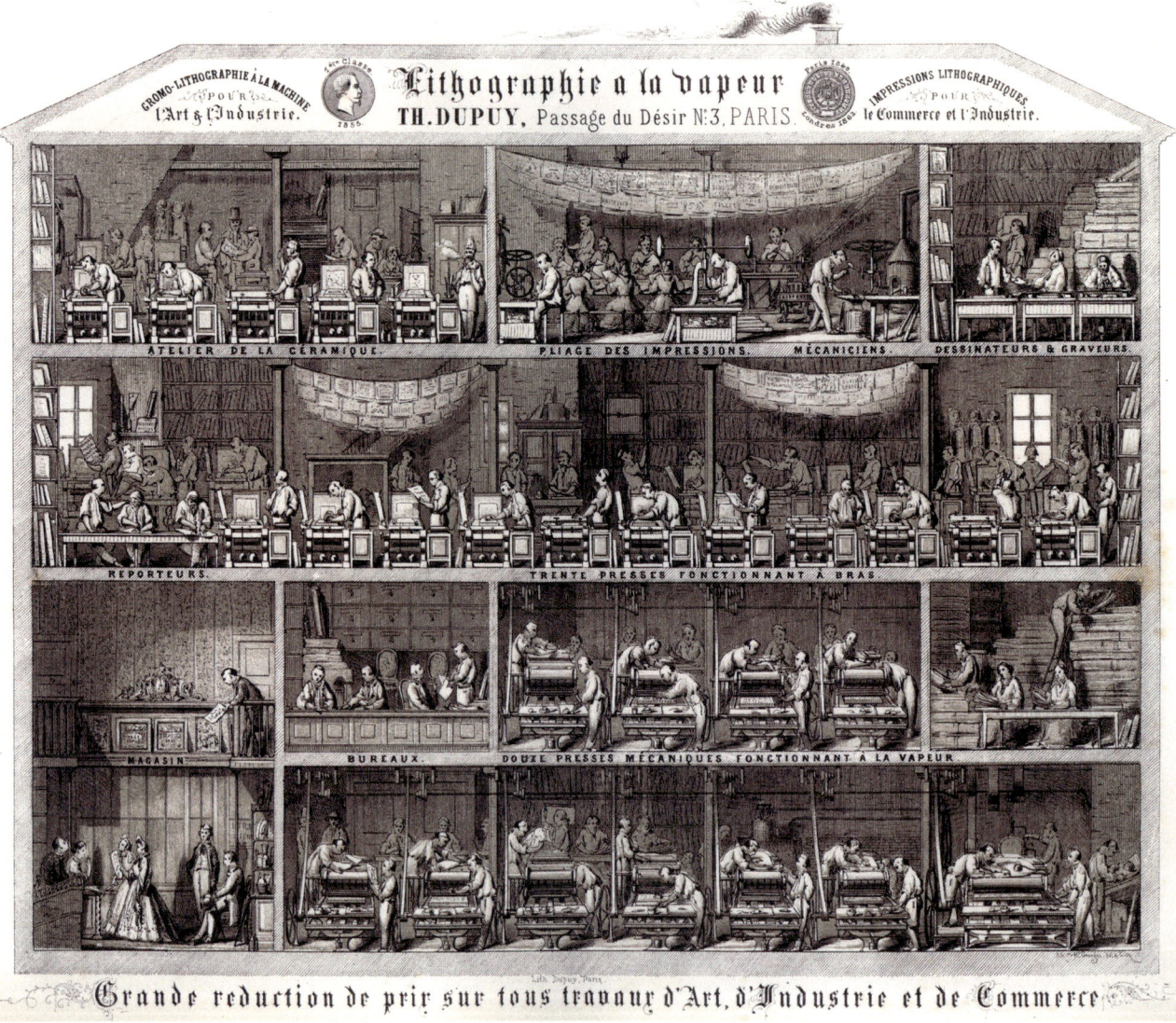

From *Album de L'industrie française commerciale, manufacturière et agricole* (Paris, 1865)

A visually complicated cross-sectional view of Théodore Dupuy's steam-powered lithography operation, in which he showed that he operated thirty manual lithographic presses and a dozen steam-powered presses, presumably of his own design. Dupuy printed all the images in this little-known *Album*, a collection of lithographed industrial advertisements.

in France, was that of Théodore Dupuy. An image of Dupuy's machine and a sample of chromolithography printed by it in eleven colors were included in the atlas of M. Knecht, *Nouveau manuel complet du dessinateur et de l'Imprimeur lithographe* (New complete manual for the designer and lithographic printer).[99]

Knecht stated that the machine cost 5,000 francs and that the steam engine to power it cost another 5,000 francs. The machine could be operated by hand crank or by steam. About the same time, Dupuy published an advertisement for his lithography business in *Album de l'industrie française, première année*, a short supplement to the second volume of *Album des célébrités industrielles contemporaines*,[100] in which he stated that he operated thirty manual presses and a dozen steam-powered lithographic presses, presumably of his own design. Most of the monochrome images in the *Album des célébrités industrielles contemporaines* were printed by Dupuy, though it is not specified whether the printing was done by steam or by hand.

Cette épreuve en quatorze couleurs été tirée publiquement au Palais L'Exposition Internationale de 1867 . . . sur la Presse Chromo–Lithographique–Mécanique Th. Dupuy (Paris, 1867) Chromolithograph, 12¾ × 10¼ in. (32.5 × 26 cm)

A broadside printed for demonstration purposes on Théodore Dupuy's Presse Chromo-Lithographique-Mécanique during the Exposition International in Paris in 1867. This very complicated chromolithograph, finely printed in fourteen colors, was clearly intended to prove the capabilities of his machine. Dupuy designated the Exposition incorrectly; it was formerly known as the Exposition Universelle.

Lithographie Artistique, J. Minot & Cie., Imprimé sur la Nouvelle Presse Lithographique Marinoni (Paris, 1889)
Chromolithograph, 3 × 4⅓ in. (7.5 × 11 cm)

A small trade card for artistic lithographers J. Minot & Cie. in Paris, printed on Marinoni's new lithographic press at the Palais de l'Exposition in 1889.

At the Exposition Internationale in Paris in 1867, Dupuy demonstrated his Presse Chromo-Lithographique-Mecanique and printed a fourteen-color chromolithographic proof demonstrating the exceptionally high-quality work that could be achieved with that machine. This is one of a few images to explicitly state that it was printed on the first generation of steam-powered lithographic presses.

Even after steam-powered lithographic presses were developed during the 1860s, many lithographers operated both manual presses and steam-powered presses, and unless they specifically stated the means by which lithographs were printed, it can be very difficult to determine whether they were printed by hand or by steam, or possibly by both kinds of presses.[101] In the United States, R. Hoe & Co. did not advertise a steam-powered lithographic press in their catalogue published in 1867, but we see their version of Dupuy's machine advertised for sale in their large-format catalogue published in 1873 and in a color poster that was probably published at about that time. Hippolyte Marinoni also produced steam-powered lithographic presses. An example of work done on one of his presses is a small trade card printed by chromolithography at the Palais de l'Exposition in 1889.

By the time of the Exposition Universelle of 1900 in Paris, Imprimeries Lemercier, the leading French lithography firm that had been founded in 1826 by Joseph-Rose Lemercier, printed a series of volumes cataloguing the exhibition. In a volume entitled *Catalogue général official, volume annex* (1900), the company devoted no less than thirty pages in the catalogue to an unusually detailed and extensively illustrated infomercial about their printing facilities. A feature of Lemercier's infomercial was the emphasis on powering their printing machines by electricity. At the end of the article is an illustration of a very large horizontal steam engine that apparently could generate enough power to run most of Lemercier's printing machines and also light their factory. Electricity was, of course, the cutting edge of innovation at the time and was featured extensively at the 1900 exposition (this was before the power grid had been built out sufficiently in Paris to supply power to Lemercier's plant from the public utility). Notably, Lemercier was still printing from stones (they illustrated their extensive stone-storage facility), as well as lithographing extensively from aluminum plates. In an image captioned "*Machine rotative tirant sur aluminum*" (Rotary press printing on aluminum), the company illustrated

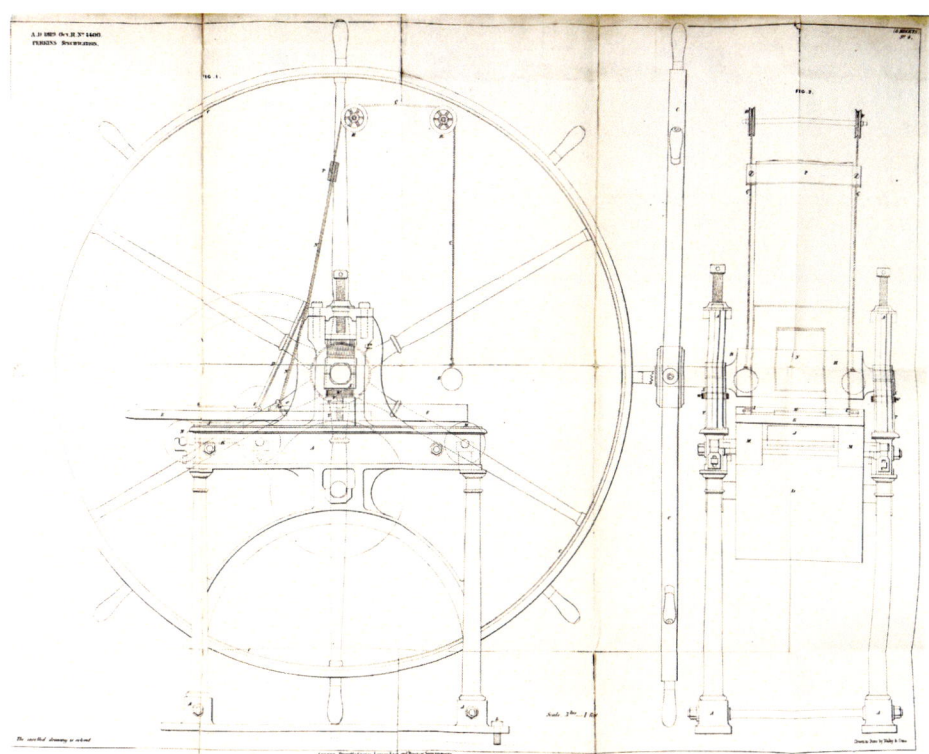

Jacob Perkins, patent no. 4400 (1819) for "Engine lathe for engraving surfaces, printing and coining presses . . ." (London, 1857)

Sheet no. 4 from Jacob Perkins's English patent of 1819 shows the outline of his steel engraving press, a key component of his system of steel engraving.

United Kingdom postage stamp (London, 1840)
Steel engraving
The Penny Black, featuring a profile of Queen Victoria

An example of the first adhesive postage stamp, first printed in England in 1840 by the steel engraving and printing process invented by Jacob Perkins. Before Perkins's method of reproducing steel engravings existed, printing the enormous number of very small, highly detailed images required for postage stamps was impossible, since only steel engravings could withstand hundreds of thousands of impressions. The total print run of the Penny Black was 286,700 sheets, containing a total of 68,808,000 stamps.

a press that appears little changed from the 1860s Dupuy design, except that it was powered by an electric motor. And they still had some handpresses for printing lithographs by hand.

The second illustration process applicable for use on printing machines, steel engraving, was a new process when it was invented by American inventor Jacob Perkins in 1792. Perkins invented steel engraving for the process of banknote printing, but he was unable to commercialize the process successfully in America. Motivated by a £20,000 prize offered by the British government for development of unforgeable banknotes, Perkins moved to England in 1818, where he and his associates set up a studio and spent months developing samples. Unfortunately for them, Sir Joseph Banks thought that "unforgeable" also implied that the inventor should be English by birth. For that reason, Banks's successors awarded future contracts to the English printing company started by English engraver Charles Heath.

In 1819, Perkins received British patent no. 4400 for his system of engraving and printing steel engravings.[102] To work around the issue of his American nationality, Perkins went into business with English engraver, currency and stamp printer, book publisher, and illustrator Charles Heath, who had the government contract. To produce steel engravings by Perkins's method, engravers such as Heath had to use special plates supplied by Perkins, which had to be printed on presses designed and provided by Perkins. Both the plates and the presses were described in Perkins's patent.

In 1840, Perkins's methods reached true mass production when they were used to print the world's first adhesive postage stamp, the Penny Black. Perkins's process, which proved the extreme durability of

Thomas Campbell, *The Pleasures of Hope* (London, 1821)

One of the first books illustrated with steel engravings. Each of the four steel engravings in Campbell's book appears to have been engraved as if it were the title page. The small images credit R. Westall as artist and C[harles] Heath as engraver. In very small type at the foot of each page, we read "Perkins, Fairman & Heath," the name of Heath's partnership with Jacob Perkins and Gideon Fairman.

steel plates compared to any other available graphic reproduction medium of the time, remained in use until 1879.

It was proved that fully 400,000 imprints could be taken from a single plate without signs of wear. Altogether, over twenty-two thousand million stamps for Great Britain and the Colonies were printed by the Perkins' process during these years.[103]

The book publisher who first recognized the aesthetic and economic advantages of steel engraving was George Longman. Beginning in 1821, Longman issued twenty books containing, altogether, around seventy steel engravings. Longman's first production using steel engravings was in an edition of Thomas Campbell's *The Pleasures of Hope*, issued on 10 January 1821. Charles Heath's four engraved illustrations for this work, including its engraved title page, were dated 1820. According to Longman's ledgers, 3,000 copies of this edition were printed, and in November 1824, a further 3,000 copies were printed from the same plates, reflecting the extreme durability of steel engravings compared to engravings from copperplates. There was also an 1822 printing; I have a copy in my collection bearing that date.[104]

The last of the widely used image-reproduction media for mechanized printing in the first half of the nineteenth century—especially of large editions—was the electrotype. Electrotyping is an electro-chemical method of forming metal parts that exactly reproduce a three-dimensional model; it was invented in 1838 by Russian electrical engineer and physicist Moritz von Jacobi of Saint Petersburg. The electrotype process had applications in a wide variety of fields, including making printing plates by electroplating, since it produced an exact facsimile of any object with an irregular surface, such as an engraved steel or copper plate, a woodcut, or a locked-up form of set type. The ability to duplicate woodcuts or wood engravings precisely also made it possible for publishers to sell electrotype copies of images to other publishers, creating a new source of revenue. The reproduction quality of electrotypes tended to be higher than the much older technology of stereotype plates. Stereotypes typically used plaster of Paris or papier-mâché (flong) for creating a mold from which a printing plate was made. These materials were prone to some degree of irregularity or shrinkage during the drying and casting process, potentially introducing slight distortions into the printing plate.

The first use of electrotyping for printing in England was in *The London Journal* of April 1840. The process was first used for illustrations in America by Joseph Alexander Adams in 1841.[105] Chapter 9 discusses the first large-scale application of electrotypes in book illustration in the United States in Harper and Brothers' *Harper's Illuminated and New Pictorial Bible*, published in parts between 1843 and 1846. Among the extraordinary features of this exceptional Bible were the 1,600 historical illustrations engraved by Joseph Alexander Adams and reproduced by electrotype. By the 1850s, electrotypes, which cost more than stereotypes to produce, were widely used in the United States and Europe to print books and magazines, especially because of their extreme durability for large print runs. Nicolas Barker stated that making electrotypes was a standard practice for Macmillan and Co. publishers from the 1850s onward:

If sales, though adequate, were likely to be of short duration, the book was printed from the type it was set in. If it had a longer life, stereotype plates were made from the type by moulding it in papier-mâché and casting hard metal in them; the plates could be faced with nickel for extra strength. If it [the book] enjoyed both large sales and a long life, the extra expense of electrotype plates was called for. Pages placed in an acid bath were covered in copper by electrolytic action, and the copper "shells" filled with hard metal. Roughly speaking, type was good for a run of 20–25,000 [copies], stereos for 50,000, and electros for anything up to 300,000; in practice, these figures were often exceeded.[106]

Numerous other methods for reproducing images on printing machines were invented in the second half of the nineteenth century, but wood engraving remained the most widely used methodology until 1880, with lithography running a close second.[107] After that date, these methods were gradually supplanted by newer technologies.

Jean-Charles Develly. Maquette for a ceramic plate (September 1831) Watercolor drawing, diameter: 5⅓ in. (13.5 cm)

This watercolor drawing, dated September 1831 on its back, may be the earliest original dated image of a printing shop running a steam-powered printing machine. The machine is a reasonably accurate artist's rendition of a double-cylinder perfecting press based on the Applegath and Cowper design. Clouds of steam from the steam engine on the left side of the image are visible. Whether the plate for which this maquette was created was ever completed is unknown. Develly did complete a plate depicting a handpress printing shop operating a Stanhope Press that is preserved at the Sèvres–Cité de la Céramique, the French national ceramics museum.

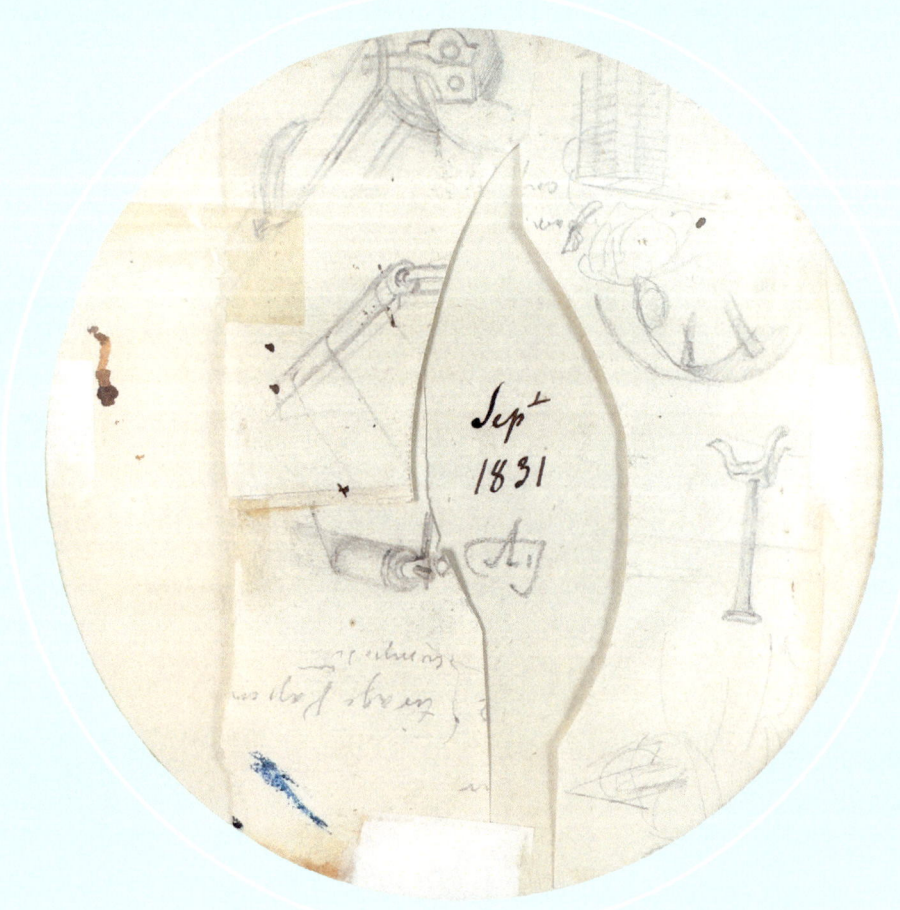

CHAPTER 7

Developments in Mechanized Book and Newspaper Production, 1800–1850

After Friedrich Koenig and Andreas Bauer departed England to establish their printing machine factory in Oberzell, Germany, printing technology continued to advance in England, and for the first three decades of the nineteenth century, English printing engineers led in the innovation and manufacture of printing machines. Some of the first major improvements were made by Edward Shickle Cowper, who invented his double-cylinder perfecting press in 1817. Cowper's press improved upon the Koenig and Bauer machine by greatly simplifying Koenig's design, reducing the number of wheels required for operation from sixty to sixteen. In 1818, Cowper received British patent no. 4194 for "Certain improvements in printing presses or machines used for printing" for his double-cylinder perfecting press. From this date forward, most of the printing machines sold by Cowper and his business partner, printer, engineer, and inventor Augustus Applegath, were perfecting cylinder presses. Presses of this design, or a modification of this design, were in wide use for decades in England and on the Continent. The early customers for the machines were primarily printers of newspapers and magazines.

Other significant innovations made by Cowper were his ink-distributing table, patented in 1818, and his "method of conveying the sheet of paper from one cylinder to another in a perfecting machine by the construction of two subsidiary 'carrying drums' between the impression cylinders, on which the sheet was carried by means of two sets of endless strings, 'each composed of two or more strings kept tight by weights or springs', the printing cylinders and carrying drums being connected by means of toothed wheels."[108] Cowper also invented a method of creating curved stereotype plates that would fit on the cylinders of rotary printing machines.[109]

A description of Cowper and Applegath's perfecting press is included in Thomas Curson Hansard's monumental *Typographia*,[110] which may be considered the most significant manual on printing after Moxon's *Mechanick Exercises* (1683-1884). Writing just a little more than a decade after Koenig's invention of the steam-powered press, Hansard took a particular interest in the latest innovations, devoting chapter VI to "Improved Manual Presses" and chapter VII to "Printing Machines." Reflective of the novelty of mechanized printing presses, Hansard called them "Printing Machines, or Engines, in which the art of the engineer is every thing, and the printer, nothing; whether the motive power be that of steam, horse, or man, the impulse to the machinery being unaided or undirected by professional judgement, or an effort of mind."[111] His statement appears to reflect his judgment that the design and operation of printing machines, as well as their output, were more representative of engineering than the traditional handpress "printer's art." In 1825, when Hansard published, printing machines were mainly being used to print newspapers and a few unillustrated books,

The Printer & the Letter Founder (Mainz and Cincinnati, c. 1850)
Hand-colored lithograph, 10⅔ × 13 in. (27 × 33 cm)

This lithograph was published in Mainz, Germany, by Joseph Scholtz, and in Cincinnati, Ohio, by Eggers and Co. It depicts a typical handpress printing office that printed on a Stanhope iron handpress during the mid-nineteenth century. The letter foundry is on the right of the image. Handpress operations like this continued to print smaller editions during the nineteenth century, while printing-machine engineers built faster and faster machines to meet the demands of large-circulation newspapers and magazines, and large-edition book printers.

Cowper, Machinist (London)
Business card, 4 × 3 in. (10.5 × 7.5 cm)

An early business card of printing engineer Edward Cowper.
His first invention was a paper cutter, patented in 1813.

Portrait of Edward Cowper (c. 1830)
5 × 4 in. (12.7 × 10.15 cm)
Courtesy of The Science Museum, London

The image shows Cowper with a small model of one of his printing machines.

and experts typically considered their results to be of less than optimal quality.

In his chapter on printing machines, Hansard reviewed the history of the machines as he knew it and described and illustrated the latest machines as they had evolved after Koenig. He also described a smaller rotary printing machine of his own invention, designed to be powered by one or two men turning a crank. Hansard seems to have developed his printing machine around the time he was writing and publishing *Typographia*; he stated that his machine, which appears to have been a modification of machines designed for newspaper production by the Scottish engineer David Napier, had been tested for six months before *Typographia* was published. Avoiding the added costs of connecting a steam engine to a printing machine, Hansard designed his machine to be operated by "two men turning a fly-wheel" rather than a steam engine. This machine, which Hansard called "the Nay-Peer" after its manufacturer, was, according to Hansard, "more likely to succeed in all its pretensions than any which has yet been offered to us; more particularly as it supersedes the necessity of steam power."[112] He illustrated his machine with a large folding plate, larger than any other plate in the book. Hansard also provided a beautiful illustration of Thomas Rutt's printing machine, the design of which was based upon the design of Friedrich Koenig's first cylindrical machine, invented in 1812. The illustration shows that Rutt's machine was driven by

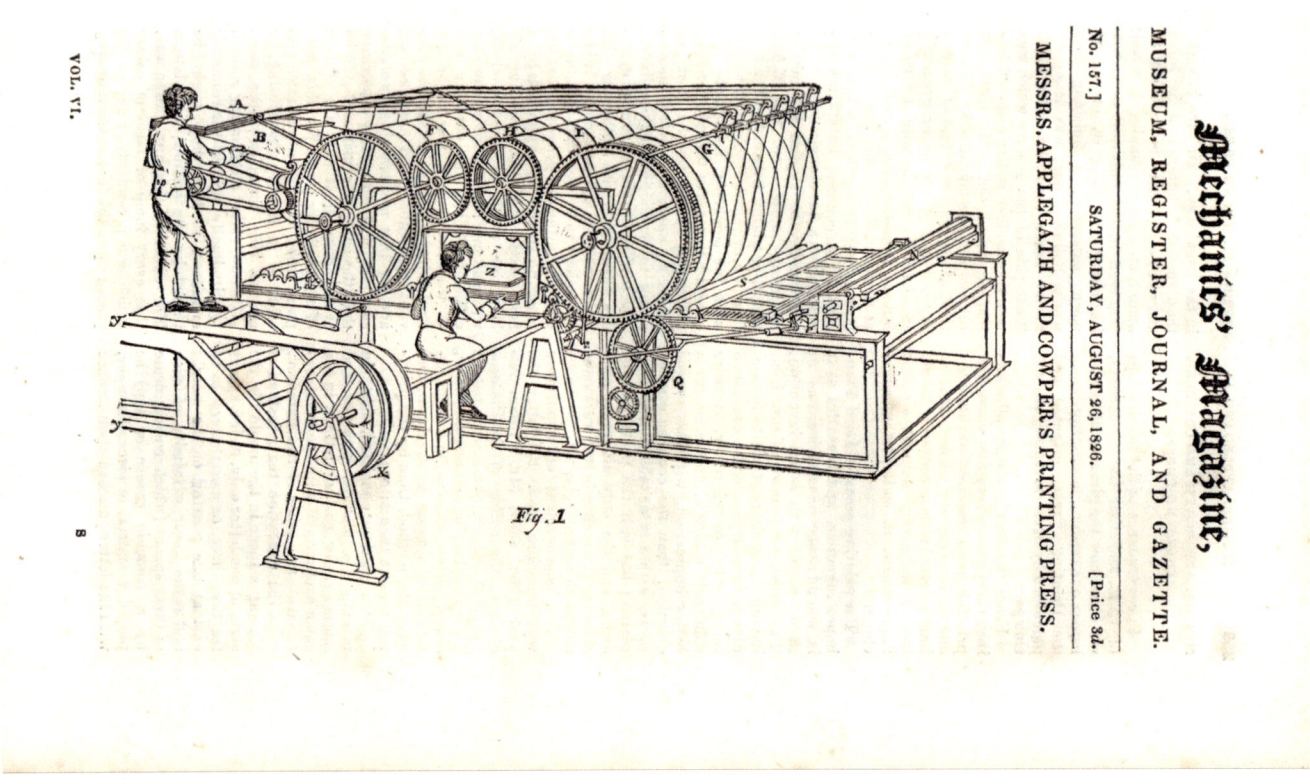

Mechanics' Magazine, no. 157 (26 August 1826)

An Applegath and Cowper double-cylinder perfecting press illustrated on the cover of *Mechanics Magazine*.

a hand crank. Hansard also republished an image of "Bensley's Printing Machine," designed by Applegath and Cowper, that had originally been published in 1822.[113] Because of the expense of steam engines, difficulties in their operation, and scarcity of the equipment in certain locations, there was a definite demand for rotary presses operated by hand cranks and flywheels during the first half of the nineteenth century and beyond, especially for mid- to low-volume printing operations.

An unusual aspect of Hansard's book were his comments upon very recent developments. Among those was his elaborate critique of "The printing machines and other inventions relative to printing of Doctor William Church,"[114] describing some of the remarkable and seemingly incredible claims that this American inventor made in his patent specification of 1822.[115] I will discuss this in chapter 11 on the history of mechanized typesetting. Another invention that Hansard mentioned was Charles Babbage's Difference Engine No. 1, announced in Babbage's *A Letter to Sir Humphry Davy, Bart. . . . on the Application of Machinery for the Purpose of Calculating and Printing Mathematical Tables*.[116] Babbage's motivation in developing his Difference Engine No. 1 was to present the errors in mathematical tables caused by their calculation by human calculators and to prevent typesetting errors by having the Difference Engine do the typesetting of the tables. To Hansard, Babbage's goals seemed more realizable than Church's.

As excellent as the machines designed by Cowper and Hansard were, they were not rapidly adopted by all printers. In 1820, there were only eight steam-driven presses in the huge printing industry of London, nearly all used by newspapers, except for those of Strahan, the King's Printer, by then owned by

From Thomas Curson Hansard, *Typographia* (London, 1825)

An illustration of Hansard's double cylinder "Nay-Peer" printing machine. Hansard's caption states that the machine was invented and manufactured by David Napier, presumably to Hansard's specifications. With crank handles on both sides of the wheel, the machine was designed to be cranked by two men.

Andrew and Robert Spottiswoode. "One of the first steps taken by the Spottiswoodes was to install steam printing. Thus in 1819 we find them purchasing from Maudslay a steam engine at a cost of £782, as well as a patent perfecting cylinder machine from Applegath for £1200, and a foundry for £785."[117]

A factor that slowed the adoption of printing machines, apart from their high cost, was that some publishers were able to print remarkably large editions on handpresses. One example is *The Mirror of Literature, Amusement, and Instruction*, which began publication in London on 2 November 1822. This

John Limbird, *The Mirror of Literature, Amusement, and Instruction*, no. 1 (2 November 1822)

For the cover of the issue launching his magazine, publisher John Limbird featured people operating a treadmill at the Brixton workhouse—undoubtedly a controversial topic at the time.

sixteen-page weekly magazine, published by John Limbird, has been characterized as England's "first long-lived cheap periodical"; priced at 2 pence per copy, it may have sold as many as 80,000 copies per week. With one woodcut per issue, it was also one of the first illustrated large-circulation magazines.

Another factor that deterred the rapid adoption of printing machines was that many books were printed in relatively small editions, which could be effi-ciently printed on handpresses. For example, Victorian novels were typically published in three-volume or "triple-decker" format in editions of around 1,000 copies, priced at a guinea and a half for the three-volume set. This price made purchase of the novels unaffordable to all but the upper class; instead, most people rented novels from commercial lending libraries such as Mudie's. This convention in publishing restricted the number of copies of novels sold but all but guaranteed publishers and the commercial lending libraries a small profit on each.

The first printer to exploit machine printing in book production at scale was William Clowes, who built a printing business in London far larger than had been possible before the availability of steam-powered printing machines.[118] In 1823, Clowes installed his first steam-powered printing press, designed by Applegath and Cowper. Clowes's facility happened to adjoin the palace of Britain's wealthiest man, the duke of Northumberland, who instituted a court action for noise and pollution abatement caused by Clowes's machines. The presses were excessively noisy, but Clowes succeeded in forcing the duke to pay the huge costs of moving Clowes's heavy machinery, and in 1827, the firm took over Applegath's premises in Duke Street, Blackfriars. Within a few years, the firm was operating twenty-five high-speed Applegath and Cowper steam-powered printing machines, twenty-eight handpresses, and six hydraulic presses, and employing over 500 workers. It is notable that the firm reached this extent before Clowes died in 1847, indicating the enormous increase in book production made possible by the new high-speed presses and the availability of machine-made paper. That the Clowes firm also operated twenty-eight handpresses during the same period reflects the continuing use of handpresses for short-run printing. Iron handpresses of the Stanhope and later designs remained in wide use throughout the nineteenth century, continuing to employ pressmen trained in traditional printing methods. For this reason, handpress printers, who chose to continue with traditional methods rather than learn the new technologies, were able to keep their jobs.

To supply his printing operation, Clowes operated a type foundry, casting up to 50,000 pieces of new type per day and holding around 500 tons of type and 2,500 tons of stereotype plates. He kept tons of type locked up for months and, during the height of his

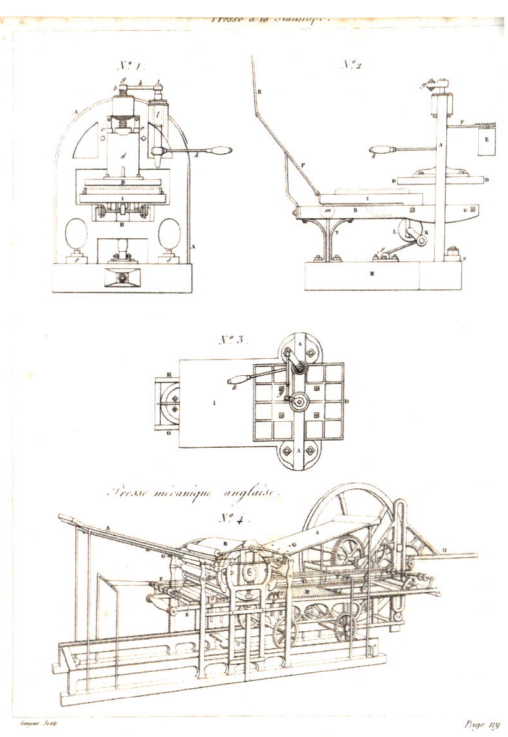

Audouin de Geronval, *Manuel de l'imprimeur* (Paris, 1826)

Below the schematics of the Stanhope Press, this engraved plate illustrates a single-cylinder printing machine, invented in England by a "Mr. Miller," that could print 2,000 copies per hour on one side of the sheet when the machine was cranked by one man "assisted by four children." At this date, French manufacturers did not yet produce printing machines or iron handpresses, so all the examples that Geronval illustrated were made in England.

career, could print a 100-page folio report in a day or night, or a thousand pages in a week from newly set type. As Samuel Smiles attests:

From his gigantic establishment were turned out not fewer than 725,000 printed sheets, or equal to 30,0000 volumes a week. Nearly 45,000 pounds of paper were printed weekly. The quantity printed on both sides per week, if laid down in a path of 22 1/3 inches [57 cm] broad, would extend 263 miles [423 km] in length.[119]

Clowes was also an innovator in terms of working practices and, in 1820, became one of the first employers to start a benevolent fund for his workforce.

Printing machines were introduced to France about a decade after they were invented in England. According to James Moran, "The first printing machines used in France were of foreign manufacture, beginning in 1823 with a two-cylinder made by Applegath for the *Bulletin des lois*, Paris, and a 'presse à gros cylindres' for Firmin Didot. Two years later the *Journal des Débats* and *Le Globe* were printed on Napier drum cylinder machines. A general printer, Fournier, of rue de Seine, ordered a large Applegath cylinder in 1827, and by that year there were twelve printing machines at work in Paris, eight of which were of English origin."[120]

Hansard's *Typographia* appears to have been widely

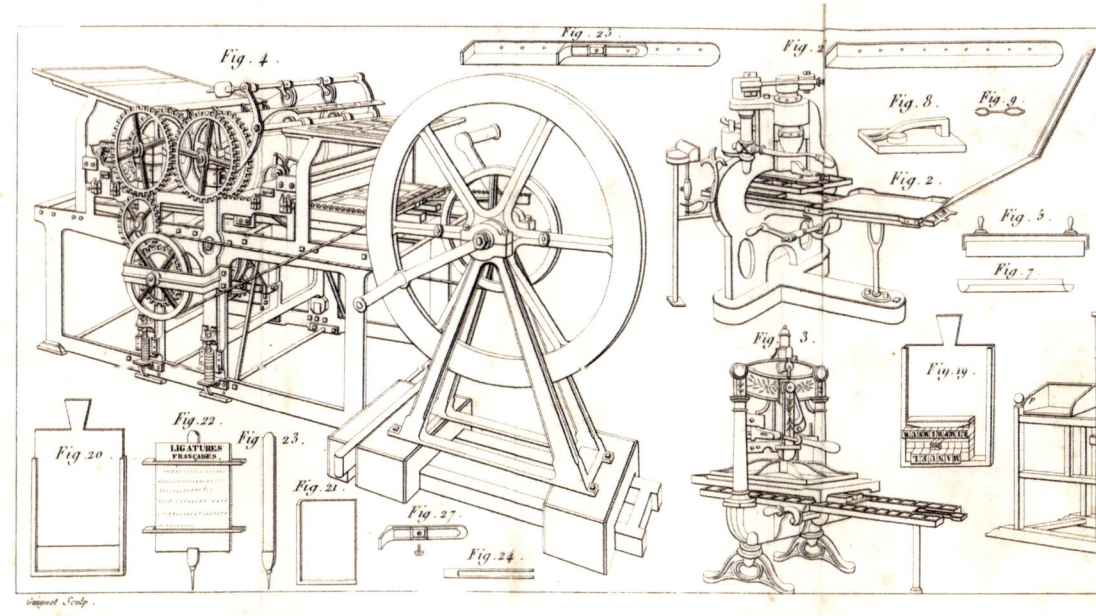

From Audouin de Geronval, *Manuel de l'imprimeur* (Paris, 1826)

The folding plate includes a reduced copy of Hansard's "Nay-Peer" machine, an example of which had been purchased to print the *Journal des débats* in Paris. That both machines featured by Geronval were powered by hand crank, rather than a steam engine, reflects the shortage of steam engines in France at the time.

read by printers on the Continent as well as in England. In 1826, the year after Hansard published *Typographia*, Maurice Ernst Audouin de Geronval issued *Manuel de l'imprimeur*,[121] the first printing manual in French to discuss machine printing. This little-known work is probably the first manual after Hansard's to discuss the design and operation of "machine presses." Audouin focused on the most efficient printing technology then available. With respect to presses, he emphasized the Stanhope handpress and a rotary machine of the Koenig type. Later in his book, he illustrated a rotary machine that he says was ordered for the printing of the *Journal des débats*, the most widely read newspaper in France of the period; the machine illustrated was designed to be driven by a large hand crank rather than a steam engine. The plate that Geronval published was an exact copy of the plate reproduced by Hansard in his *Typographia*; from this, we may assume that the *Journal des débats* had ordered from David Napier one of Hansard's "Nay-peer" machines. In the first part of the nineteenth century, France, like Germany, lagged behind England in the development of steam technology, and it is likely that the *Journal* wished to experiment with machine printing without the complications of a steam engine.

Opposition to the introduction of printing machines in France may not have been immediate. We have evidence of it during the three-day July Revolution, also called the Second French Revolution or *Trois Glorieuses*, in which the Bourbon monarch Charles X was overthrown and replaced by his cousin

Camille Hyacinthe Odilon-Barrot, *Habitans de Paris!* (Paris, 1830)
Poster, 11⅔ × 17½ in. (29.5 × 44.5 cm)

At the end of the July 1830 revolution in France, after damage had been done to printing machines at the Imprimerie Royale, politician Camille Hyacinthe Odilon-Barrot published this poster imploring the public not to destroy printing presses.

Louis Philippe, duke of Orleans. A factor motivating that revolution was the decree of Charles X, on 25 July 1830, suspending freedom of the press. On 29 July 1830, during the July Revolution, thirty printers broke into the Imprimerie Royale and destroyed or damaged the five printing machines that had been installed at the end of 1829. This was both a political statement and an act of resistance against mechanization. Printers felt threatened by the installation of printing machines, and typesetters opposed the introduction of stereotyping, to which they attributed loss of employment.

On 30 July, the end of the *Trois Glorieuses*, French politician Camille Hyacinthe Odilon-Barrot, active in the revolution, implored the public not to destroy or damage any other printing presses. A poster that he had printed for posting on the streets of Paris may be translated as follows:

RESIDENTS OF PARIS!

The conquest of our liberty has cost us blood, some very precious blood, since it was the blood of our friends, but thank God it has not been defiled by any disorder so far. It is our honor to all who have fought for the most beautiful and just cause, to prevent any property from being violated. But of all the properties which we hold most sacred those which we must protect most effectively are the Presses, which were the first instruments of our deliverance and on which despotism first exerted its violence. Shame on us if, imitating despotism in this way, a single one of our presses is damaged! The Commission Municipale de Paris places under the safeguard of all the citizens all the presses, public and private.

Made at City Hall, this 30 July 1830.
For the Municipal Commission,
Odilon-Barrot

In September 1830, two months after the destruction of the Imprimerie Royale's printing machines, certain printers and typesetters of Paris organized to boycott government printing projects. The workers involved in the boycott were supported by Parisian printers, such as Louis-Armand-Jean Fain and François-Jean Baudouin, and publishers like Würtz, of Treuttel and Würtz, and Pierre-François Ladvocat also defended them, urging the government to further increase its financial support. According to Ladvocat, only 300 out of 2,000 typographers/typesetters in Paris were employed full-time in September 1830, and if the crisis continued, bookbinders, papermakers, and other workers in the publishing industry would be similarly affected. Ladvocat reminded the government of the action of the Parisian printing workers during the *Trois Glorieuses*, especially of their key role in the mobilization of Parisian artisans against the ordinances of Charles X. At the end of September, the *Tribunal correctionnel de Paris* acquitted typographers accused of forming an illegal association.[122]

The first book published in German to provide an early history of the development of machine printing was Benjamin Krebs's *Handbuch der Buchdruckerkunst* (Handbook of the art of printing), issued anonymously in Frankfurt in 1827. Krebs, who referred frequently to Hansard's *Typographia* in his text, devoted nearly 100 pages to the development of high-speed printing machinery (*Schnellpresse*) from its conceptualization by Nicholson up to the time of writing. As one would expect, he devoted considerable space to the contributions of Friedrich Koenig and Andreas Bauer at Oberzell, but he also discussed available early printing machines from other manufacturers.

While Krebs was up-to-date on high-speed printing machines, the printing shop, type foundry, and bookstore that he operated, the Andreäschen Buchhandlung,[123] used smaller presses. In the beautiful font catalogue that he published in 1834, he added a remarkable image of its printing office in operation. This is probably the only illustration to show a Ruthven press and a Stanhope Press in use, along with the even more obscure Cogger press.[124] It is also one of the only illustrations from the 1830s of a printing office in operation.

At this time, English printing engineers were the printing technology leaders. In 1827, Augustus Applegath invented a four-feeder printing machine that could print between 4,200 and 5,000 copies per hour. The illustration, titled "The 'Times' Printing Machine," shows the press that was operational in 1828. About 13 feet high and 14 feet long (4 × 4.25 m), it remained operational at *The Times* until about 1850.[125]

Applegath's business partner, Edward Cowper never wrote a book, but he did publish a significant article on printing machines in 1828, and he followed up with another summary article in 1850. Cowper's

Andreäschen Buchhandlung, *Proben aus der Schriftgiesserey der Andreäischen Buchhandlung* (1834; facsimile ed., Pinneberg: Verlag Renate Raecke, 1984)

In 1834, the Andreäschen Buchhandlung book-printing office in Frankfurt am Main showed off their three different English presses in operation. They had a Ruthven press, a Stanhope Press, and a Cogger press. At this time, none of their presses were made in Germany. The only printing machine company in Germany, Koenig and Bauer at Oberzell, was producing rotary presses for newspapers, rather than machines for book printing.

The "Times" Newspaper Printing Machine (London, 12 October 1833)

The image, printed as a negative, shows eight men operating the four-feeder machine designed by Augustus Applegath.

Illustrirte Zeitung, vol. 2, no. 45 (4 May 1844)

Augustus Applegath employed engineer Thomas Middleton for assistance in building his four-feeder cylinder press for *The Times*. In 1843, Middleton built machines based on Applegath's design for other newspapers, including the two machines illustrated here for *The Illustrated London News*, which first published this woodcut in its issue of 2 December 1843. The two separate double-cylinder machines next to one another in the image each printed one side of sheets of the paper. Each machine had two paper feeders and two "takers-off." This is one of the only images that shows how the fanned-out sheets of paper were hand-fed into the press, how the printed sheets came out at the top level, and how men carried sheets of paper to or from the press on their shoulders.

Andrew Ure, *A Dictionary of Arts, Manufactures, and Mines* (London, 1839)

This image of a pressroom shows a single-cylinder rotary printing machine being driven by a man turning a hand crank.

Andrew Ure, *A Dictionary of Arts, Manufactures, and Mines* (London, 1839)

This image of a pressroom shows a double-cylinder printing machine credited in small letters to "Cowper" (i.e., Edward Cowper) on the center base of the drawing of the machine. This is one of the only images of a printing machine in operation that actually shows the table steam engine driving a flywheel powering the machine. The boiler driving the table engine would have been kept outside for safety reasons.

1828 article, "On the Recent Improvements in the Art of Printing,"[126] laid out the key developments in the history of printing machines up to the time of writing, presenting his unique perspective on the significant details. He described nuances of the differences in the paper paths through the cylinders of the different machines; tracing the paper path was one of the best ways to distinguish the operation of the different machines. He also described variations in the inking mechanisms and in the ways that stereotype plates were mounted in the machines. He pointed out that an image reproduced in the *Literary Gazette* and reproduced again in Cowper's article was deceptively labeled as "Bentley's Machine," as if Bentley had invented it, rather than owning it. At the end of his paper, Cowper stated that he and Applegath had sold sixty printing machines to date, citing the newspapers that had acquired them.

One of the people who read Cowper's paper most carefully was Andrew Ure, who summarized Cowper's analysis and illustrated the machine printing process in considerable detail in his *A Dictionary of Arts, Manufactures, and Mines: A Containing a Clear Exposition of their Principles and Practice*,[127] which contains some of the very best illustrations of printing machines being powered by hand crank and steam engine on pp. 1036–1038.

Two decades later, Edward Cowper summarized the progress in high-speed printing that had been made during his career in a paper titled "On Printing Machines, Especially Those Used in Printing 'The Times' Newspaper."[128] The paper was, to a certain extent, the summation of his career, as Cowper, who was born in 1790 in the era of handpress printing, died two years after this paper was published. Cowper's role in the advance of printing from 200 to 250 copies per hour on a handpress to 10,000 copies per hour when he published this paper was, of course, monumental.

Cowper began his paper with a recap of the paper he published in 1828. He then discussed in considerable detail his partner Augustus Applegath's eight-cylinder vertical machine, which, when fed by eight boys at sufficient speed, could print 10,000 impressions per hour. One of the main drawbacks of this system was that it required hand feeding of paper and was thus dependent upon the speed of the humans feeding it. Hand feeding of sheets of paper was required until the invention of the web press, which printed from a continuous roll of paper.

At the end of Cowper's paper, he mentioned that printing at *The Times* was "performed" by four of Applegath and Cowper's four-cylinder horizontal machines, each "producing five thousand sheets per hour, and two of Applegath's new eight-cylinder vertical machines, each producing ten thousand sheets per hour." Twenty-five pressmen operated the printing machines at *The Times*; the vast amount of hand typesetting required by *The Times* was then accomplished by no less than 110 compositors.

Obviously, Cowper, Hansard, and other printing-machine designers and builders were enthusiasts for the new technology. What did book printers, who did not require such high production speed, think of printing machines? In 1841, almost thirty years after John Walter II introduced mechanized printing by printing *The Times* on Koenig's printing machine, printer William Savage commented about "Machines" in *A Dictionary of the Art of Printing*.[129] By machines, Savage basically meant "cylindrical printing" or printing on rotary presses. He acknowledged that, in addition to the obvious speed advantage, cylinder or machine presses also provided the very significant advantage of being able to print on sheets of paper of much larger size than could be printed on any handpress. As late as 1841, decades after Koenig developed the cylinder press, Savage regarded machine presses as very novel, and this was certainly the case as they represented the most significant advance in speed of

Fig. 15. Applegath's "Times" Vertical Printing Machine.

Edward Cowper, "On Printing Machines, Especially Those Used in Printing 'The Times' Newspaper," *Institution of Civil Engineers, Minutes of Proceedings*, vol. IX (London, 1850)

Applegath's eight-cylinder vertical printing machine built for *The Times* newspaper. Applegath built a similar machine for *The Illustrated London News*. That magazine operated their vertical printing machine at the Great Exhibition in 1851 and illustrated it in operation in their issue of 31 May 1851. See chapter 13.

production since Gutenberg's invention of the handpress circa 1450.

After a long discussion of Nicholson's early conceptualization of the cylindrical press and Koenig's development of the machine or cylinder press, Savage compared the quality of machine output with handpress or "press" output. His remarks are strikingly analogous to comparisons sometimes made between offset and letterpress printing today:

With respect to the comparative merits of the cylindrical method of printing and those of the press, the manufacturers of machines as well as most master printers, not content with the real superiority of properties which the machine does certainly possess, attribute to it properties which it does not possess, and which are incompatible with it, namely, those of producing the finest work, and printing the finest impressions from highly finished engravings on wood at the rate of eight hundred or one thousand per hour.[130]

In spite of the limited printing quality available from cylinder presses during their early decades, Savage appreciated their ability to produce consistent impressions with uniformity of color. He wrote,

Another advantage in machine printing is, the regularity and uniformity of colour through any number of impressions, as it can be regulated with the greatest nicety to any shade; in this instance it is superior to the press for the production of common work, in the uniformity of colour, but only superior to common work in its rivalry with the press.[131]

By "colour," in this instance Savage may have been referring to the subtle tones of black ink, since at this time hardly any color printing was being done on printing machines.

Charles H. Timperley, a contemporary of Savage, appears to have taken a more conservative attitude toward printing machines than Savage in his *The Printer's Manual* (1838), a book written nearly entirely about handpress printing. Timperley offered these respectful, but cautious, comments about what he perceived as the limited applications of the new machines built by Cowper, Napier, and others, beyond printing newspapers. He emphasized that the "serious expense of a printing" machine could only be recovered by printing large editions.

But those of which limited numbers are printed; those also requiring a superior description of press-work with fine ink; fine and large paper copies, with alterations of margin; and many other peculiar circumstances which are continually occurring, will always require a judicious choice of men and materials, for the old mode of working, varied as circumstances may at the moment require. Half-sheet work, or jobs printed on one side only, are either impracticable or disadvantageous at a perfecting machine.[132]

Thirty-four years later, in his *A Dictionary of Typography and Its Accessory Arts*,[133] John Southward clarified some of the points Savage had made earlier. By this time, mechanized printing was well established, and all the advertisements at the back of Southward's book concerned mechanized presses or "machines" rather than handpresses. Southward summarized the accomplishments in the mechanization of printing and book production that had occurred by 1850:

The invention of machines has given an impetus to the progress of the art of printing, and has thereby accelerated the diffusion of knowledge to an extent which cannot be contemplated without a feeling of amazement. By the use of machines, sheets of paper can be printed of a size which could not possibly be obtained on a press worked by hand, and at a speed which, compared with that of the hand press, is that of the express train to the tortoise.[134]

From *Parley's Visit to the Printing Office* (detail). See p. 128.

CHAPTER 8

Charles Dickens and His Imitators Exploit the New Technologies with Great Success

One of the first novelists to profit from new high-speed, steam-powered presses was Charles Dickens. Indeed, Dickens's publishers, Chapman and Hall, would not have been able to supply as many copies of the monthly parts of his novels as they did without the existence of the new printing machines, though when Dickens began experimenting with serialization, no one expected *The Posthumous Papers of the Pickwick Club* to sell anywhere near the number of copies eventually sold.

Between April 1836 and November 1837, *Pickwick* was published in twenty-five separate parts or fascicules. The work was printed in London by Bradbury and Evans for the London publishers Chapman and Hall, who sold each part for 1 shilling. Written for publication as a serial, Dickens's novel, a series of loosely related adventures, is widely considered to have established the serialized format for literature. Serialization in parts enabled a wider range of customers to purchase a novel by spreading the cost over the individual parts. Serialization also allowed the publishers to charge advertisers for placing advertisements inside each part. After serialization was complete, the publishers Chapman and Hall issued the novel in bound-book form without the advertisements.

Prior to *Pickwick*, Victorian novels were typically 150,000–200,000 words in length and published in three volumes, sometimes called the three-decker or triple-decker format. They were typically printed in editions of around 1,000 copies, priced at one and a half guineas or 10 shillings and sixpence a volume. This price was equivalent to about half the weekly income of a middle-class household, and it made purchase of the novels affordable mostly to the upper class. Instead of purchasing novels, most readers borrowed them from commercial circulating libraries, such as that owned by Charles Edward Mudie. This business model restricted the number of copies of novels that an author and publisher could sell, but it almost guaranteed a modest profit on each.

Following the traditional formula of printing 1,000 bound copies of a novel, Chapman and Hall ordered 1,000 copies of the first part or fascicule of *Pickwick*. Even that number of parts was difficult to sell, so the print run of the second part was reduced to 500 copies, making the first printing of the second part by far the hardest to collect today. By the third part, the print run was restored to 1,000 copies, after which sales of each succeeding part increased; by the time the final part was issued, in November 1837, sales had reached 40,000—a remarkably large number of copies for the first edition of a Victorian novel. As sales of the later parts increased, Chapman and Hall had to reprint the earlier parts twenty times, making the bibliographical history of the complete set of twenty-five parts and their related advertisements one of the most complex of any Victorian publication. Without a high-speed printing machine for production,

Charles Dickens, *The Posthumous Papers of the Pickwick Club* (London, 1836)

A facsimile of the cover of the first part of Dickens's *Posthumous Papers of the Pickwick Club* (1836).

Charles Dickens, "Extraordinary Gazette," *Bentley's Miscellany* (London, 1837)

Portrait of the young Charles Dickens by Hablot Knight Browne ("Phiz") on the cover of the pamphlet entitled *Extraordinary Gazette*, issued with *Bentley's Miscellany* magazine. Dickens edited that magazine while he was writing *Pickwick* and *Oliver Twist*.

printing enough copies to keep up with the rapidly increasing sales of the separate parts would have been much more difficult, if not impossible.

With *Pickwick*, Dickens and publishers Chapman and Hall had happened upon a way of issuing fiction that revolutionized publishing, distribution, bookselling, author-publisher relations, copyright provisions, and the writing of fiction. "The thirty-two-page, two-illustration part became a standard to which the public rapidly became accustomed."[135]

Production of Dickens's serial parts required a large crew at Bradbury and Evans. According to the elaborately detailed account by Jane E. Chadwick,[136] Bradbury and Evans used six printing machines, including an Applegath four-feeder printing machine similar to what Applegath developed for *The Times* newspaper. That machine could print over 4,000 copies per hour.

Writing rapidly to meet his production obligations, Dickens did not make the typesetting process easy for the compositors:

One former Compositor who worked at Bradbury and Evans for over fifty years, Charles Cawte (1832–1907), reminisced about working from Dickens's copy: "Well I remember the thick, spluttering, blue ink, quill-

penned manuscript. After getting over the first few lines the copy would not have been called 'bad'—that is, from a compositor's point of view."[137]

At this early stage in his career, Dickens was juggling several jobs. On 4 November 1836, the publisher Richard Bentley hired Dickens (who was using the pen name "Boz" at the time) to serve as the editor of *Bentley's Miscellany*, a new monthly literary and humor magazine. Dickens had already achieved extraordinary fame for *The Pickwick Papers*, and Bentley was banking on Dickens's popularity to help *Bentley's Miscellany* tap into the same "non-partisan, sentimental, comic market"[138] that had made *Pickwick* such a hit. Dickens edited *Bentley's Miscellany* for a little over two years, quitting the post in February 1839.

Under the initial terms of his hiring agreement with Bentley, Dickens was responsible for "recruiting, selecting, editing, and proofing the contents of a ninety-six-page magazine,"[139] as well as supplying original content of his own for every number. Dickens began serializing *Oliver Twist* in the second number, published on 1 February 1837; the final chapter of the novel appeared in the March 1839 issue of the *Miscellany*. In keeping with the terms of his agreement, Dickens contributed other humorous pieces throughout his tenure at the magazine, including the famous "Extraordinary Gazette" at the end of the March 1837 number; this four-page leaflet, written "in a style parodying a Royal speech,"[140] announced the coming of *Oliver Twist*. The masthead vignette by George Cruikshank ("Phiz") contains the earliest drawing of Dickens, recorded in William Glyde Wilkins's *Dickens in Cartoon and Caricature* (1924). The version illustrated is the larger-format first printing of the "Gazette," intended for insertion into the third number of the *Miscellany*; a smaller-format second printing with four additional advertising pages was issued separately.[141]

The relationship between Bentley and Dickens began cordially, but soon devolved into protracted legal wrangles over money, editorial control, and contract obligations. In the twenty-eight months between Dickens's hiring and departure, Bentley's attorneys drew up no fewer than five agreements regarding Dickens's obligations as editor and writer for the *Miscellany*, each more complicated and restrictive than the last, until the final one that terminated Dickens's editorship. The ongoing friction between Bentley and Dickens was in no small part due to the lack of an adequate legal framework defining author-publisher relations and international copyright; the law had not yet caught up with the enormous changes—such as mechanized printing and mass distribution—that were revolutionizing the nineteenth-century publishing industry.

As his success increased, Dickens faced increasing problems from copyright infringement and plagiarists. Among his many problems in this regard was the fact that English copyrights did not transfer to other countries, so that American publishers routinely published large editions of his novels without paying him any royalites. In England in 1843, Chartist leader, printer, and publisher John Cleave published *Parley's Visit to the Printing Office; with a Familiar Account of the Steam Engine, the Printing Machine, and the Arts of Composition, Engraving and Stereotyping*. This small pamphlet was "Presented Gratis, to all regular Subscribers, with No. 25 of 'Parley's Penny Library'"; it was, for all intents and purposes, an advertisement for Cleave's adaptive plagiarisms or piracies of Dickens. It is also one of the scarcest ephemera of book production in the Industrial Revolution, with one of the very best illustrations of a printing-machine room in operation powered by a table steam engine. Signs on the wall of the machine room depicted in the illustration read *Barnaby Rudge*, *The Old Curiosity Shop*, and *Parley's Penny Library*. Cleave issued adaptive plagiarisms of both of those Dickens novels in *Parley's Penny Library*, numbers 1-7 and numbers 8-12.[142] By publishing adaptations of Dickens's writings that were changed enough to avoid being infringements of copyright, Cleave was able to plagiarize elements of Dickens's works very profitably. Cleave eventually issued 108 numbers of *Parley's Penny Library; or, Treasury of Knowledge, Entertainment and Delight*, collected in nine volumes.

Dickens also had to deal with plagiarisms or piracies of *A Christmas Carol*. Dickens's celebrated Christmas tale was published on 19 December 1843. Just over two weeks later, the first installment of "A Christmas Ghost Story,"[143] a "re-originated" abridgement of the *Carol* by the hack writer Henry Hewitt, appeared in *Parley's Illuminated Library*, an illustrated twopenny weekly that published unauthorized knockoffs of popular works of the day. It is estimated that the 6 January number, like the other numbers of *Parley's*

🔖 *Parley's Visit to the Printing Office* (London, 1843)

Chartist leader, printer, and publisher John Cleave presented this small, thirty-six-page pamphlet, measuring 5½ × 3½ in. (14 × 9 cm), to the subscribers to *Parley's Penny Library*.

🔖 From *Parley's Visit to the Printing Office* (London, 1843)
Enlarged from the original woodcut, which measures 3½ × 5 in. (9 × 12.5 cm)

This crudely executed woodcut showing the pressroom in *Parley's Visit to the Printing Office* is possibly the most accurate representation published during the nineteenth century showing how a table steam engine powered book-printing machines. The belt-drive mechanism is depicted. Note the posters advertising *Barnaby Rudge* and *The Old Curiosity Shop*, along with *Parley's Penny Library*, on the wall of the printing office.

🔖 From [Henry Hewitt,] "A Christmas Ghost Story," *Parley's Illuminated Library* (London, 1843)

The opening of Hewitt's plagiaristic adaptation of Dickens's *A Christmas Carol*. Note the unusual textual border surrounding each page.

🔖 *Parley's Illuminated Library* (London, 1843)

This book contains Henry Hewitt's "re-originated" abridgment of Dickens's *A Christmas Carol*, entitled "A Christmas Ghost Story."

Illuminated Library, had a print run of about 50,000 copies; despite this large number, no single copy of issue no. 16 has survived.[144]

Apart from the copy illustrated, the only recorded copy of "A Christmas Ghost Story" appears to be in the bound copy of Vol. I of *Parley's Illuminated Library* that Iona and Peter Opie donated to the Bodleian Library, Oxford, in 1988 as part of the Opie collection of children's literature. Another copy of the volume, in the University of Chicago Library, lacks the first installment of Hewitt's abridgement, but contains a later pirated version of *A Christmas Carol* titled *The Miser's Dream: A Story of Christmas Eve.* OCLC cites only the University of Chicago copy and microform copies of the Opie copy.

Dickens had previously suffered the travesties of *Parley's* literary piracies, which he regarded as copyright infringements, but hesitated to take any legal action against them until "A Christmas Ghost Story," which was similar enough to the original *Carol* that he felt confident of success. Shortly after the first installment's appearance, Dickens brought a lawsuit in the Court of Chancery against both Hewitt and *Parley's* publishers, Richard Lee and John Haddock. After much effort, he ended up winning the case, but the victory was a Pyrrhic one—Lee and Haddock avoided paying damages by declaring bankruptcy, and Dickens was stuck with £700 in court costs. Dickens's disgust and bitterness over this experience found later expression in his novel *Bleak House,* which satirizes the Court of Chancery as a place where "lawyers grow fat, and justice is never served."[145]

Dickens did manage to stop the publication of the second installment of "A Christmas Ghost Story," which was supposed to have appeared in the next issue of *Parley's Illuminated Library.* Instead, this issue has a 16-page gap (pages 257–272) where the installment would have been printed, and includes a piece titled "A Genuine Ghost Story. (Not by Charles Dickens)," which makes sly reference to the lawsuit:

> *The world can no longer get a ghost story, either for love or money; and even should, we venture upon a spiritual ebullition, for the behoof of our illuminated readers, an injunction plays the very* dickens *with us, for "colourably imitating" Parley's Ghost of Christmas . . . Since we have not been suffered to see or hear of Marley's*

[Thomas Peckett Prest,] *The Penny Pickwick* [London, 1837–1839]

One of the weekly parts of *The Penny Pickwick,* written by Thomas Peckett Prest and published by Edward Lloyd between 1837 and 1839. Lloyd was careful to modify Dickens's stories enough to avoid being legally prosecuted for plagiarism. He was also careful to underprice Dickens's selling price for the individually published parts of *Pickwick.*

THE

𝔓𝔬𝔰𝔱𝔥𝔲𝔪𝔬𝔲𝔰 𝔓𝔞𝔭𝔢𝔯𝔰

OF

THE PICKWICK CLUB.

BY CHARLES DICKENS.

WITH A FRONTISPIECE.
FROM A DESIGN BY C. R. LESLIE, ESQ., R.A.
ENGRAVED BY J. THOMPSON.

LONDON:
CHAPMAN AND HALL, 186, STRAND.
MDCCCXLVII.

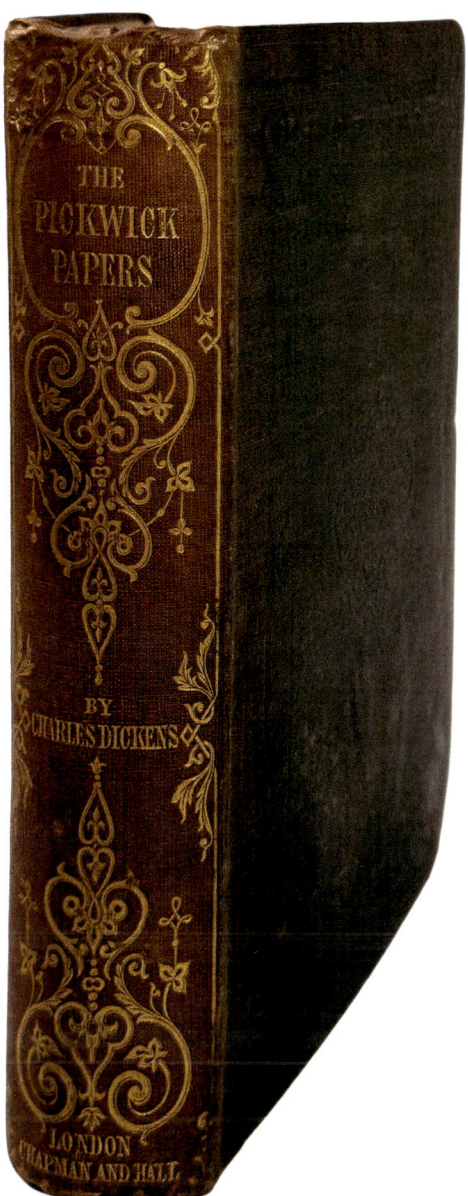

Charles Dickens, *The Posthumous Papers of the Pickwick Club* (London, 1847)

Title page and original cloth binding on the "Cheap Edition" of *Pickwick*, issued as part of a "Cheap Edition" of Dickens's collected works.

Charles Dickens, autograph letter signed to Edward Chapman, Chester Place (6 April 1847)

Dickens's letter to his publisher Edward Chapman expressed hope that sales of the "Cheap Edition" of his works might reach 100,000 copies—an unheard-of number for a large set of books at that time. Dickens's enormous commercial success in selling so many copies of printed books within a relatively short time was made possible by the development of high-speed printing machines in the second printing revolution.

ghost, a visitation at this time from that of Parley may have the same effect.[146]

A remarkable feature of *Parley's Illuminated Library* is the text layout, in which nearly every page is framed "with printed maxims or poetic quotations at the top and bottom and along the sides. The maxims stand alone in the top or bottom margins, or in arbitrary pairs in a side margin; the verse quotations continued discursively from top to bottom, or side to side, across several pages."[147] *Parley's* appropriated this unusual design from Samuel Maunder's *The Treasury of Knowledge* (1830) and similar works intended for the juvenile market.

Another plagiarist of Pickwick was the English hack writer Thomas Peckett Prest, who penned *The Penny Pickwick* "Edited by 'Bos'" (instead of Boz), which was published by Edward Lloyd in weekly parts from 1837 to 1839. Prest and Lloyd published a whole series of parodies of Dickens novels. Lloyd once bragged that an issue of his *The Penny Pickwick* sold 50,000 copies, outselling Dickens's original. Each issue of *The Penny Pickwick* was a pamphlet of eight pages cheaply printed on cheap paper with crudely cut illustrations.

When he was not pursuing plagiarists, Dickens naturally wanted to promote sales of his writings. At the end of March 1847, to further promote and increase sales and make his writings more affordable, Dickens initiated the "Cheap Edition" of his works to be sold in parts, with smaller page size, type set in two columns, and with different and fewer illustrations. The first of Dickens's works published as a "Cheap Edition" was *Pickwick*, issued between April and September; it was issued as a complete volume on 8 October 1847 in wrappers for 4 shillings 5 pence and cloth for 5 shillings. The copy illustrated is in the original cloth binding.

On 26 April 1847, shortly after the launch of the "Cheap Edition," Dickens wrote a letter to his publisher Edward Chapman. The letter refers to the "Cheap Edition" parts publication of *Pickwick*, which Dickens hoped might reach 100,000 copies, as well as the first issue of the complete book. Such an enormous volume of sales could not have been imagined before Dickens and Chapman and Hall used the new printing technology to expand the market for fiction. The actual sales of the "Cheap Edition" of *Pickwick* in parts ranged from 62,238 copies for the first part to 31,263 for the last.[148]

Dickens's letter, with his witty reference to accounting in the final paragraph, reads as follows:

Chester Place
Twenty Sixth April 1847
My Dear Sir,

I think it will be a great thing to advance the parts as you propose. I am sanguine for our getting up to the hundred thousand, including the first issue of the complete book.

You shall hear from me about the Frontispiece, very shortly. I will take care of it, without loss of time.

It is not worth while, I think, to send any presentation copies of the parts. But I shall be glad to have a couple myself, always, regularly, and promptly.

Will you mention to your book-keeper that in case he should meet a fair copy of our accounts to last Christmas, walking about anywhere, I should be glad if he will give her my compliments, and say she may rely upon a welcome, whenever she is disposed to come towards this part of the town.

Faithfully yours always,
[Signed with a flourish] Charles Dickens[149]

New-York

(FOR THE COUNTRY.)

VOL. VI....No. 604] NEW-YORK, WEDNESDAY, JANUARY 4, 1826.

PRINTED FOR THE PROPRIETORS
BY J. M. ELLIOTT,
AT No. 3 NASSAU-STREET, NEW-YORK.

☞ THE AMERICAN FOR THE COUNTRY is published every TUESDAY and FRIDAY, at No. 3 NASSAU-STREET, between Wall and Pine streets, New-York, at FIVE DOLLARS per annum, (or *Four Dollars, if paid in advance,*) and regularly sent by mail, agreeably to direction, to any part of the United States. All letters or communications must be addressed to the " *Editors of the American, No. 3 Nassau-street, New-York.*"

** No subscriptions discontinued till all arrearages are paid up, unless at the discretion of the proprietors.

NEW-YORK AMERICAN.

FRIDAY EVENING, DECEMBER 30, 1825.

Verplanck on Contracts.—This work, which has just appeared, is a very original and beautiful speculation upon an important title of law—that of misrepresentation and concealment in the contract of sale and other commercial bargains. It discusses a practical question of no little difficulty, with great clearness; and where we do not assent to the conclusions of the writer, we are struck with the acuteness and extensive learning employed in the investigation, and with the lofty tone of morals asserted throughout.—The author appears to have had his attention drawn to the subject by the controversy which has frequently arisen, how far merchants and other men of business have a right to take advantage of their knowledge of such circumstances as may influence the market price of commodities in dealing with others who are not on a footing of equality in that respect. He has investigated it, not only with this view, but also with respect to the law of warranty in sales, of inadequacy of price, and other similar questions which arise in the Courts of law and equity upon sales of real and personal property. These he has fully examined, both on original principles, and as they are settled by the rules of positive law. For this purpose, he has compared the institutions of different nations, and has, we think, satisfactorily shown the present state of this branch of law in this country, to be very unsettled, and to call loudly for the interposition of the

difficulty. The writer of this article has seen 2400 sheets worked upon a similar machine within the hour in London; and the Editor of the London Courier, whose paper is worked upon the Napier Press, announced to his readers on the 14th of November, 1823, that his press in one instance produced at the rate of 2880 sheets within the hour. But the better to enable the public to contrast the difference between the Napier machine and the common press, we will put down the number of impressions (each newspaper requiring two impressions) given in each week by the offices of the Daily Advertiser and American in their daily and country papers. They amount to *sixty-four thousand four hundred and sixteen*. To perform this work at the common press at 360 impressions per hour, would require within a few minutes of *one hundred and seventy-nine hours, or, seven days and nights and seven hours, without allowing a moment for rest.* The same quantity of work is now performed by the two offices in *forty-six hours; less than two days and nights,* and this with one third of the labour formerly required. The fact is, the only real labour about the machine is that of turning the fly wheel, and even this bears no comparison with that of sawing wood, or many occupations in which thousands are daily employed. The machine, it is true, is expensive compared with other Printing presses; but the early dissemination of intelligence in the country, and the timely delivery of papers both morning and evening in the city, was of too much importance to the public and the proprietors, to let the price stand in the way of the great advantages which the machine promised, and which it performs to our entire satisfaction.

Having said thus much of the advantages of it, we will now endeavor to give the public a brief description of the machine itself. It is called THE DOUBLE IMPERIAL PRINTING MACHINE, and is the invention of D. Napier, Engineer, of London, a Scotchman, not less distinguished for his mechanical ingenuity than for his gentlemanly and upright character.

The machine weighs about two tons, and is composed of a great variety of parts, but so compact and beautiful in all its movements, that it occupies no greater space than the common press. It is ten feet in length, five in breadth, and five in height. The motion is given to a fly-wheel by manual labour, and a stout man will turn it with ease. This supercedes the necessity of steam, and is superior to horse power, because it is less expensive, less troublesome, and more regular. The impression is given by two Cylinders. The bed, or carriage upon which the type or form is placed, runs upon friction rollers, which move from one end of the press to the other on ways or ribs which extend the full length of the machine. The carriage with the form of type passes under the Cylinders by a power given by half a dozen cog wheels all moved by the fly-wheel. At each end of the press there is a supplying board upon which the paper is laid by tokens, (ten quires at a time,) and at each board a man or boy is stationed merely to move down the blank sheets three or four inches, to a gauge; when by a beautiful movement of a part of the machine, called the feeding bar, which unlocks and falls at a

an institution, was sinking." Nor is it obvious how any one could possibly construct an argument, unless of ignorance and prejudice, to the contrary.

Perhaps I shall be told, that under this same defective system were formed those illustrious commanders of our Navy, whose names must ever adorn the pages of its annals, with a lustre resplendent and inextinguishable as the stars which emblazon its banners.—But if this were even true, it would not be decisive. If those officers whose conduct and valor retrieved the fortunes and sustained the honor of their country, and drew on them the eyes of every nation in Europe, and filled the naval world with the fame and admiration of their victories; if they were indeed formed *under* this system, I might still aver that they were never formed *by* it. They formed themselves in spite of it; or, at any rate, without its aid. They became great naval officers, because they were great men; because, like the men who led forth the armies or presided in the councils of our revolutionary struggle, they were as little to be vanquished by the discouragements of adversity, as by the foes of their country; because, being animated with an ardent and disinterested patriotism, and with a lofty and generous emulation of renown, that instinct of noble minds, which has ever been the sister, if not rather the parent, both of patriotism and gallantry, they were capable of an elevation and expansion of soul, equal to the greatness and emergency of the occasions on which they were called to act; and in those dark and dubious conjunctures of impending disaster, at the approach of which "the feebler faint" and are thrown into despondency, could

"Feel, to the rising bosom's inmost core,
"Its hope awaken, and its spirit soar."

To men of qualities like these, capable of *their* enterprize and determination, and inflamed at the same time, with a like enthusiastic devotion to their profession, no want of advantages, no combination of difficulties with which the entrance to that profession was beset and obstructed, could prove invincible. They seem always to have felt the surrounding perplexities and perils of every crisis, merely as serving to compress and concentrate their powers into a more intense and elastic vigour of reaction. Such qualities, exerted in a cause worthy of them, were almost enough to assure their possessors of conquering even the vicissitudes of fortune; and to render the ulti-

try have celebrated as the sound of the "church-going bell," should not be found any where but in the city of New-York? Ages have rolled away, during which all Christian nations have thus been accustomed to mark the "holy hours of prayer." The crowded cities, and the lonely villages of our own country, resound with these calls to assemble at the temples of the Most High—proclaiming the religious sense of the community—reminding the careless and indifferent of their duty. The "great nuisance" of this practice, so general in every age and country, that it may be considered as the dictate of a powerful religious feeling, has been no where discovered except in this city. Is it painful to the "sick and the dying?"—But an argument which proves too much, proves nothing. The clattering of pleasure carriages, and the heavy thunderings of carts, are often seriously annoying not only to the *sick,* but to the *well.* Why not put a stop to these, and require that no carriage or cart shall be driven faster than on a walk? In cases of peculiar nervous irritability in sick persons, the noise of carriages is guarded against, by strewing the street with tan: and in every instance, where requested as necessary to the comfort of a sick person, the ringing of a neighbouring bell has either been entirely stopped, or diminished in frequency and loudness.

It really seems to me wonderful that only in this city should there be that exquisite sensibility which is so much agonised at the ringing of bells. This in the country from which we are descended, and in some places in our own country, is frequently resorted to as an *amusement*. With this view solely Christ-Church bells in *Philadelphia* where the quiet habits of the *Friends* prevail, sound a merry peal for an hour or more, two evenings in every week.

The usage of *all* religious denominations to summon their people to Church on *Sundays,* and of *some,* on the *festival* or *prayer-days of the week,* has been of so long standing and is so sanctioned by the immemorial practise of all christian nations, that it may be considered as a religious right. I confess I cannot for a moment think that in this *free* country, where legislation is managed with so scrupulous a regard to the religious feelings and habits of the people, the Corporation of the city will depart from the wise maxim of not "governing too much," and will attempt to invade a right which is respected in the strong monarchy of England, and in the despotic governments of the old word. *A Friend to Old Customs.*

TO THE EDITORS OF THE AMERICAN.

Gentlemen,—Mr. Waterman, of the Greenwich-street Hotel, will present the debtors in gaol with a dinner; and Mr. Delamater, proprietor of the new hotel near the Park Theatre, with a suitable quantity of whiskey punch, on New Year's Day. The debtors on Thursday received five loads of wood and ten blankets, for which they return thanks.

New-York American, vol. VI, no. 601, New York, 4 January 1826 (detail). See p. 238.

This issue of the *New-York American* newspaper published a long article on its first page regarding the paper's import from London and first use of a Napier Imperial printing machine, the first printing machine designed to print newspapers that was used in North America. The publishers emphasized that they acquired a machine that could be hand-cranked, since appropriate steam engines were not available in New York, and human power was more efficient than having the machine powered by a horse.

CHAPTER 9

The Development of Mechanized Printing in America: Daniel Treadwell, the American Bible Society, R. Hoe & Co., Isaac Adams, the Harper Brothers

The history of the mechanization of newspaper, magazine, and book production in the United States is quite distinct from the history of mechanization in England, France, or Germany. Far fewer printing machines from England were exported to America than were sold in England or the Continent of Europe, providing an opportunity for American inventors to create and build their own printing machines. The first printing machine invented in the United States was Daniel Treadwell's mechanized platen press, called Treadwell's Power Press, invented in 1821; the first books produced on printing machines in North America were printed on Treadwell machines. Within the decade, Isaac Adams improved upon Treadwell's Power Press with the Adams Power Press, widely used to print books in America from about 1830 to 1880. The first American newspaper printed on a mechanized press was the *New-York American*, which advertised the fact in its issue of 4 January 1826; because Treadwell's press was not advantageous for printing newspapers, the *New-York American* was produced on an imported hand-cranked machine built for that purpose by the Scotsman David Napier in London. In 1829, Robert Hoe of New York City, who learned about printing machines by working on the Napier at the *New-York American*, began building printing machines for the rapidly growing American newspaper industry; he was followed by his son Richard March Hoe, and his grandson Robert Hoe III, in building the largest nineteenth-century American printing-machine design and manufacturing company, which flourished until it eventually closed after World War II. Beginning in 1828, the Harper brothers pioneered the use of American printing machines in the production of books and large-circulation magazines.

As in England, stereotyping preceded the introduction of mechanization of book and newspaper production in America. Implementation of the Stanhope improved method of stereotyping in the United States lagged its implementation in England by seven years. The first book printed from stereotype plates in North America was an edition of the Bible published by the Bible Society at Philadelphia in 1812;[150] the stereotype plates for this edition were made in London by T. Rutt. The following year, the first book printed from stereotype plates made in North America, an edition of the Presbyterian Church's *The Larger Catechism*, was produced using plates made by J. Watts and Co. in New York; the title page of the work advertises it in very small type as "the first book ever stereotyped in America."[151]

In 1819, Daniel Treadwell traveled to London, where he saw the Koenig printing machine in operation. In 1821, after returning to Boston, he invented the Treadwell Power Press, the first American mechanized press.[152] Treadwell developed his power press at a time when there were few steam engines in America and even fewer steam engines small enough to power something the size of a printing machine; Treadwell thus used horsepower to operate his press.

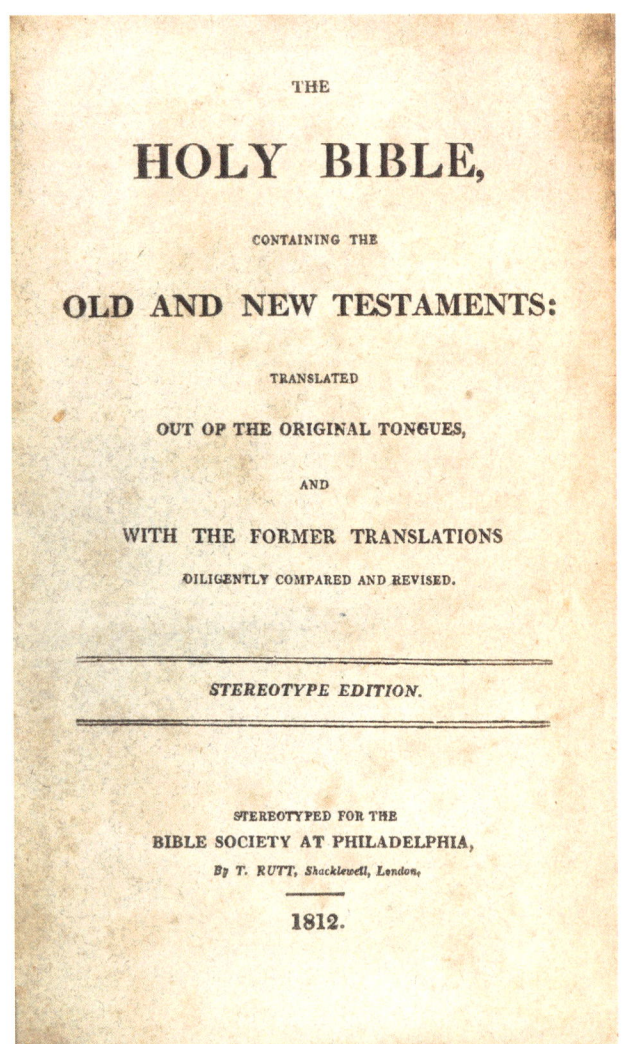

The Holy Bible. Stereotype Edition (Philadelphia, 1812)

The stereotyped edition of the Bible, published by the Bible Society of Philadelphia in 1812, was the first book in the United States that was printed from stereotype plates made from the improved Stanhope process of stereotyping. The plates were made in London by T. Rutt.

Alexander McLeod, ed., *The Larger Catechism . . .* (New York, 1813)

This was the first book printed in America from stereotype plates produced in America.

Not finding any printers who would buy his press, Treadwell went into the printing business himself, printing several books and employing women to operate the press. This was probably the first time that women were known to operate a printing press of any kind in North America—something strongly opposed by the printing trades in Boston.

In a brief autobiography written in 1854, Treadwell told how he chose to produce a platen press rather than a rotary press since the intention was to print books, and he concluded that platen presses were superior for "book work." Treadwell designed his press to be powered by horse or water or human power since steam engines were in very short supply in America at the time. His press was operational by horse in 1821, but he did not patent it until 1826.[153]

From the start, Treadwell employed women to operate his power presses. Though he did not mention the cost differential, it was widely assumed that women would be paid roughly half the cost of men,

Portrait of Daniel Treadwell
Frontispiece from Morrill Wyman, "Memoir of Daniel Treadwell," *Memoirs of the American Academy of Arts and Sciences*, Centennial Volume, vol. XI, pp. 325–524 (Cambridge, Massachusetts, 1888)

Treadwell invented the first American printing machine, the Treadwell Power Press.

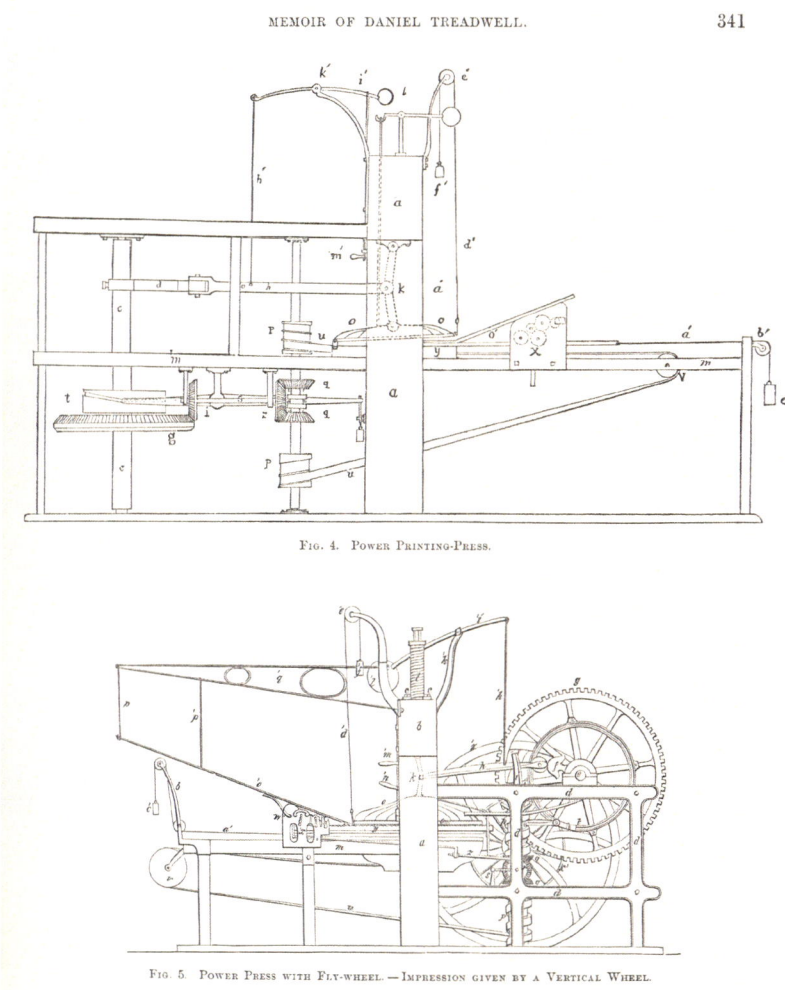

Morrill Wyman, "Memoir of Daniel Treadwell," p. 341

Schematic of the Treadwell Power Press, the first American printing machine. Intending his press primarily for book printing, Treadwell invented a bed-and-platen press and designed it to be powered by horse, water, or humans rather than a steam engine, since steam engines were in very short supply in America during the 1820s.

therefore giving Treadwell's Power Presses a cost advantage in their operation. Between 1821 and 1822, Treadwell printed many books with his presses, including an edition of the New Testament printed from stereotype plates in 1822. Treadwell sold his Boston printshop in 1822. Unfortunately, the establishment operating the business in Massachusetts burned down in 1826.[154]

Books printed on the Treadwell Power Press included small items like a sixteen-page pamphlet titled *Constitution of the Widow's Society*,[155] as well as William Mitford's eight-volume *The History of Greece*;[156] this last was probably the first large multi-volume set of books to be printed by a mechanized press rather than a handpress. On the back of each volume's title page, Treadwell printed an advertisement for his press, reading (in black letters) "Treadwell's Power Press, Boston." Thomas Brown's *Lectures on the Philosophy of the Human Mind*, a two-volume work published in Boston by S. T. Armstrong in 1826, is identified on

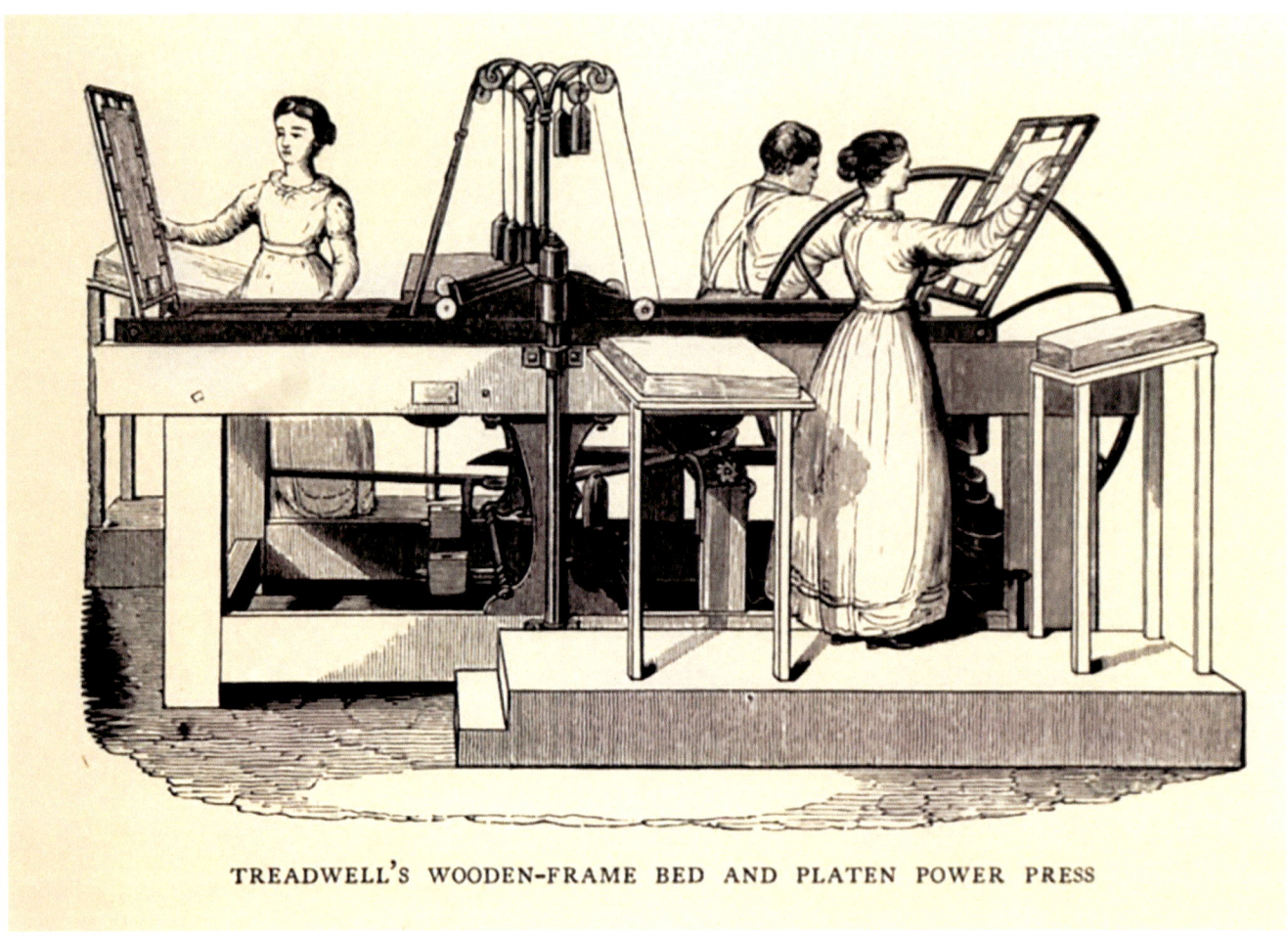

TREADWELL'S WOODEN-FRAME BED AND PLATEN POWER PRESS

Robert Hoe III, *A Short History of the Printing Press and of the Improvements in Printing Machinery from the Time of Gutenberg up to the Present Day* (New York, 1902)

A drawing of Treadwell's Power Press in operation. From the original introduction of his power press, Treadwell employed women to operate his printing machines, mainly with the intention of lowering labor costs, since women were typically paid half or one-third the wages of men. This tradition continued after Treadwell's machines were replaced by Isaac Adam's bed-and-platen power presses, with women often depicted operating Adams's presses in America throughout the nineteenth century.

its title pages as having been printed on "Treadwell's Power Press—J. G. Rogers & Co." None of these three items was illustrated.

Unlike in England, where Oxford and Cambridge Universities and the King's Printer held a monopoly on Bible printing, any publisher in the United States could issue an edition of the Bible. In 1826, Boston publishers Hilliard, Gray, Little and about ten other publishers issued an edition of the complete *Holy Bible, Containing the Old and New Testaments*, stating on the title page that it was printed on Treadwell's Power Press. Since a few copies have survived, the edition was presumably printed before the press burned down in 1826. The edition was printed from stereotype plates made at the "Boston Type and Stereotype Foundry, Late T. H. Carter & Co." Because Oxford University Press and Cambridge University Press did not introduce printing machines until the 1830s, this Bible printed on the Treadwell Power Press was the first Bible printed on a printing machine anywhere.

Treadwell's first large customer for his printing machines was the American Bible Society (ABS), founded in New York in April 1816 by Elias Boudinot. Initially, the society's main function was the pro-

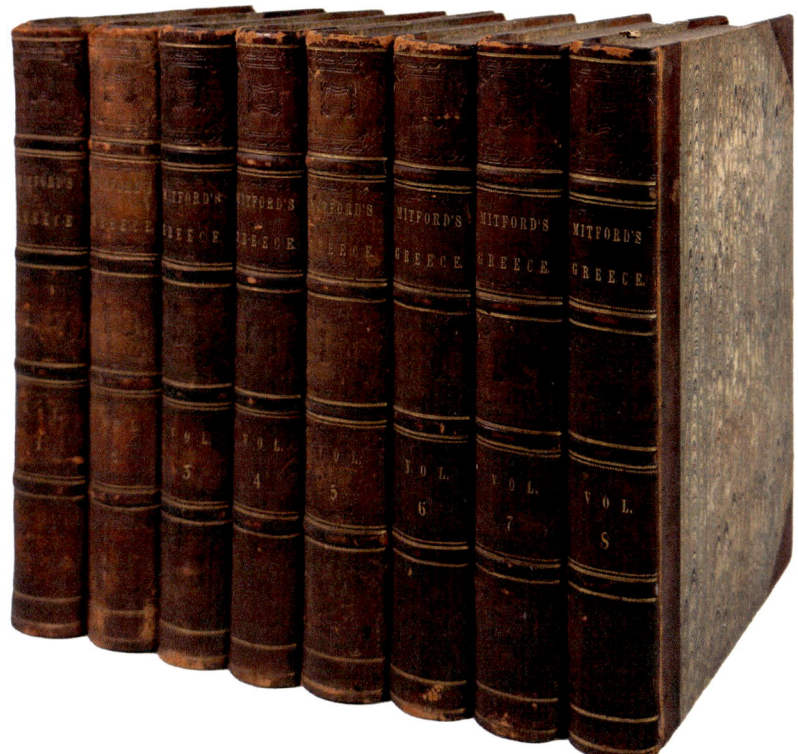

Treadwell's Power Press, Boston.

🔖 William Mitford, *The History of Greece*, 8 vols. (Boston, 1823)

This set, printed by Treadwell's Power Press, was probably the first multi-volume set of books printed on a printing machine anywhere. Treadwell advertised his press on the verso of each title page.

🔖 *The Holy Bible* (Boston, 1826)

This edition of the Old and New Testaments printed on Treadwell's Power Press in Boston in 1826, as stated just above the imprint date, was the first edition of both testaments ever printed on a printing machine.

duction of stereotype plates of the Bible that could be used to produce Bibles by local constituent Bible societies around the country, but they rapidly got into the publishing business themselves, issuing quality books at very small markups over cost. The organization issued its first annual report on 8 May 1817; the report's final three leaves reproduced sample pages of the ABS's *Stereotype Long-Primer Bible, Octavo*, its *Stereotype Brevier Bible, Duodecimo*, and its *Stereotype Minion Bible, Duodecimo*. In 1816, the society issued 10,000 copies of the Bible in brevier type, and by 1819 they had printed 100,000 copies.

Besides adopting stereotypes for Bible printing, the American Bible Society was also an early adopter of machine printing. The society opened negotiations with Daniel Treadwell in 1823 and, by 1829, had installed sixteen Treadwell Power Presses, probably built by Robert Hoe of New York under a franchise arrangement with Treadwell. Printer Daniel Fanshaw operated the Treadwell Power Presses for the American Bible Society.

Initially, the Treadwell presses at Fanshaw's establishment were powered by donkeys on the top floor of his building. By 1830, the American Bible Society had introduced a steam engine to power their printing ma-

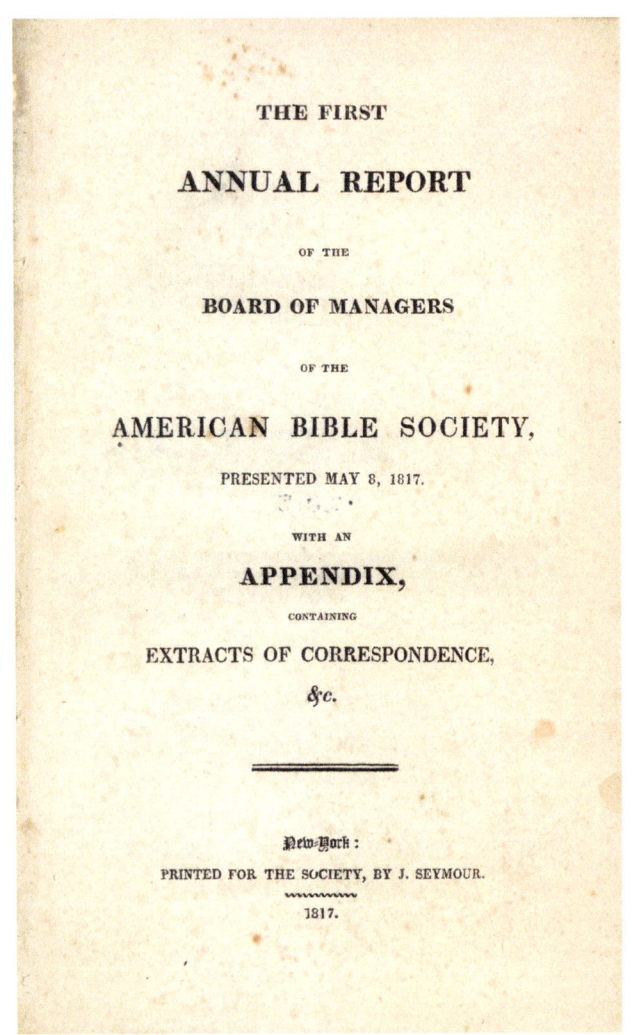

The First Annual Report . . . of the American Bible Society (New York, 1817)

The first annual report issued by the American Bible Society contained sample pages of the various stereotyped Bibles published by the society. The distinction between the different editions was primarily related to the size of type in which the Bible text was set.

chines; they were probably the first book publishers in North America to do so. According to *An Abstract of the American Bible Society* issued that year,[157] by 1830, the Society had issued over one million Bibles and Testaments. Regarding the printing facilities of the ABS, the *Abstract* stated that they were operating sixteen steam-powered presses and six handpresses.

In 1828, New York publishers J. and J. Harper began printing books on a Treadwell Power Press in their Cliff Street offices. Initially, that press was powered by a horse:

In the early days at Cliff Street, at least by 1828, horsepower was added to manpower through the purchase of one of the new presses invented and manufactured by Daniel Treadwell of Boston. However, hand presses were not discarded. As late as 1837 twenty-four hand presses were still in use. The Treadwell press was connected by a system of gears to a vertical shaft which extended to the basement. A horizontal beam was then connected to the bottom of this shaft and the far end of the beam a white draft horse was harnessed. As the horse walked steadily in a circular path, the shaft turned the gears upstairs and the presses flapped away as fast as the sheets could be fed to them. In 1833 the new Adams steam press was installed, and the horse was retired to Father Harper's farm on Long Island. For a while he frolicked around the pasture enjoying his new freedom; then old habits asserted themselves. When a seven-o'clock whistle blew he emerged from the barn, went to a solitary tree in the pasture and walked steadily around it till the twelve-o'clock whistle told him he might rest till one, when again he took up his solitary and circular tramp till the six-o'clock whistle released him for the night.[158]

In 1833, the Harper brothers replaced their Treadwell press with a steam-powered R. Hoe & Co. press, becoming possibly the first American book printers to do so. This terrified Harper's chief foreman:

"They are going to do something dreadful at Harpers," he told his wife. "It will take the bread out of the mouths of so many of the boys. It is terrible to think of how much misery it will cause."

"Why, what can it be?" she replied. "Have they failed or lost money, and must they cut down the force?"

"No, it is not that," he answered, "but they have decided to put in a steam engine and get presses that run by steam. There is no telling how many of us will be thrown out."[159]

The first American to improve upon the Treadwell Power Press was inventor and politician Isaac Adams, who had worked with Treadwell. Adams invented the Adams Power Press between 1827 and 1830; this machine, a platen press operating under power, rev-

Portrait of Isaac Adams
Image from Wikipedia

Portrait of American inventor and politician Isaac Adams, inventor of the Adams Power Press, a bed-and-platen printing machine widely used for book production especially in America, but also elsewhere, between 1827 and 1880.

[Jacob Abbott,] *The Harper Establishment . . .* (New York, 1855)

A woman operating an Adams Power Press at Harper Brothers publishers in New York.

olutionized the printing industry, especially book production in America. Introduced in 1830, by 1836 the Adams Power Press had become the leading machine used in book printing in America and remained so for much of the nineteenth century. It was sold (with updates) as late as 1881 by R. Hoe & Co. and was distributed worldwide. The Adams Power Press substantially reduced the cost of book production, making books more widely available. One Adams press could print 1,000 impressions per hour—four times the speed of a handpress. It also handled sheets of paper twice as large as a handpress; however, it was a platen press operating under power, with all the limitations of platen printing. It is estimated that, between 1830 and 1880, some 90 percent of American book printing was done on Adams presses.[160]

Harper and Brothers were among the first publishers to adopt the Adams Power Press. In 1855, Jacob Abbott, author of books for young people, issued *The Harper Establishment, or How the Story Books Are Made*,[161] an illustrated account of the new Harper and Brothers publishing facility in New York. Even though it was written for a youthful readership, Abbott's book remains perhaps the most detailed illustrated account of book production in America during the mid-nineteenth century. Page 120 in Abbott's work illustrates Harper and Brothers' Adams Power Press, operated by both men and women; other illustrations show customers visiting the large new "fireproof" establishment, which also served as a retail facility. Bookbinding, which at this time remained mostly a hand process, is also featured, as well as the process of marbling paper.

Harper and Brothers was proud of introducing the best and most advanced book production processes in its operations. They were among the first to adopt both the Treadwell and Adams power presses, and were also the first in the United States to publish electrotype illustrations on a large scale.[162] Invented in 1838 by Russian electrical engineer and physicist Moritz von Jacobi, electrotype was an electro-chemical process that could be used to make printing plates by coating stereotype, wood-block, or intaglio plates with a thin layer of copper, thereby strengthening

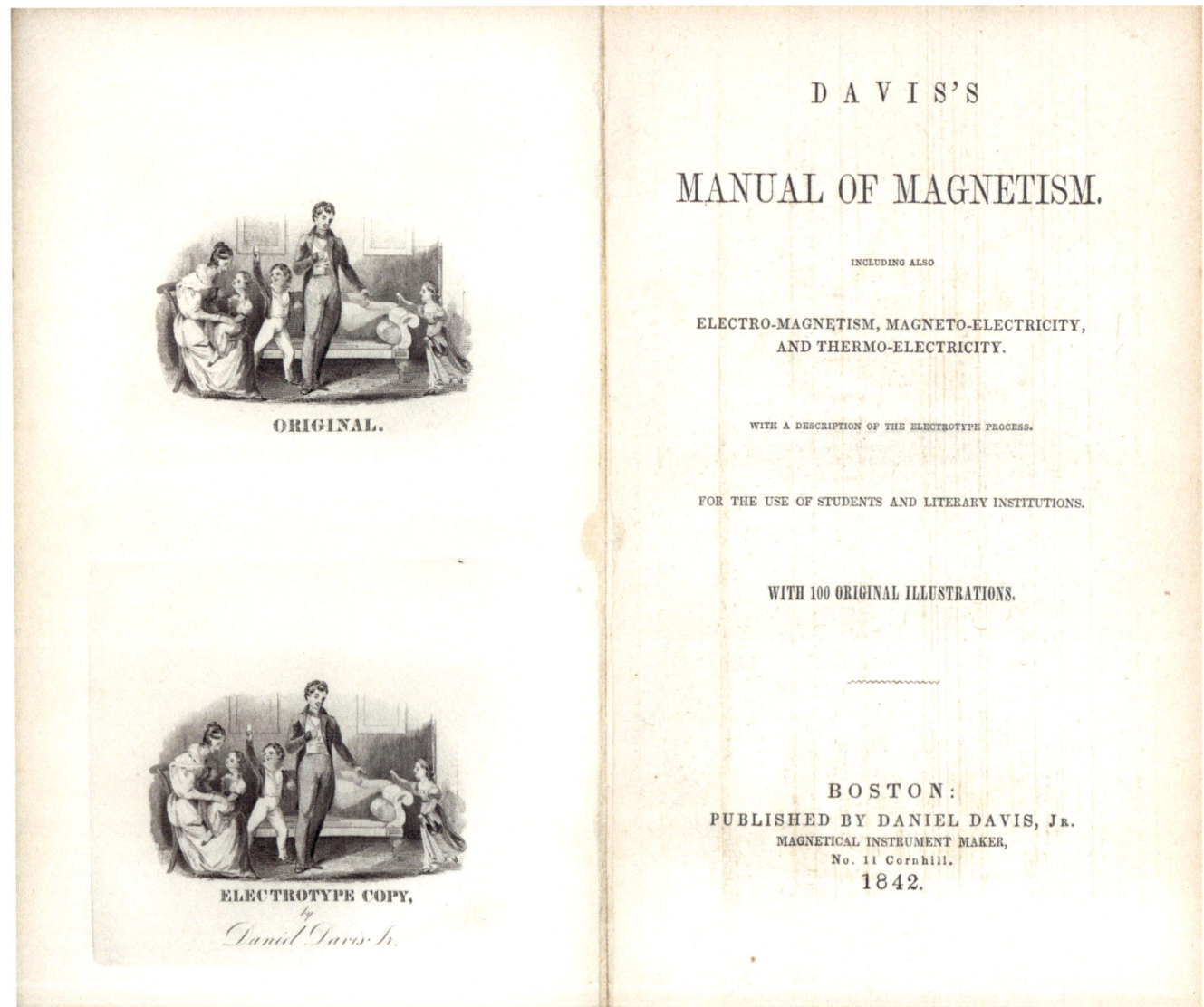

Daniel Davis, *Davis's Manual of Magnetism* (Boston, 1842)

The frontispiece of Daniel Davis's book showed that electrotype copies could not be distinguished from the originals.

them for use in high-speed, high-pressure presses; the process allowed for large print runs of extremely fine-quality text and pictures. The first in America to publish an electrotype illustration was engraver Joseph Alexander Adams, whose electrotype copy of a wood-engraved image was published in the April 1841 issue of *The American Repertory of Arts, Sciences, and Manufactures*.[163] The following year, American photographer Daniel Davis Jr. reproduced an engraved image and an electrotype copy of it as the frontispiece to his *Davis' Manual of Magnetism, Including also Electro-Magnetism, Magneto-Electricity, and Thermo-Electricity. With a Description of the Electrotype Process*.[164]

The first large-scale application of electrotypes in book illustration in the United States was Harper and Brothers *Harper's Illuminated and New Pictorial Bible*, published in parts between 1843 to 1846; it is considered to be the finest book produced in the United States up to that date. Among the extraordinary fea-

Prospectus for *Harper's Illuminated and New Pictorial Bible* (New York, 1843)
12¼ × 9 in. (31 × 23 cm)

The prospectus indicated that Harper's Bible would include 1,600 historical engravings, plus an illuminated letter for each chapter, by Joseph Alexander Adams.

CHAPTER 9

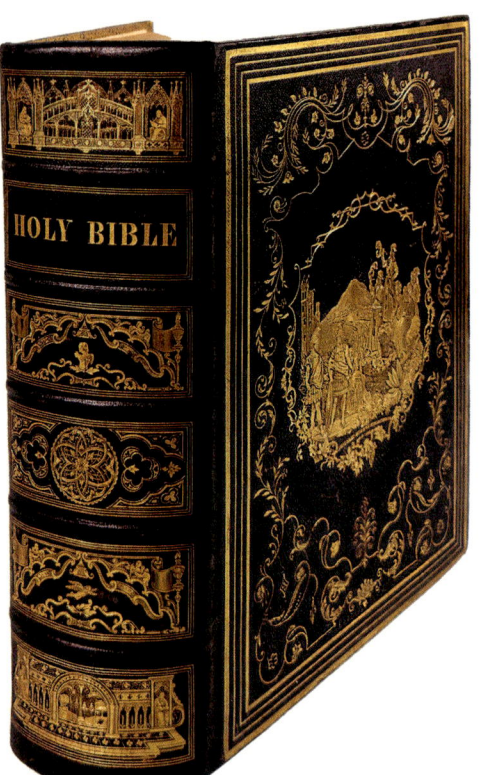
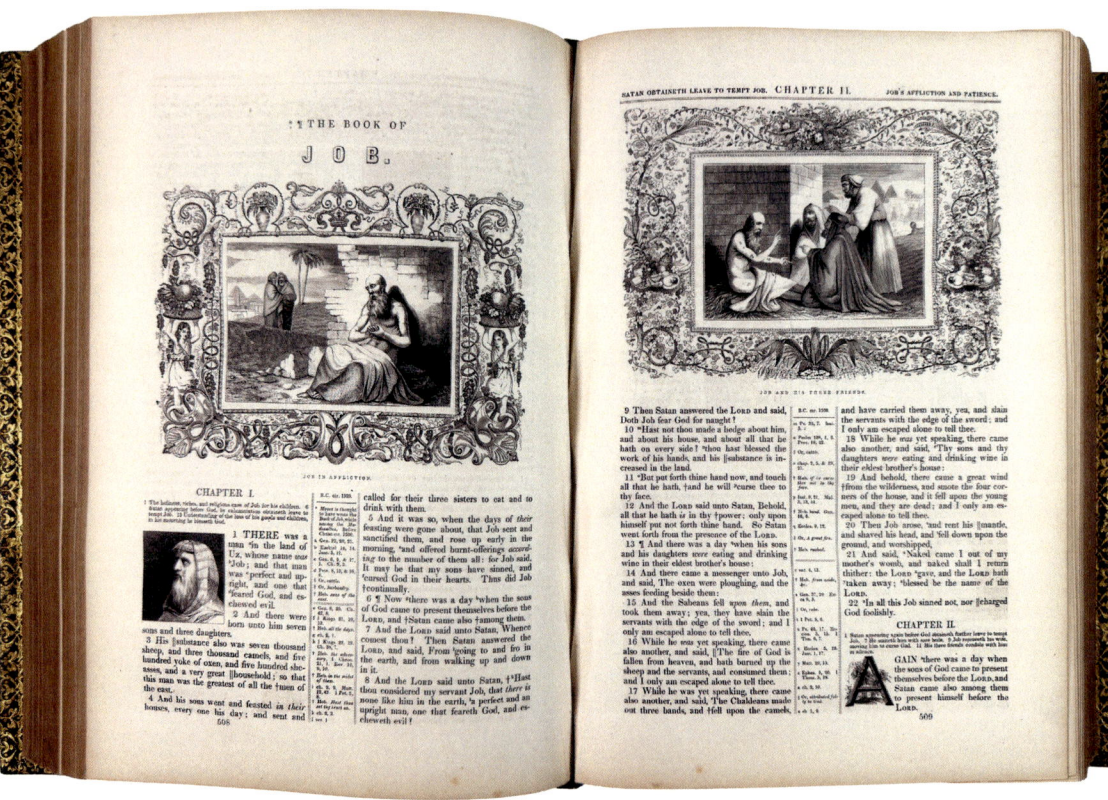

Harper's Illuminated and New Pictorial Bible (New York, 1846)

👉 A very elaborately gilt morocco binding on a copy of the book.

👉 A page opening showing the ornately framed engravings and historiated initial letters printed from electrotype plates.

tures of this exceptional Bible were the 1,600 historical illustrations engraved by Joseph Alexander Adams and reproduced by electrotype; in 1844, Harper and Brothers ordered a new set of presses specially designed to facilitate the electrotyping printing process. The complete work, printed on high-quality paper, weighed over thirteen pounds when bound. Even though it was a large and expensive book, Harper and Brothers sold over 75,000 copies of the *New Pictorial Bible*, reflective of the high demand for Bibles in mid-nineteenth-century America. After the sample published in Daniel Davis's manual, *Harper's Illuminated and New Pictorial Bible* was the first book printed in the United States using this technology.

The idea of publishing such a Bible came from Joseph Alexander Adams, a local printer and engraver, who came to the Harpers with the proposal to produce the grandest Bible ever published in the United States. To make the Bible exceptionally special, Adams proposed illustrating it with 1,600 illustrations, printed mostly on the same pages as their reference in the text instead of on separate illustrated pages, and using electrotype plates to reproduce the illustrations. Adams emphasized that the electrotypes were exceptionally durable and could withstand a huge number of impressions.[165]

To make the *Harper's Illuminated and New Pictorial Bible* affordable to the widest number of people, Harper and Brothers decided to print the edition in 54 parts, ranging from 25 to 60 pages and costing 25 cents each. Optimistic about sales via this method, they ordered an initial press run of 50,000 copies per installment; subscribers could purchase the installments as they appeared and then have them bound upon the completion of the book in 1846. To make these installments more enticing, Harper and Broth-

144

Harper's Illuminated and New Pictorial Bible, 54 parts (New York, 1843–1846)

To make the expensive Bible affordable, Harpers offered it for sale in fifty-four separate parts costing 25 cents each.

ers decided to print some pages in two colors. *Harper's Illuminated and New Pictorial Bible* was an immediate success: the initial press run of 50,000 copies quickly sold out, and Harper and Brothers reprinted 25,000 copies in 1846. Sales of the *Illuminated Bible* remained strong enough that over the next two decades the firm issued reprints in both 1859 and 1866.

At the end of 2018, I was fortunate enough to add to my collection a complete set of all 54 original parts of *Harper's Illuminated and New Pictorial Bible*. Very few complete sets of parts of this work have survived as, considering how awkward it is to use the parts, nearly all copies would have been bound upon completion. Besides this factor, any remaining issues of the parts that Harpers had kept would most probably have been destroyed in the fire that burned down the Harpers warehouse in 1853.

In his children's book *The Harper Establishment*, Abbott seems to have deliberately excluded discussion of the large-circulation *Harper's New Monthly Magazine* that the firm had launched in 1850 (that magazine was, of course, intended for an adult readership). By 1855, the magazine's circulation was 50,000 copies, requiring production facilities different from books, which typically were issued in only a thousand or a few thousand copies at a time. *Harper's Magazine* continues to be published today.

Beginning with 7,500 copies, *Harper's New Monthly Magazine* reached a circulation of 50,000 within six months. The December 1865 issue (vol. XXXII, no. CLXIIIVII), published fifteen years after the magazine's inception, opened with a thirty-one-page illustrated article titled "Making the Magazine," an updated expansion, intended for an adult audience, of Abbott's 1855 *The Harper Establishment*. The 1865 article made use of some of the same images Abbott had

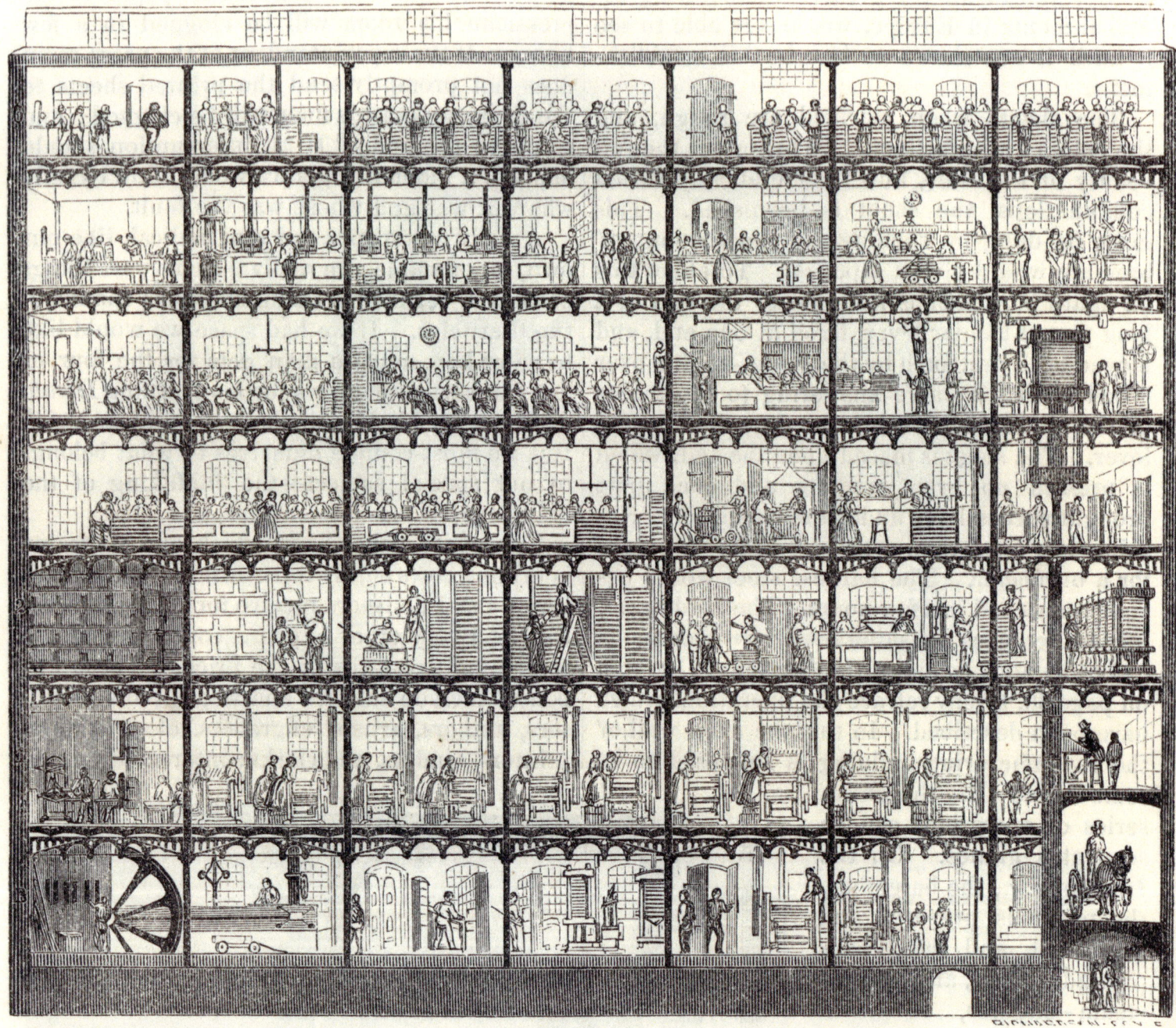

SECTION OF MANUFACTORY.

🔥 "Making the Magazine," *Harper's New Monthly Magazine*, vol. 31, no. 187 (New York, December 1865)

A cross section of the Harpers' six-story building, enlarged to show detail. On the ground floor, we see the steam engine on the far left and printing machines on the far right operated by men. On the second floor, there is a handpress room on the far left, and then the remainder of that floor shows a series of Adams Power Presses being operated by women. The floor above that appears mainly to be book storage. The next three floors up are the bindery, with many women employed. The top floor, which would have had the best light from windows and skylights, is occupied by typesetters, who appear to be entirely men.

🔥🔥 "Making the Magazine," *Harper's New Monthly Magazine*, vol. 31, no. 187 (New York, December 1865)

The R. Hoe & Co. Four Cylinder Type-Revolving Printing Machine, used to print the 50,000 copies of *Harper's New Monthly Magazine* in 1865, was illustrated in the magazine.

THE HOE ROTARY PRESS.

The type-plates are fastened upon the large cylinder, or "drum," in the centre, covering its whole length, and about one-fifth of its circumference. On each side, near the top and bottom, where the four "feeders" are standing, are the four "impression-cylinders;" between these are the "inking-rollers." A feeder places a sheet so that the fingers of the impression-roller can catch it; this is pressed against the type-plates, which have been inked by passing under the inking-rollers just above. The sheets thus printed are taken by a series of "endless tapes" to a "fly," which deposits them in a regular pile. There is a separate fly for each impression-cylinder. The one at the left lower corner is most clearly shown in the cut. It has just laid down a sheet on the pile below, and is coming up for another, which it will receive from the tapes just above the feeder's head. The fly above, which is now lying flat upon its pile, receives its sheets from the tapes at the top. The arrangement at the other end of the press, which is not here shown, is precisely similar.

used in 1855 and added several new ones. The 1865 article was ostensibly about producing the *Monthly*, but it also discussed production of Harper's other magazine, *Harper's Weekly: A Journal of Civilization*, which Harper and Brothers began issuing in 1857. Production of these magazines required facilities rather different from the Harpers' relatively short-run book-production facilities.

Besides production of the magazines, the article discussed and illustrated the bindery facilities at Harper's, as they sold many copies of the magazines as bound volumes with a wide variety of binding styles offered. At least in the early decades, Harper's printed the *New Monthly Magazine* on paper of relatively high quality, with the result that paper was most expensive component in their production process.

To print the magazines, Harper's used three Taylor Cylinder Presses, which could print 1,200 sheets per hour. They also used what R. Hoe & Co. called a Four-Cylinder Type-Revolving Printing Machine. It is probable that this press, which became available in 1847, made it possible for the Harpers to increase the circulation of *Harper's New Monthly Magazine* to 50,000 copies and the circulation of *Harper's Weekly* to 200,000 copies. The anonymous author of the article pointed out that Harper's required higher-quality printing for their magazine than was required by newspapers, which were the typical customers for these presses, so Harpers operated their Hoe "lightning" press at half its maximum speed—5,000 impressions per hour.

American book printers in the United States con-

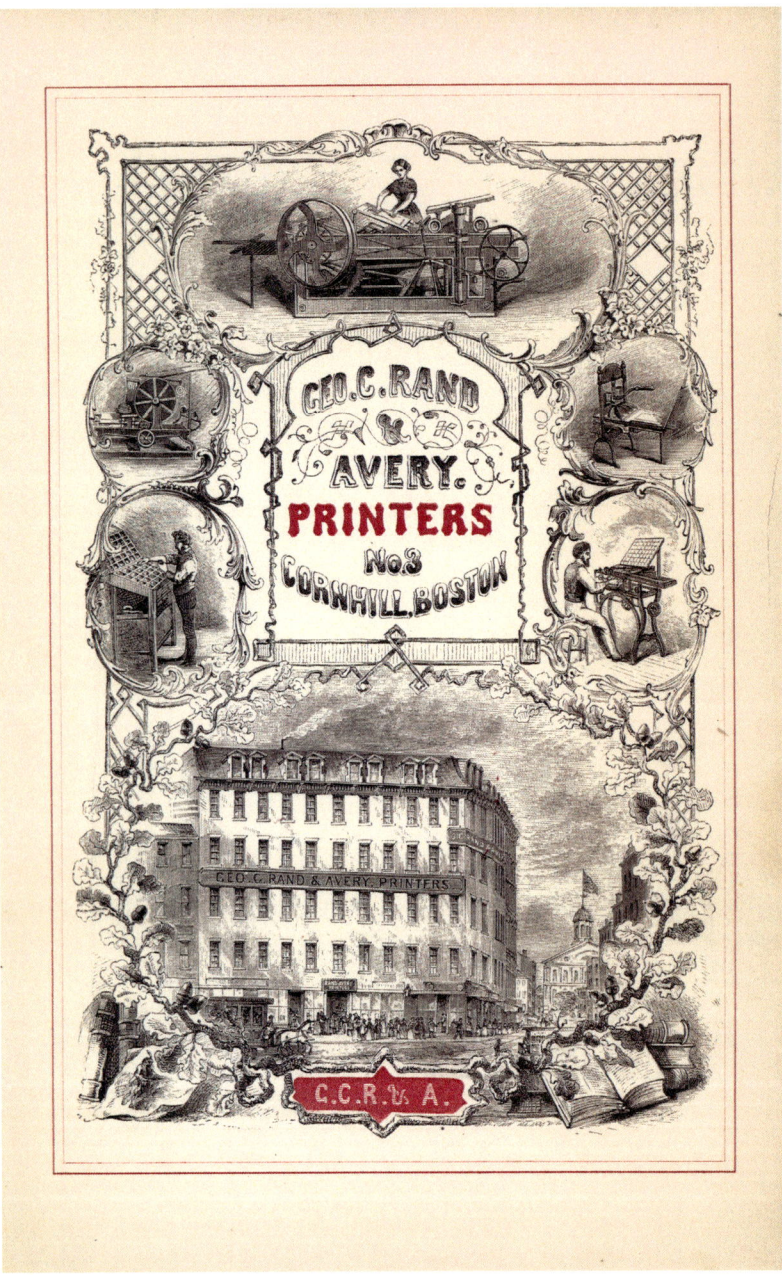

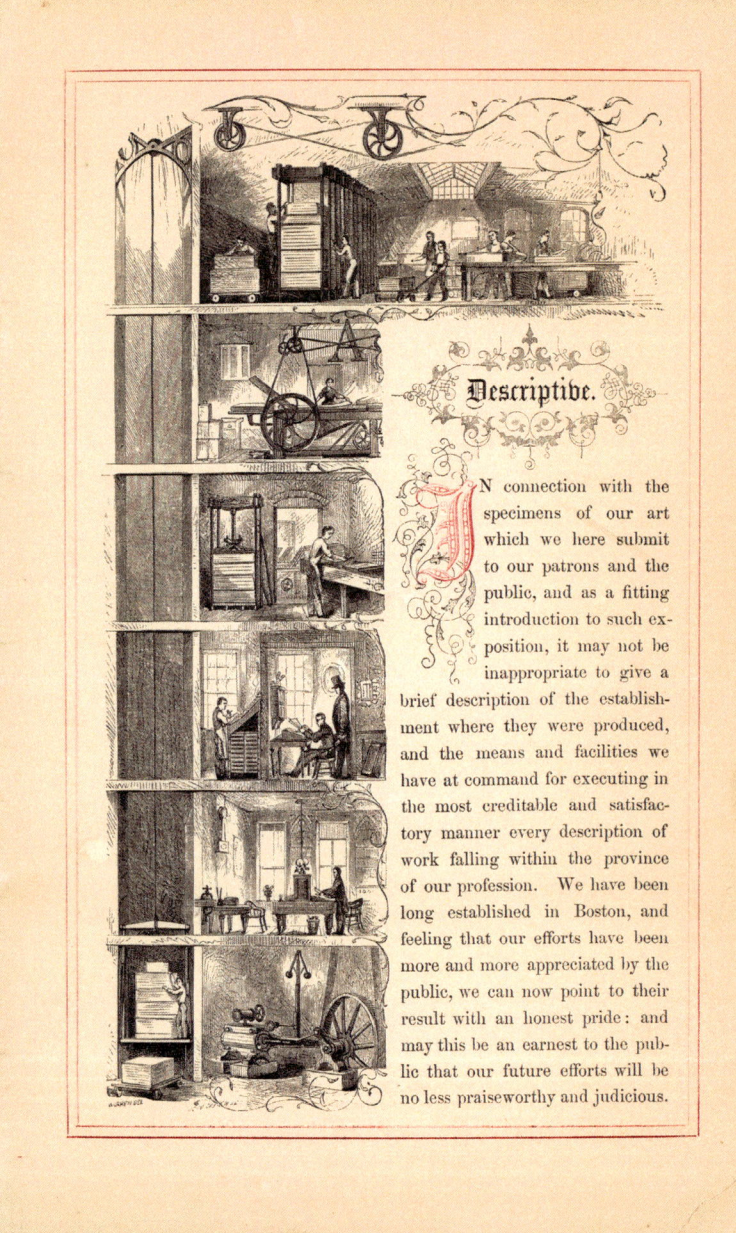

Geo. C. Rand and Avery Printers, *Specimens* (Boston, c. 1865)

The title page of the Rand and Avery specimen book depicted their building and the main features of their typesetting and printing establishment, including the Adams Power Press, being operated by a woman, as shown at the top of the page.

This cross section of the Rand and Avery building showed their Corliss steam engine on the ground floor, the elevator, and the arrangement of the different departments on the various floors in the building.

tinued to use versions of the Adams Power Press throughout most of the nineteenth century. Around 1865, Geo. C. Rand & Avery Printers, the largest book printers in New England at the time, published a type specimen book[166] that opens with a beautifully illustrated account of their Boston facility, undoubtedly one of the most advanced in the United States in the Civil War period. Though the book was undated, we know that it was published before 1867, when the name of the firm was changed to Rand, Avery & Frye. Regrettably, their outstanding and large printing establishment was destroyed by the Great Boston Fire on 8 November 1872.

Along with the elegant engraved, two-color title page for this volume, which contains depictions in miniature of many of the significant features of their facility, Rand and Avery devoted a page in their volume to the Corliss steam engine powering their plant, a machine about which they were undoubtedly proud, as it was absolutely "state of the art." In the caption to the image of their steam engine, they wrote, "We had, at the outset, designed to describe more fully the various machines in use in our Establishment; but we now find how entirely inadequate a book of this size is to any satisfactory description of the many Printing and Hydraulic Presses, Paper Folders, Cutting Machines, &c., which we employ in our business. However, our curious friends,—and, indeed, all who wish to know how books are made—will find our office open to their inspection at all seasonable hours, and skilled workmen to explain what is mysterious in the art of Printing."

Gleason's Pictorial Drawing-Room Companion, vol. 2, no. 22 (detail). See p. 159.

CHAPTER 10

The Role of Women in Eighteenth- & Nineteenth-Century Book Production: Emily Faithfull

Beginning quite early in the history of printing, from time to time women may have been employed as typesetters. The first press known to have employed women in this capacity was the press of the convent of San Jacopo di Ripoli, which employed nuns as compositors; the press was in operation in Tuscany from 1476 until 1484. However, it is probable that the San Jacopo di Ripoli convent experience was the great exception. Until the nineteenth century, documentation of the role of women in book production is scarce. As discussed in chapter 2, we know that women traditionally played a very active role in papermaking by hand, which was often a family affair with women working alongside men. Lalande's *L'Art de faire le papier* (1761) includes three plates showing women performing tasks in papermaking.[167] Perhaps out of tradition, employment of women continued to a certain extent after the mechanization of papermaking.

The first evidence that we have of an actual training school for women in any aspect of the printing trades occurred after the French Revolution. In 1794, in the spirit of social reform, a printer named Deltusso published *Petition à la Convention nationale pour l'école typographique des femmes* (Petition to the national convention for the women's typographic school), printed by female students at the school. Deltusso's seven-page pamphlet promoted the Typographic School for Women that had been proposed in 1791 by a Madame Bastide; the school, which opened in 1793 or earlier

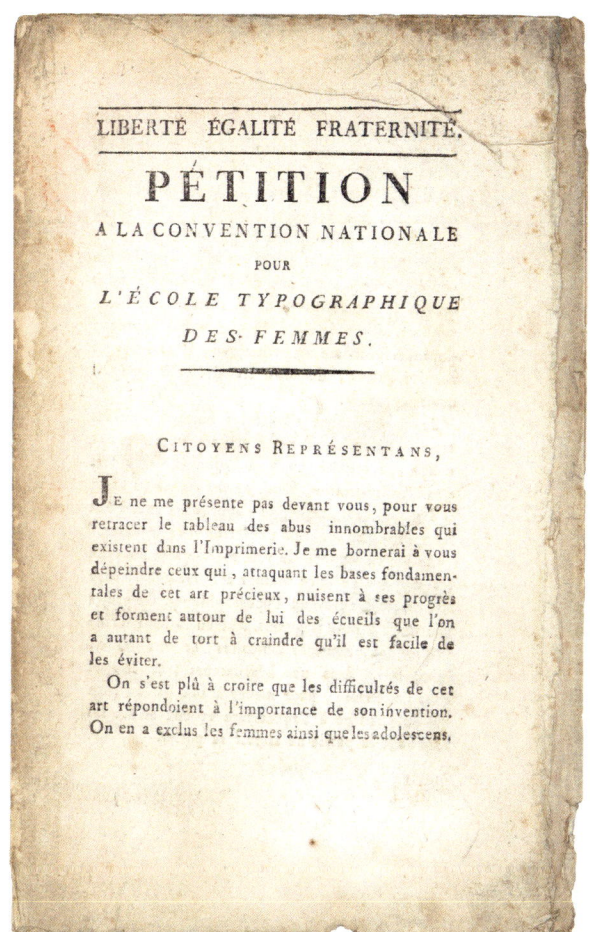

[C. Deltusso], *Petition à la Convention national pour l'école typographique des femmes* [Paris, 1794]

Deltusso's seven-page petition to the National Convention promoting the École Typographique des Femmes was printed in 1794 by women students at the school.

151

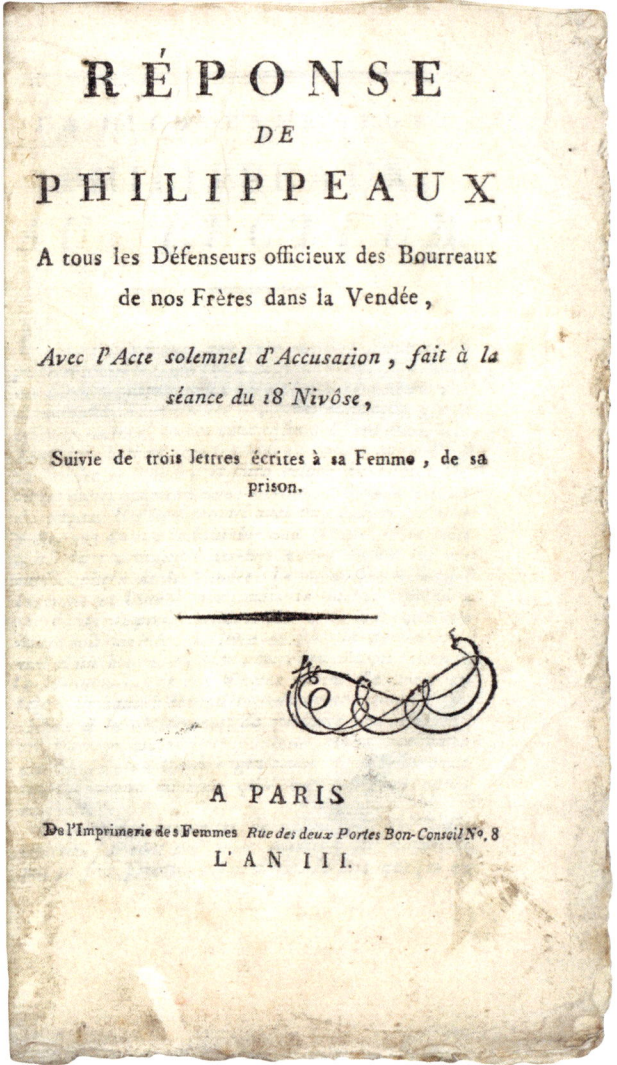

A. C. Thibaudeau, *Histoire du terrorisme dans le département de la Vienne* (Paris, [1795])

At the foot of the title page we see that the booklet was printed by the Imprimerie des Femmes at the École Typographique des Femmes.

Pierre Philippeaux, *Réponse de Philippeaux a tous les défenseurs officieux des Bourreaux . . .* (Paris, an III [1795])

Printed by the Imprimerie des Femmes at the École Typographique des Femmes.

under Deltusso's direction, also operated as a printing press. The school's typography apprenticeship lasted six months and cost 400 livres; students had to be at least twelve years old, and enrollment was limited to sixty. The school operated a press that published several works. The goal of the typography school for women was to provide female citizens with professional skills so that "the most worthy half of the human race" was not "reduced to dependence on the other half."[168]

However, even in the new spirit of social reform, the idea of women working as typesetters threatened the male typesetting establishment, who understandably regarded women working at lower wages as competition. It appears that the school may have existed for only two years, since the last known imprint from the school is dated 1795.[169]

In addition to Deltusso's petition, the *École Typographique des Femmes* issued at least two other works, both in 1795: Antoine-Claire Thibaudeau's *Histoire du terrorisme dans le département de la Vienne* and *Réponse de Philippeaux à tous les défenseurs officieux des Bour-*

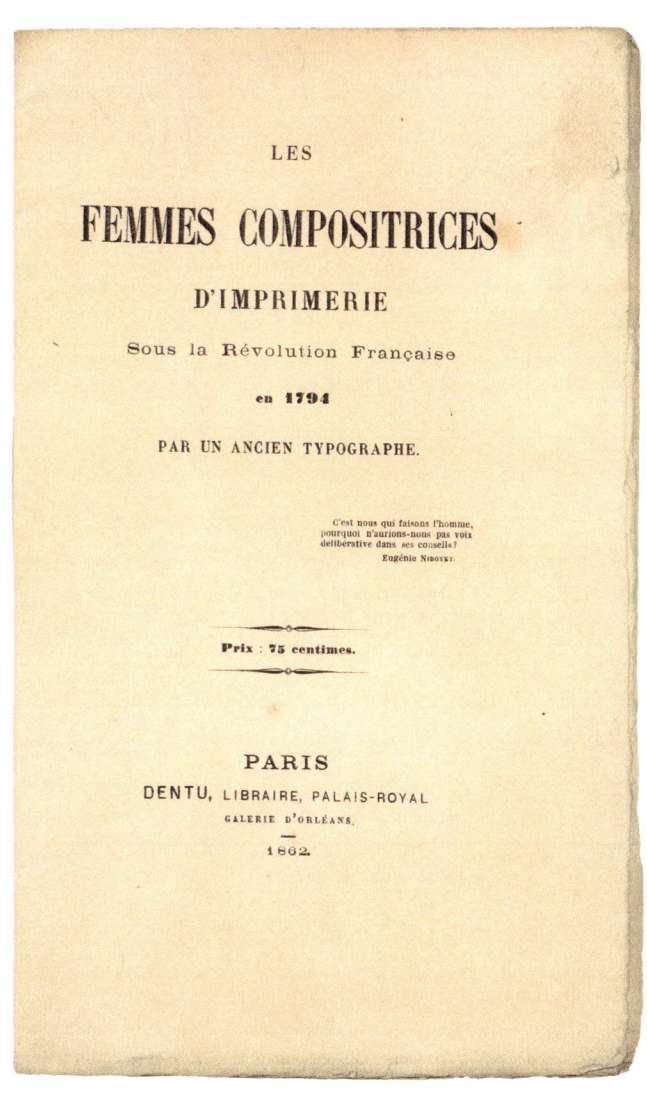

[Alphonse Alkan,] *Les Femmes compositrices d'imprimerie* (Paris, 1862)

The question of whether women should work as typesetters became a controversial issue in France and England during the 1860s. Alphonse Alkan's pamphlet on women typesetters reprinted Deltusso's petition on the subject, originally published in 1794.

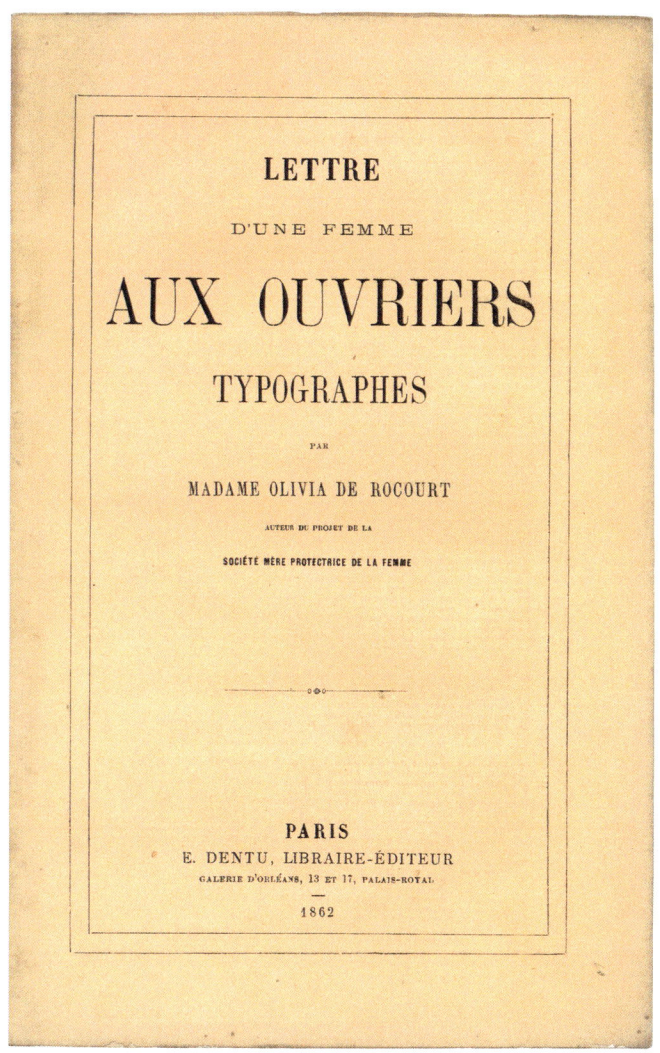

Olivia de Rocourt, *Lettre d'une femme aux ouvriers typographes* (Paris, 1862)

In 1862, female activist Olivia de Rocourt published a pamphlet supportive of women typesetters.

reaux de nos frères dans la Vendée, avec l'acte solemnel d'accusations, fait à la séance du 18 Nivôse, suivie de trois lettres écrites à sa femme, de sa prison.

The issue of women typesetters seems to have continued quietly in France until the 1860s. In 1862, Olivia de Rocourt published a pamphlet supportive of women typesetters: *Lettre d'une femme aux ouvriers typographes* (Paris: E. Dentu, 1862). In the same year, Dentu published a pamphlet by typographer Alphonse Alkan titled *Les Femmes compositrices d'im-* *primerie sous la Révolution française en 1794 par un ancient typographe,* which included a reprint of Deltusso's 1794 pamphlet.

While in Paris in September 2012, I acquired an album or scrapbook in quarto format, bound in nineteenth-century blind-stamped red cloth and labeled *Receuil de journaux* on the spine. On all 184 pages of the album, a former owner had pasted newspaper clippings from printing-trade journals and other souvenirs of the Parisian printing trade published be-

L'Univers illustre, vol. 8, no. 485 (Paris, 15 November 1865)

The woodcut published on the first page of this French magazine shows women typesetting and making up pages at Emily Faithfull's Victoria Press in London. The magazine did not take a stand on the issue. This illustration was originally published in the 15 June 1861 issue of *The Illustrated London News*.

15 CENTIMES LE NUMÉRO
20 centimes par la poste.

PRIX DE L'ABONNEMENT
DANS TOUTE LA FRANCE.
         PARIS. DÉPARTM.
Un an . . . . 15 fr. » 17 fr.
Six mois . . 8 fr. » 9 fr.
Trois mois . . 4 fr. 50 5 fr.
Le numéro : 15 cent.

LA COLLECTION DU JOURNAL
14 VOLUMES BROCHÉS
AVEC TOUTES LES PRIMES
65 fr. au lieu de 85
LA MÊME COLLECTION
reliée en 8 volumes
85 fr. au lieu de 107 fr.
(Envoi franco)

Bureaux d'abonnement, rédaction et administration :
Passage Colbert, 24, près du Palais-Royal.
Toutes les lettres doivent être affranchies.

8ᵉ ANNÉE. — N° 485.
Mercredi 15 Novembre 1865.

Vente au numéro et abonnements :
MICHEL LÉVY FRÈRES, éditeurs, rue Vivienne, 2 bis
et à la LIBRAIRIE NOUVELLE, boulevard des Italiens, 15.

SOMMAIRE

Chronique, par M. GÉRÔME. — Bulletin, par M. TH. DE LANGEAC. — Les Femmes typographes, par M. X. DACHÈRES. — Paris nouveau, par M. HENRI MULLER. — Les Bourgeois de Darlingen (suite), par M. HENRI CONSCIENCE. — Épisode d'une chasse au renard, par M. H. VERNOY. — Causerie scientifique, par M. S. HENRY BERTHOUD. — Marguerite en prison, par M. R. BRYON. — Courrier des Modes, par Mᵐᵉ ALICE DE SAVIGNY. — L'Homme depuis cinq mille ans, par M. A. DARLET.

CHRONIQUE.

Ah! quel bonheur d'être fusilier! — Les almanachs pour 1866. — Luxe effréné des almanachs. — Le *Double Liégeois*. — Son rang dans la société. — Le *Sapeur* de Thérésa. — Malheur aux vaincus! — Retour des poltrons. — Fuite des frères Davenport et du choléra. — Les Parisiens et leurs médecins. — Souvenirs de 1849 à propos des internes décrétés. — La croissance électrique au Cirque Napoléon. — Ici on redresse les hommes. — Le moteur Lenoir. — La vapeur en chambre. — M. de Bismark. — Un portrait qui nous coûterait 50,000 francs. — Un premier ministre des arts. — La *Bible* illustrée, par Gustave Doré. — Quelques détails sur l'artiste. — Ses journées et ses soirées. — Les dimanches de la rue Saint-Dominique. — Ici l'on fume et l'on chante. — Doré violoniste, chanteur et escamoteur. — De quoi l'on cause. — La vieille servante de comédie. — Une bonne qui tutoie son maître. — Ce que la vieille servante pense de Gustave Doré.

Évidemment, si j'avais à recommencer ma vie, je choisirais une carrière autre que celle de journaliste!

En ce moment, par exemple, je regrette de ne pas avoir l'honneur d'appartenir au régiment des fusiliers de Poméranie, dont la musique a un très-grand succès à Paris.

Tout homme qui n'est pas quelque peu fusilier de profession et musicien poméranien par vocation doit renoncer à occuper l'attention publique. Les Parisiens ne veulent plus entendre parler des braves gens qui n'ont que le seul défaut de ne pas porter de casque orné d'un aigle prussien. On s'arrache positivement ces braves gens qui soufflent, au Cirque de l'Impératrice, dans les instruments en cuivre inconnus dans les rangs de nos musiques militaires; on fête ces Prussiens à Saint-Cloud, tantôt ils fraternisent avec les musiciens de la gendarmerie impériale, tantôt encore on les invite à dîner chez des banquiers allemands et des financiers prussiens.

L'histoire de la semaine peut se résumer en quelques mots: musique des fusiliers, un duel entre un journaliste et un étranger, et l'ouverture du magasin de nouveautés, créé au boulevard Malesherbes par celle qui fut Mˡˡᵉ Figeac au Théâtre-Français. Ajoutez à cela l'arrivée des almanachs de toutes espèces et vous en saurez, sur l'histoire de la semaine, tout aussi long que votre très-humble chroniqueur.

Or, quand les almanachs arrivent, le tambour de la garde nationale n'est pas loin; c'est un premier avertissement donné à l'homme pour lui dire que le jour néfaste approche où la moitié de Paris demande des étrennes à l'autre moitié.

Il ne m'est plus permis de douter de l'approche du jour de l'an. Depuis que les nouveaux almanachs sont arrivés, déjà ils se disputent la succession de 1865, comme les neveux partagent l'héritage dans l'antichambre du moribond. Les voilà devant moi sur une table tous les enfants précoces de l'année à venir, tous étalent sur leurs couvertures le chiffre

1866

Comme s'ils voulaient nous dire :

UN ATELIER DE FEMMES TYPOGRAPHES, dessin communiqué (Voir page 723).

tween 1865 and 1867. The articles emphasize social issues, especially in typesetting, particularly the issue, which was then highly controversial in the French printing trades, of whether or not women should be employed as typesetters—one of the few roles in the French printing industry where women were sometimes employed at the time (women were also sometimes given the job of feeding paper into printing machines). Included in the scrapbook is a series of articles published in August 1865 recording considerable debate in newspapers over the role of women as typesetters. The chief complaint from men was the standard one—that since women were paid much less than men, they tended to depress the salaries of male typesetters. Several of the articles stated that it was very difficult for male typesetters to make a good living and that competition from women in typesetting jobs was very unwelcome; besides, women were meant to stay home with children, etc., etc. In an article published on 15 August 1865, the author, A. Bazin, who characterized himself as a typesetter, stated that 5,000 male typesetters were competing against 500 women typesetters in Paris.

There is no identification of ownership in the album except an unusual French, diamond-shaped bookplate reflecting a serious interest in the history of printing. This contains a monogram that may be read PLM or LMP or some other combination of the letters. Images in the album show women doing typesetting, a lithography plant, a bank-note printing plant, and some unusual ephemera. Whoever assembled the album went to the great effort of preparing what appears to be a complete manuscript index to people and places mentioned, making this an unusually valuable reference source for two years in the history of printing in Paris.

The topic was of wide enough general interest that the 15 November 1865 issue of *L'Univers illustré* published a brief article on the topic and reproduced a woodcut of women typesetters at the Victoria Press in London (originally published on 15 June 1861 in *The Illustrated London News*) on its upper cover, without identifying the source of the image or the Victoria Press. The article stated that *L'Univers illustré* would not take a side in this dispute. Here is a translation of the conclusion of the article:

In publishing the sketch of a workshop today where women typographers are employed, we will only add that the idea took off in England and America much faster than in our country. In those two countries, there are quite a large number of printing houses which employ men and women.

In France, the only important printing house, we believe, where women regularly work in typesetting, proofreading, and as paper feeders for printing machines is that of Paul Dupont, at Clichy. Her Majesty the Empress visited this establishment some time ago and expressed her august sympathy for the lot of poor women in modern society.[170]

In England, women were long involved in the bookbinding industry, folding, sewing, and performing other non-strenuous tasks; their employment in bookbinding continued through the nineteenth century.[171] The first evidence we have in England of women participating in the printing trades is in the 1840s and 1850s, as operators of early typesetting machines that were designed to be operated by women. The inventors of these machines deliberately promoted their use by women to show cost savings resulting from mechanization, since women were paid roughly half the wages of men. Typesetting by hand or by machine became a career for women in certain printing facilities during the second half of the nineteenth century, but employing women as typesetters was often controversial and nearly always faced aggressive resistance from men, primarily because women typesetters were paid less and their presence as typesetters put a damper on men's pay.

In 1860, coinciding with the controversy over women typesetters in Paris, Emily Faithfull founded the Victoria Press for the employment of women in London. In founding the Victoria Press, Faithfull was influenced by the Society for Promoting the Employment of Women, established in 1859, for which she served as secretary. At the Victoria Press, Faithfull employed women as compositors; she also employed men to do some of the presswork and heavy lifting, necessitated by the strength required to pull a handpress and the need to lift the very heavy locked-up forms of type.

One of the first books printed at the Victoria Press was *Trade Societies and Strikes: Report of the Commit-*

Stereoscopic Co.
Photograph of Emily Faithfull (London, c. 1860)

This carte-de-visite photograph was signed by Emily Faithfull in the lower margin.

tee on *Trades' Societies, Appointed by the National Association for the Promotion of Social Science.*[172] The book was an early production of the National Association for the Promotion of Social Science, a reformist organization formed in 1859; Faithfull was one of its members. The 651-page book included "Account of Printers' Strikes and Trades' Unions since January, 1845," by J. W. Crompton, and "Account of the London Consolidated Society of Bookbinders," by bookbinder and trade unionist Thomas Joseph Dunning. Its year of publication coincided with the formation of the London Trades Council that united London's trade unionists in 1860.

This work is one of the few conveniently organized sources of information on eighteenth- and early-nineteenth-century labor organizing and strike activity in England. Because strikes and formal union organizing were illegal in England until 1871, the earlier history discussed in this book mostly involves informal trade disturbances, until the 1842 general strike in England and the government repression that followed. Even though strike activity was illegal, there were actual attempts at what the government defined as strike activity, and those attempts sometimes resulted in imprisonment for participants.

In October 1860, Faithfull, who was the best-known promoter of women in the printing trades in the nineteenth century, published an article titled "Victoria Press" in *The English Woman's Journal*, which she began publishing in 1860.[173] The journal published articles mainly about female employment and equality issues. Issues with printing that Faithfull raised included potential problems associated with printing that were rarely mentioned at the time, including occupational health hazards from typesetters' extended contact with lead and antimony through touch and inhalation. Before opening her press, Faithfull conducted experiments to see if women could be taught to be successful typesetters and to ascertain the aspects of printing that were beyond the strength of most women, such as lifting a form onto the press.

By the time Faithfull wrote her article, she was employing sixteen female compositors, one of whom had been trained as a printer by her father. She mentioned that four of the other compositors were very young, under fifteen years old. Regarding the health of printers, Faithfull stated that the annual reports of the Widows' Metropolitan Typographical Fund for the previous ten years (1850–1860) indicated that the average age of death of printers at that time was forty-eight years. Lead poisoning and tuberculosis were known causes.

Besides advocating for women as compositors, Faithfull urged highly educated women to apprentice as readers and correctors for the press. These were among the highest paid professions in printing, requiring both technical knowledge of printing and a high level of skill in languages, grammar, etc.

One of the most elaborate books printed and published by the Victoria Press was *The Victoria Regia: A Volume of Original Contributions in Poetry and Prose*.[174]

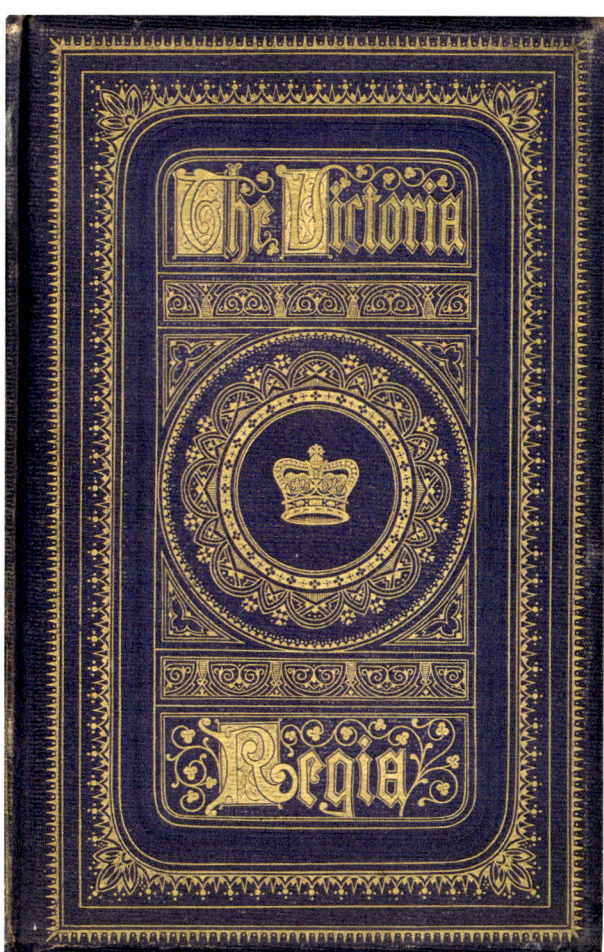

Adelaide A. Procter, ed., *The Victoria Regia* (London, 1861)

The elaborately stamped publisher's cloth binding on the Victoria Press edition of the poetry collection entitled *The Victoria Regia*. The imprint on the title page reads "Printed and Published by Emily Faithfull and Co., Victoria Press (for the Employment of Women)."

This collection of poetry, which included poems by Tennyson and Thackeray, was issued in ornately stamped cloth bindings or bindings with a similar design in leather. In appreciation for this book and in honor of her work creating jobs for women in the printing trades, Faithfull was appointed printer and publisher in ordinary to Queen Victoria.

On 15 June 1861, *The Illustrated London News* published an article about Faithfull's Victoria Press, along with an illustration of work at the press that was frequently reproduced or copied without attribution. Considering the traditional opposition of the predominantly male printing trades to the entrance of women, I was surprised by the supportive nature of this review:

> *The introduction of women into the printing trade can now claim the consideration due to a successful experiment. The Victoria Press was opened in March, 1860, and has therefore stood the test of one year's work, and has triumphed over all the preliminary difficulties with which such an undertaking could not fail to be attending. The originators of this enterprise believed that former failures were to be traced to causes altogether apart from the nature of the work. They believed that these failures arose from the small scale of the experiments, which from the nature of the work and the considerable outlay of capital required before economical success can be secured, carried risk of a larger experiment than any hitherto tried, feeling that the time was ripe for it. The public mind, long either strongly opposed or utterly indifferent to such innovations, was in enlightened quarters beginning to feel the necessity for extending the industrial employment of women.... If the Victoria Press continues to obtain the public support which has hitherto been extended to it there is no doubt of its ultimate entire success even as a commercial undertaking, and this is earnestly to be desired, as the only ground on which the extension of the experiment can be justly urged.*[175]

As noted above, typesetters in England typically resisted women typesetters for the obvious reason that they were paid less than men. On 5 August 1867, a Manchester printer who signed himself "W.H.W." published this paragraph in a letter to *The Printers' Journal and Typographical Magazine*, written after a newspaper attempted to recruit women typesetters:

> *Wanted, Sixty well-educated Girls to learn to set type for a newspaper in Manchester.—Address M53, at the printers.*
>
> *A number of applications were made, and inquiries were put by those friendly disposed to our profession with a view of ascertaining what was the intention of the advertiser. The sequel has proved that the advertisement answered the required purpose in obtaining an idea of the number and character of aspirants for typographical honours in our city among the ranks of the fair sex. Female labour has been fairly tried within*

twenty miles of our city some years since, and it has been weighed in the balance and found wanting, and if half what I hear of the working of Miss Faithful's [sic] establishment in London be correct, the day is far distant when such labour can hope to supersede our own. There are natural impediments in their path which neither time nor patience can overcome. Physical labour and intellectual power they will never possess in sufficient force to enable them effectually to cope with the difficulties and exigencies of our profession. The legitimate function of women is not yet thoroughly understood and appreciated. The extra income derived from the results of female labour can never compensate for the negligence of home duties and the comforts resulting therefrom. Females so employed become independent of parental control, and they step by step, imperceptibly, but none the less surely and certainly, deviate from the paths of moral rectitude, and eventually add to the growing throng of that class of unfortunates who stain the fair page of our social history with a foul and loathsome blot.[176]

In response to this attack on women compositors, William Wilfred Head[177] of the Victoria Press published this letter written by Matilda Alexander, a former partner in the press, in the 19 August issue of the same journal:

To the Editor of the Printers' Journal:

Sir,—in reference to a paragraph contained in your journal of August 5, reflecting, I may say, not very flattering terms on the efforts made by this firm and others for promoting the employment of women in the printing trade, your correspondent appears quite to have mistaken the object, at least of the heads of this establishment, when he triumphantly refers to the day being "far distant when such labour can hope to supersede our own."

The promoters of the innovation, if I may so call it, in the interest of women, or, as your correspondent would probably call them, the usurpers of the rights of man, never dreamed of ever being able to accomplish anything so futile as the exclusive labour of women in the art of printing. But they thought, and think still, that whatever labour adapts itself to female handicraft that women, who have a livelihood to seek as well as men, are equally entitled with men to resort to any employment for which their nature and intellect are fitted.

This capacity of fitness your correspondent utterly denies that women possess. It would be needless, I am sure for me to attempt to show that the girl who has received a tolerable education—the future wife, it may be, of a printer, and the mother of a family—should not to the extent required in our business be endowed by nature with faculties equivalent to such a demand. . . .

The following testimonial on my relinquishing partnership with Miss Faithfull will, I think prove to you the appreciation of the females in my employment of the work they have had to perform and the treatment to which they have been subjected:—"The females employed at the Victoria Press cannot allow the occasion for the change in the firm to pass without expressing their deep sense of obligation to Mr. Head, for the liberal and honourable manner in which has redeemed his promise on entering the concern to give printing by female hands a fair trial. . . . Will Mr. Head accept the thanks of the employees of the Victoria Press, and their best wishes that he will carry out in the future this good work as satisfactorily to himself as he hitherto has to them? Signed on behalf of the employees,—Matilda Alexander.—August 5, 1867."[178]

In 1876, Emma Paterson, a former bookbinder's apprentice and assistant secretary to the Workmen's Club and Institute Union, founded the Women's Printing Society (WPS) under the aegis of the Women's Protective and Provident League. This organization, which initially included some alumna from the Victoria Press, remained in business for several decades. At the WPL, women compositors set type and proofread their work. "Later they carried out every stage of the compositor's work, including imposition (i.e. dividing the galleys of type into 'pages' in a wooden frame or chase and arranging them in a 'forme' so that the pages are printed in the correct order for folding the sheets of printed material) and distributing the type."[179] One of the many books that they published was *Sunny Memories. Containing Personal Recollections of Some Celebrated Characters*, by M. L. [Mary Lloyd].[180]

It is fair to say that opportunities for women in the printing trades during the nineteenth century were generally limited, except for paper manufacturing and bookbinding, and some exceptional situations in typesetting, either manual or by machine, some of which are enumerated in the chapter on typesetting.

Gleason's Pictorial Drawing-Room Companion, vol. 2, no. 22 (29 May 1852), p. 352

Women operating steam-powered printing machines at the printing facility for *Gleason's Pictorial Drawing-Room Companion*. Only outlines of the presses appear in the image. The presses appear to be of the Adams type.

When women were hired, it was usually for cost savings since it was almost universally accepted in the nineteenth century that women were paid around half the wages of men.

Women were generally excluded from operating handpresses because of the strength typically required to pull a handpress. Though printing machines reduced some of the strength requirements, their operation was almost entirely the province of men except in the United States. From the beginning of mechanized printing in America, Daniel Treadwell hired women to operate the Treadwell Power Press. It appears that the concept of women operating a printing machine was accepted without resistance in the United States. When the Adams Power Press was introduced, to replace the Treadwell Power Press, virtually every image of that press in operation for book production, especially from the 1850s onward, shows it being operated by women. All the images of that press in operation at Harper and Brothers in New York show it being operated by women. About the same time, images of printing machines printing *Gleason's Pictorial Drawing-Room Companion*, which was published under that name in New York from 1851 to 1855, show their presses, including a Taylor Cylinder Press, being operated by women.

From *Gleason's Pictorial Drawing-Room Companion* [no date]

Women operating a steam-powered Taylor Cylinder Press, from an undated page of *Gleason's Pictorial Drawing-Room Companion*. Alva Burr Taylor, a blacksmith by trade, worked with R. Hoe & Company of New York from 1822 until 1842, when he formed his own company. Taylor produced cylinder presses, Washington handpresses, and a few jobbing presses, as well as steam engines. From images like this, we may draw the tentative conclusion that women were frequent operators of power bed-and-platen presses in nineteenth-century America.

Considering the documentation that we have of women operating printing machines in mid-nineteenth-century America, it is possible that women had greater opportunities in the printing trades in the United States than in England or Europe. An indication of this in the field of typesetting appeared in *Gleason's Pictorial Drawing-Room Companion* in the issue of 15 October 1853. On p. 153, the editor, Maturin M. Baillou, wrote under the heading "Female-Type-Setters":

We observe by our southern and western exchanges, that it is thought something very peculiar to employ girls for type-setters in printing offices, with them. Here in Boston it is very common, and we may add that half of the type-setting done in our office has been performed by girls for several years. There is no good reason why this should not be very general. Females are better adapted to this sort of work than men, they are fully as intelligent, quite as quick and correct as men, while the character of the work, in all respects, seems to be more properly in their province than belonging to strong and healthy men, whose natural occupation would seem to be of a character to call into action the thews and sinews that Providence has given them. We observe that the plan of employing female type-setters is being very largely adopted in New York, Philadelphia, and other large ciies, and we most heartily approve of the object. It is an intelligent, pleasant and profitable employment.

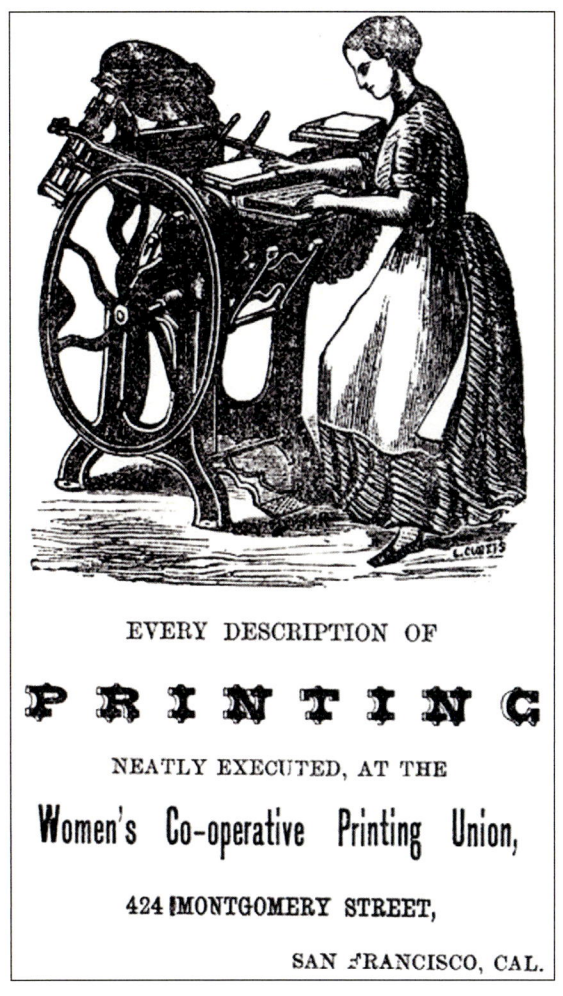

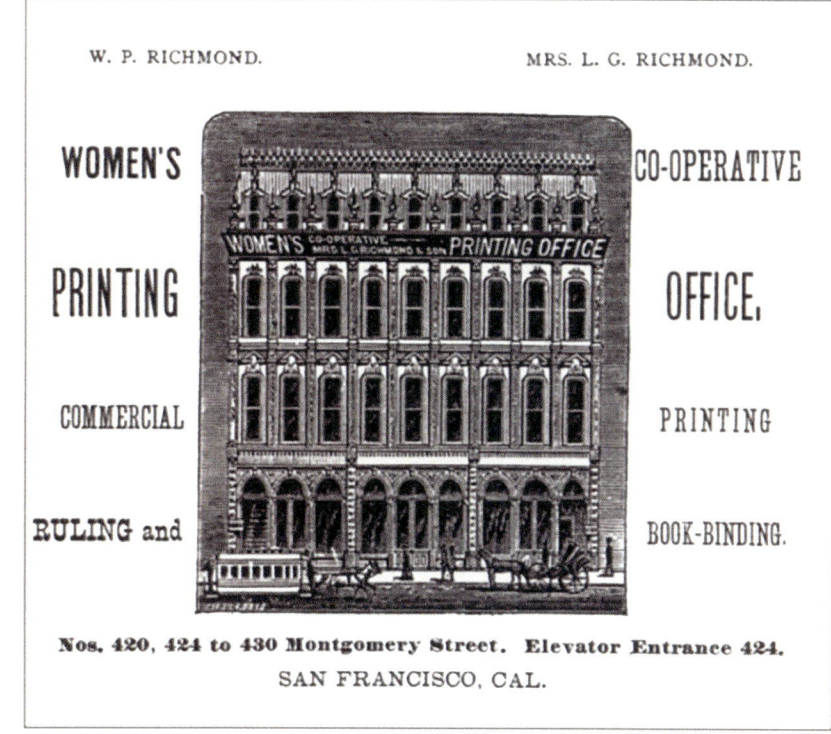

Advertisements for the Women's Co-Operative Printing Unions founded in San Francisco 1868 by Agnes B. Peterson. Image from HathiTrust.

What Baillou did not mention was the likely cost savings involved with hiring women typesetters, since women were typically paid about half the wages of men.

Another example of opportunities in printing for women is what appears to be the first woman-owned printing business in California, the Women's Co-Operative Printing Union, founded by Agnes B. Peterson in San Francisco in 1868. An advertisement published in the early 1870s gives the business's address as nos. 420 and 424 to 430 Montgomery St., Elevator Entrance 424; at the time this advertisement was published, the business was owned by L. P. Richmond and Mrs. L. G. Richmond. Besides printing, the company produced account books and operated a bindery.[181] However, the employment of women as typesetters in nineteenth-century America was usually very controversial. One of the best discussions of this is Mary Biggs's "Neither Printer's Wife nor Widow: American Women in Typesetting, 1830–1950."[182] Biggs pointed out that male typesetters often complained, without valid evidence, that working in a printing shop was too arduous for women's "delicate health" and limited physical strength, or that women were simply not as competent for the task as men. It is difficult to measure what percentage of typesetting in the United States was undertaken by women during the nineteenth century; however, Biggs cites a statement, published on 12 May 1870 in the feminist newspaper *The Revolution* (founded by women's rights advocates Elizabeth Cady Stanton and Susan B. Anthony), that women printers had "'for years [been] successfully employed by the Harpers and . . . in nearly all the book and paper offices in the city' (excluding the morning papers) and that their work equalled men's 'both in quantity and quality'."[183]

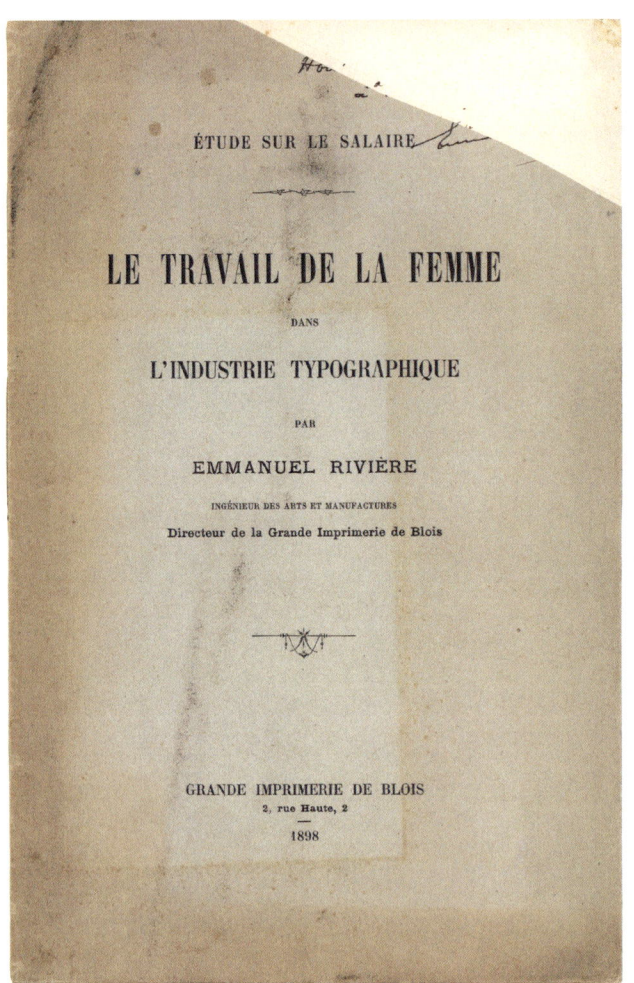

Emmanuel Rivière, *Le Travail de la femme dans l'industrie typographique* (Blois, 1898)

This study was critical of women working as typesetters in France.

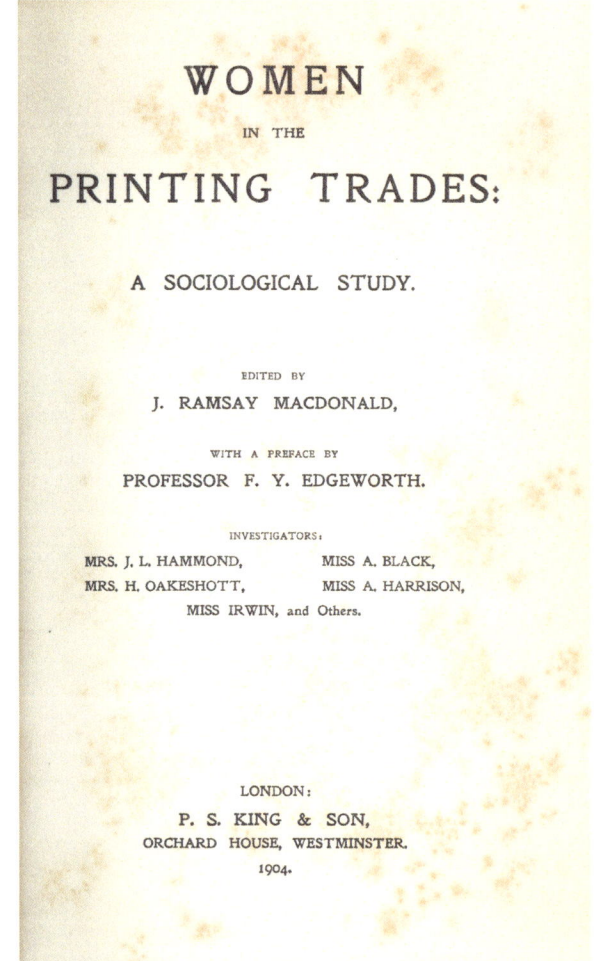

J. Ramsay MacDonald, ed., *Women in the Printing Trades: A Sociological Study* (London, 1904)

This relatively comprehensive survey of women's work in the printing trades credits five women on the title page for conducting the research.

In his article on "Printing the Magazine," published in *The Century Illustrated Monthly Magazine* in November 1890,[184] Theodore Low De Vinne stated that the composition of the magazine was "done by young women, whose work is as accurate and acceptable as that done by men. The women are paid the same rates as men." Undoubtedly, De Vinne's "equal pay for equal work" policy for women typesetters was exceptional for his time, or he would not have mentioned it. De Vinne, whose firm printed the magazine, also wrote about the role of women in the bindery of the magazine.

In France, the controversy over the employment of women as typesetters continued in 1898 during the period in which a considerable amount of typesetting was becoming mechanized by Linotype, Monotype, and other methods of machine typesetting. Printer and conservative activist Emmanuel Rivière, director of La Grande Imprimerie de Blois, self-published a pamphlet, the title of which may be translated as *A Study on Wages: Women's Work in the Typographic Industry*.[185] In his pamphlet, Rivière blamed women for the economic structure that deprived women of equal pay for equal work. He criticized the employment of women as compositors on a global basis, stating that even though they were paid about half the salary of men (from 2 to 3.50 francs for 10,000 letters versus 4.50 francs paid to men for the same amount of work), women's work was inferior and whatever cost savings were involved were lost in extra time wasted in corrections. Rivière wrote that women, by accepting less pay than men and taking men's jobs, also deprived men of earning a salary that could support a family. He also criticized women for working, stating that, in doing so, they neglected their children and housekeeping, implying also that they neglected their husbands. He acknowledged that single women or orphans or prisoners were justified in holding a job.

In 1904, union activist and politician J. Ramsay MacDonald issued *Women in the Printing Trades: A Sociological Study*, an analysis of women's work in the printing trades in England and Scotland; five women were credited as investigators on the title page.[186] The book summarized the situation for women in the British printing trades as it had evolved during the second half of the nineteenth century, suggesting that there had been little progress for women in the printing trades. Despite the pioneering work of Emily Faithfull, printing unions in Great Britain resisted the employment of women, though women continued to be employed, especially as typesetters and in bookbinderies, usually at lower salaries than men. By accepting less pay for equal work as typesetters, women were sometimes regarded as union busters, used to break strikes or to defeat the goals of the unions representing male typesetters. Regarding the situation in Scotland,

The few attempts made to organise women in the printing trades have failed. Women have been introduced into these trades at times of trouble with the men's Unions, and are consequently not likely to form organisations of their own. Their work has been so precarious and so largely confined to the mechanical and lower grades of labour, that they have had no incentive to aspire to high standards of wages or other industrial conditions. The women employed in the actual printing processes do not seem to have regarded their work as their permanent means of livelihood to the same extent as folders [in bookbinderies], for instance have done, and have been less interested, consequently, in improving their trade conditions; and finally, the men's Societies, for various reasons, some well-founded and some groundless, have regarded women printers as a form of cheap labour—'undercutters'—and have looked upon them as dangerous intruders.[187]

Postage stamp (1991)
Deutsche Bundespost

A postage stamp illustrating women working in a printing shop at the Lette-Verein trade school for women founded by Wilhelm Adolf Lette in Berlin in 1866.

Appendix III of MacDonald's book, titled "General Glasgow Report. (A.) Letterpress Printing. Machine Feeding and Flying. (B.) Lithographic Printing. Machine Feeding. (C.) Letterpress Printing. Type-setting,"[188] reports that in Glasgow the chief role women played in printing remained feeding paper into printing machines, much as it had been during the early decades of the nineteenth century when printing machines were first introduced. Criticism of women as typesetters included "(a) Irregularity of the women's attendance at work. (b) Their shorter hours. (c) Marriage. (d) The introduction of the Linotype machine ... This was urged as the most important cause of the change back to men."[189] It is worth noting that typographic unions, controlled by men, dominated the operation of Linotype machines when they were introduced, preventing women from gaining employment as Linotype operators. However, for companies that resisted acquiring Linotype machines, employment of women may have been a temporary solution: "The cheaper type-setting of women is needed in order to compete successfully with the Lintotype machine. There is no doubt that to a certain extent the comparative low price of women's labour tends to retard the introduction of machinery."[190]

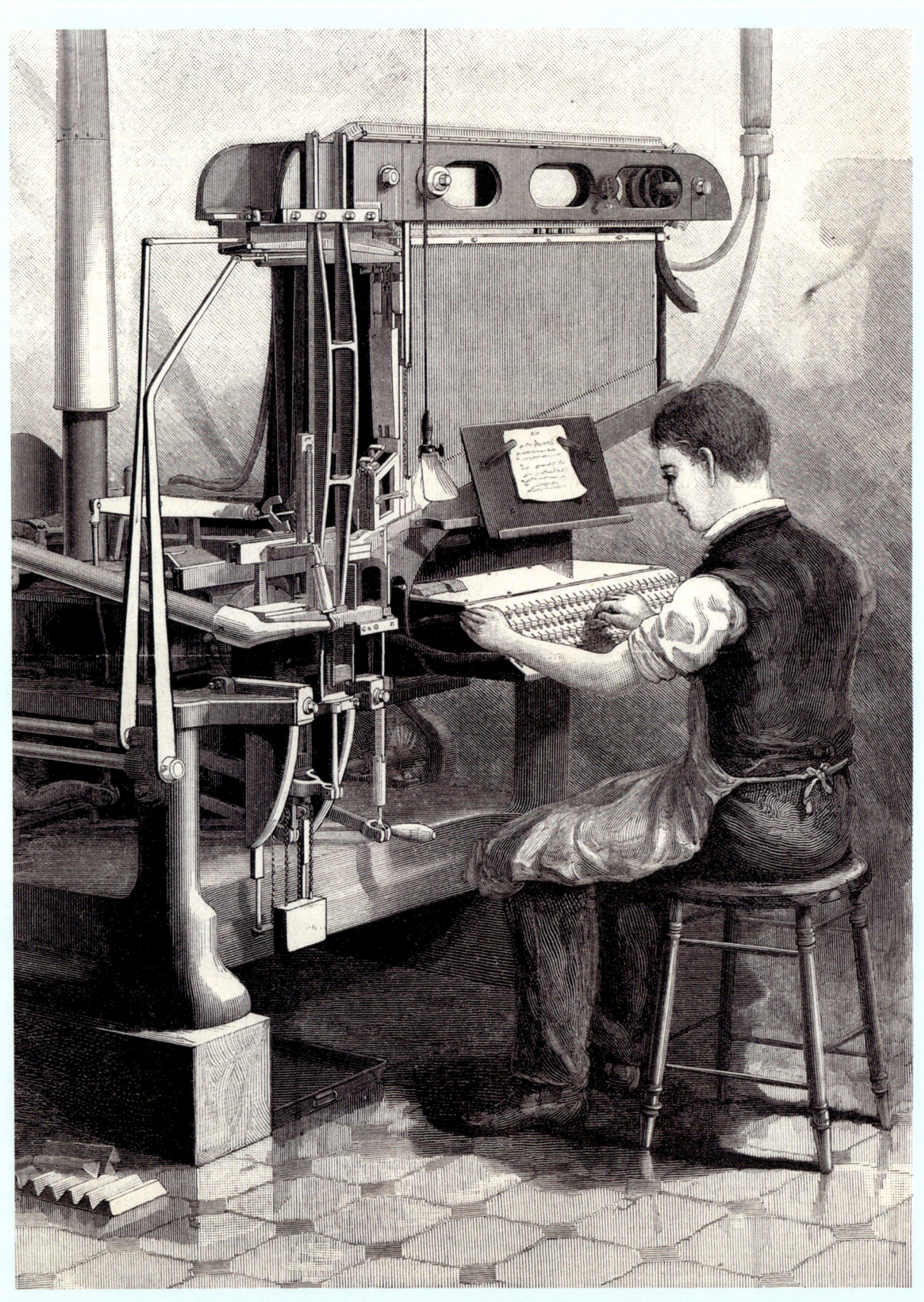

From *Scientific American*, vol. 60, no. 10 (detail). See p. 183.

CHAPTER 11

Mechanizing Typesetting & Type Distribution from William Church to Young and Delcambre's Pianotyp, and to Linotype and Monotype

Typesetting and type distribution (returning lead type to the case after printing) were the most labor-intensive and costly elements in the production of many books, magazines, and newspapers during the nineteenth century. To provide a method of cost saving, mechanizing typesetting and type distribution were the goals of numerous inventors from about 1820 onward. The two separate processes—setting lead type and distributing the type—were exceptionally complex and more difficult to mechanize than papermaking, printing, or bookbinding. As a result, during the nineteenth century, typesetting and type distribution generated a remarkable variety of often ingenious, but less than completely satisfactory mechanical solutions. Of those we will discuss only a few.[191]

All the earliest mechanical typesetting and distribution systems were relatively inefficient. Apart from mechanizing a laborious process, a primary attraction of these systems for printers was that they could be operated by women, who were typically paid half the wages of men. The cost saving from hiring women to operate the machines was intended to offset the high cost of purchasing the machines. Hiring women as typesetters, whether they set type by mechanical

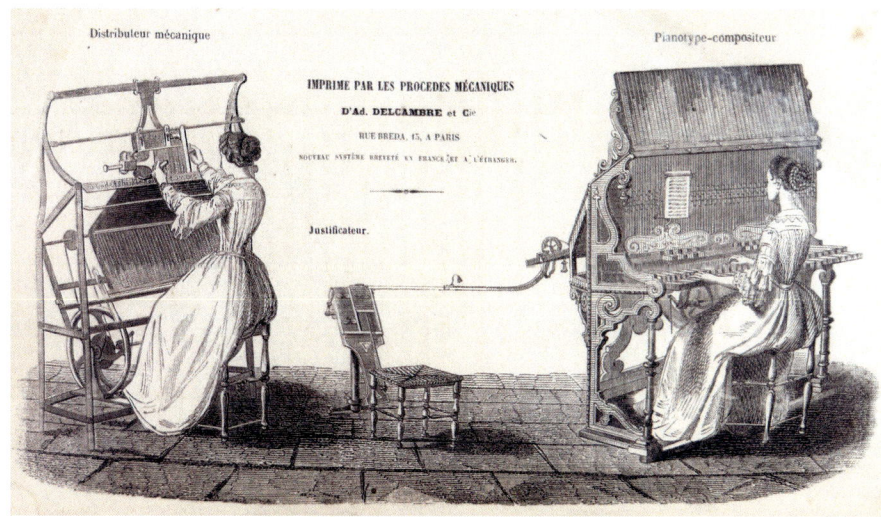

From J.-G Barthe, *Le Canada reconquis par la France* (Paris, 1855)

An advertisement for the Young and Delcambre typesetting machine (right) and type-distribution machine (left). A coinventor of the Young and Delcambre equipment, Adrien Delcambre was one of the few nineteenth-century typesetters and printers who advertised his mechanical typesetting system within some of the books that were typeset by his machines. Showing women operating the machines was controversial; it also meant cost savings to printers, since women were typically paid half to one-third the wages of male manual typesetters.

This Typographic Specimen to the Memory of William Caxton, Wynkyn de Worde, Richard Pynson, and Their Successors . . . (London, 1826)
Poster, type and brass rule, 24 × 15¾ in. (61 × 40 cm)

Paralleling the first efforts to mechanize typesetting in the 1820s, the art of setting type manually may have also reached its peak at that time. In 1826, John Johnson of the Apollo Press in London published a poster honoring the first English printer, William Caxton, and his successors. At the foot of the poster, Johnson boasted in the caption that the design, which measured 19 × 13¾ inches (48 × 35 cm), was "composed of type and brass rule, containing upwards of 60,000 movable pieces of metal, and above 150 different patterns of [printers'] flowers." It is doubtful that any more complex design was ever set in lead type by hand or by typesetting machine.

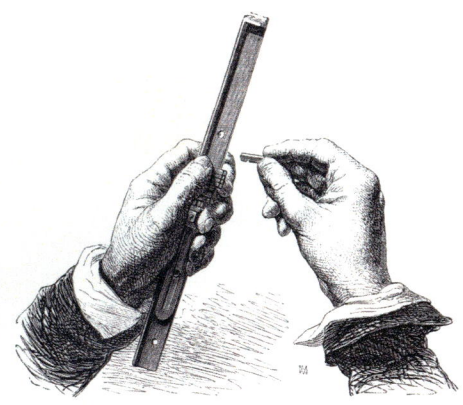

From Théotiste Lefevre, *Guide pratique du compositeur d'imprimerie* (Paris, 1873)

Manual typesetting exploited the typesetter's precise sense of touch. Gloves were not recommended. This image from Lefevre's book illustrates the precise way to hold type and the composing stick with bare hands. Throughout the nineteenth century, most typesetters continued to set type by hand, even after numerous typesetting machines were available.

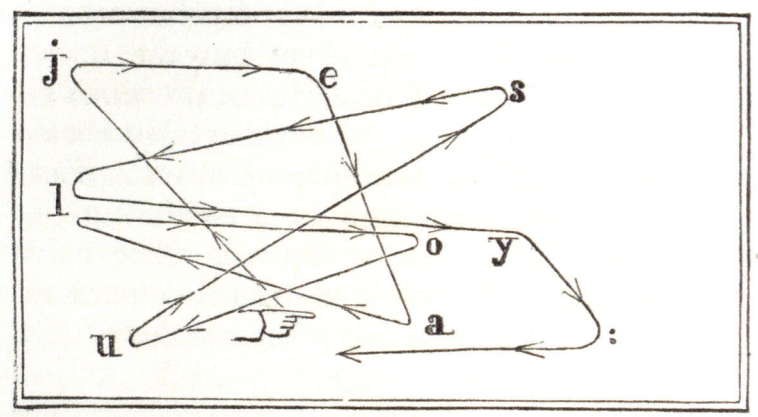

From "Making the Magazine," *Harper's New Monthly Magazine*, vol. 32, no. 187 (December 1865)

A very unusual diagram of the track of a compositor's hand distributing the letters of the single word "jealousy" back to the type case, suggesting the complexity of that less-discussed process of manual typesetting.

or traditional manual means, remained highly controversial throughout the nineteenth century. This was one reason that none of the nineteenth-century attempts to mechanize typesetting were widely accepted until the invention of Linotype and Monotype eliminated the need for type distribution at the end of the century, at which point men dominated the operation of Linotype and Monotype machines.

Paralleling early efforts to mechanize typesetting, the art of setting type manually probably reached its peak at about the same time. In 1826, John Johnson of the Apollo Press in London published a poster honoring the first English printer, William Caxton. In the caption at the foot of the poster, Johnson boasted that the design, which measured 19 × 13¾ in. (48 × 35 cm), was "composed of type and brass rule, containing upwards of 60,000 movable pieces of metal, and above 150 different patterns of [printer's] flowers." It is doubtful that any more complex design was ever set in lead type by hand or by machine.

Manual typesetting persisted throughout the nineteenth century and remained the dominant method, employing many thousands of people. Gaubert, inventor of a French mechanical system, estimated in 1843 that for France alone there were ten thousand manual typesetters, setting fifty million characters per day for ten million readers.[192] In view of the complexity of manual typesetting and type distribution, and the enormous amount of labor-intensive work involved, we should first consider how the manual processes were accomplished and compensated, and the occupational health hazards involved.

Because of the complexity of manual typesetting and the difficulty of performing the necessary tasks efficiently, manual typesetting was typically compensated by piecework rates during the nineteenth century. This system rewarded those who were able to typeset rapidly, while placing less efficient typesetters under constant financial pressure. The technique of manual typesetting is described in Théotiste Lefevre's *Guide pratique du compositeur d'imprimerie* (Practical guide for the printer compositor),[193] the best of many nineteenth-century books on manual typesetting. This comprehensive book, which included chapters on typesetting in Greek, Russian, Hebrew, and Arabic, as well as mathematical formulae and even music, was reprinted in 1878, with a second volume that included material on printing by both handpress and machine, confirming the continuing strong demand for skilled manual typesetters at that date. Numerous images of typesetting departments published in the late nineteenth century confirm that most typesetting was done by manual methods until the invention and gradual adoption of Linotype and Monotype.

Lefevre's work provided some of the very best illustrations of the fine nuances of the manual typesetting of lead types. Because of the sense of touch required in distinguishing between and handling small pieces of metal, the work was done with bare hands, with the compositor's hands in near constant contact with lead type. Such continuous contact with lead for many hours per day undoubtedly caused gradual absorption of lead through the skin, and, if by chance, the compositors handled their food with hands smeared with lead, they could have spread lead on their food and absorbed lead by mouth. They might have also breathed in lead dust.

In the second edition of his *The Effects of Arts, Trades, and Professions, and Civic States and Habits of Living, on Health and Longevity* . . . (1832), English physician Charles Turner Thackrah described some of these hazards:

Compositors are often subjected to injury from the types. These, a compound of lead and antimony, emit, when heated, a fume which affects respiration, and are said also to produce partial palsy of the hands. Among the printers, however, of whom we have inquired, care is generally taken to avoid composing till the types are cold, and thus no injury is sustained. The constant application of the eyes to minute objects gradually enfeebles these organs. . . . We can scarcely find or hear of any compositor above the age of 50. In many cases printers are intemperate.[194]

Nearly contemporaneous with Thackrah, the French physician Louis Tanquerel des Planches published a comprehensive book on lead poisoning in 1839, which was translated into English in America in 1848. In that we read, "Lastly, that trade without which all others would be valueless, the art and trade of printing, has given its victims to the disease, so inseparable from all forms of using lead. Among printers, the pressmen are never attacked; but some of the compositors, especially if they practice holding types in the mouth while correcting a typographical error, are not infrequently attacked. The continual handling of lead alloy, the setting up of types detaches lead particles; and the cases of types are often found covered with a light lead dust, which is removed sometimes by the bellows, thus filling the air with poisonous particles."[195]

Since there is no known safe amount of lead absorption, manual typesetters risked lead colic and all the other known symptoms of lead poisoning. These risks appear to have been generally ignored by typesetters; Lefevre made no mention of them in his book.

One unexpected source of information on the financial compensation of manual typesetters in the mid-nineteenth century is an obscure book titled *Five Black Arts: A Popular Account of the History, Processes of Manufacture, and Uses of Printing, Gas-Light, Pottery, Glass, Iron*, edited by William Turner Coggeshall.[196] This was a reissue, with other treatises, of a collection of Thomas Curson Hansard Jr.'s *Treatises on Printing and Type-Founding from the Seventh Edition of the Encyclopaedia Britannica* that had originally been published in Edinburgh in 1841. At the time, the processes involved had not effectively changed from decade to decade.

The explanations published in *Five Black Arts* for how typesetting was compensated by piecework are complex, but easier for the layman to understand than the technical discussions of the economics of typesetting published for typesetters. I personally find it difficult to imagine how people could accomplish an average manual typesetting speed of 2,000 characters per hour, backward, and often in very small fonts, on top of the other tasks that were necessary for them to perform, such as type distribution, which typically occupied 25 percent of their time. Most probably, only a few could achieve that speed, especially for extended periods of time, making it extremely hard for typesetters to earn a good wage. And perhaps only very near-sighted people could set the exceptionally small and difficult-to-read fonts such as nonpareil and diamond.

Since manual typesetting was challenging and it was difficult to calculate payment for it by piecework, it is understandable that inventors attempted to mechanize the complex processes of typesetting and type distribution in early years of the mechanization of book production. Probably the best retrospective summary of these nineteenth-century attempts is in Lucien Alphonse Legros and John Cameron Grant's *Typographical Printing-Surfaces: The Technology and*

Mechanism of Their Production.[197] The authors devoted many chapters to the wide variety of machines for type composition, distribution, and type casting that had been developed, various esoteric models of which may have remained in operation in 1916, the year the book was published. Most of the different machines built were probably used to typeset certain publications, but retrospective identification of those specific works typeset by specific machines tends to be difficult. By the time of the publication of Legros and Grant's book, manual typesetting had for the most part been replaced by mechanical systems, with the Linotype and Monotype systems generally standardized among printers, though other systems remained in operation to a lesser extent. The authors concluded their book with what they called a conservative estimate of the number of mechanical typesetting systems in operation in England, Europe, and the United States in 1916: 44,000 machines. At this time, by far the most popular machines were Linotype: circa 33,000, and Monotype: circa 5,000. They also listed "Monoline: circa 2,000" and "Typograph: circa 4,000."[198]

The first person to claim to have invented a mechanical typesetting machine was the American physician, inventor, and civil engineer William Church. In 1822, Dr. Church, who was living in England at the time, received British patent no. 4664 for "An Improved Apparatus for Printing." This patent, illustrated with eight large, folding engineering drawings, consisted of three parts: "First, a machine for casting the printing types, and also of arranging them in boxes of letters, so that the types of the same denomination are placed side by side in ranges, ready to be transferred to the composing machinery. The second part of the apparatus consists of a machine, by which the individual types are selected and composed into words and sentences. The third part of the apparatus is a press for printing and delivering the sheets into a pile."[199]

Church's invention, which may never have been built, is known almost entirely from his patent, which showed a "rapid" typecasting machine, a mechanical typesetter, and a newly designed bed-and-platen printing machine, in that order. Two decades later, the basic gravity-driven principle of Church's typesetting machine was adopted by inventors such as James Hadden Young and Adrien Delcambre, John Clay and Frederick Rosenborg, and Etienne Robert Gaubert.

Scientific American, vol. 88, no. 7 (14 February 1903)

The cover of this magazine issue provided one of the clearest images of the first mechanical typesetting machine, patented by William Church in 1822. Church probably never built his machine, but in the 1840s, his gravity-driven concept of typesetting, described in his British patent, inspired the inventors of the first commercial typesetting machines.

[Pierre Leroux,] *Nouveau procédé typographique qui réunit les avantages de l'imprimerie mobile et du stéréotypage* [Paris, 1822]

Pierre Leroux's privately printed pamphlet speculated on the development of a composing machine that he called a "Pianotyp" twenty years before Young and Delcambre adopted that name for their machine.

That Church's machine was well known among printers in England is confirmed by Thomas Curson Hansard's "The Printing Machines and other Inventions relative to Printing of Doctor William Church," a long, critical, and almost incredulous discussion of Church's invention included in Hansard's *Typographia*.[200] In 1963, the catalogue of the famous *Printing and the Mind of Man* exhibition stated, "While there is no evidence that a composing machine was built, the design included features which were embodied in later inventions. The type was stored in inclined channels, from which it was released by the operation of a keyboard. The released type fell into a horizontal race where it was assembled by rocking arms into a continuous line. Like other early composing machines, Church's did not provide for justification of the lines, leaving that to be done by hand. Power was provided by a clock-work mechanism."[201]

The same year that Church's patent was granted in England, French philosopher, political economist, and printer Pierre Leroux, while apprenticed to the Didots, privately issued a sixteen-page pamphlet titled *Nouveau procédé typographique qui réunit les avantages de l'imprimerie mobile et du stéréotypage*

(New typographic process that combines the advantages of mobile printing and stereotyping). In this very rare pamphlet, Leroux speculated on the development of a composing machine that would combine the process of polyamatype, invented by Henri Didot in 1817, with the ability to write documents. Henri Didot's process of polyamatype involved casting 140 letters into a stereotype plate at one time rather than typesetting each of these characters individually and then casting them into a stereotype plate. The machine that Leroux wanted to build would combine the writing directly with the typesetting and stereotype plate making, giving the writer control over the publishing process. Like other inventors, Leroux called his machine a "Pianotyp." However, Leroux mentioned few details as to how his theoretical machine would work.

Was it a coincidence that Leroux published his pamphlet the same year that William Church patented the first machine for type composition? It is unlikely that Leroux, in France, could have read Church's English patent shortly after it was issued, but Leroux might have read a reference to Church's patent in a magazine or newspaper.

In 1843, when various mechanical typesetting machines such as those by Young and Delcambre, Clay and Rosenborg, and Gaubert were being publicized, Leroux published a letter to François Arago, perpetual secretary of the French Academy of Sciences concerning the "new typography."[202] In this letter, he retold his history of conceptualizing a typesetting machine and reprinted the text of his original privately printed pamphlet. Fourteen years later, journalist and publisher Victor Meunier wrote in glowing terms about Leroux's still-unfulfilled notion of unifying the typesetting and printing-plate production process in an article on "Mechanical Typography," published in November 1857 in his newspaper, *L'Ami des sciences*:

The introduction of cylindrical presses, which performed a day's work in one hour, had made typographic art dependent on mechanics, but it was not known that the machine was to extend its influence further over all other departments of printing: the distribution of type, composition, justification. A more noble prerogative, intelligence, seemed to reign supreme, but, emboldened by its triumphs, mechanics does not fear to attack intelligence as such. After substituting for man in the works which mechanics operated only as a physical force it is invading areas in which it operates as almost a sentient being, when it offers itself in the function of a compositor-typographer.[203]

Of the early mechanical systems, Young and Delcambre's Pianotyp, patented in 1840, is the first machine for which at least some of the machine's actual work is documented within the publications it produced. The Pianotyp, especially as developed in France by Adrien Delcambre and his son Isidore, was adopted by several printers and publishers between the 1840s and the 1860s. Several books typeset by this machine in England and France are known. James Hadden Young and Adrien Delcambre, both of whom were residents of Lille, France, received British patent no. 8428 for "An Improved Mode of Setting up Printing Types" in 1840. Young and Delcambre's machine, as improved and made functional by English engineer and inventor Henry Bessemer, was the first typesetting machine known to have been used in a printing office. However, their machine was not a great improvement conceptually over William Church's invention of 1822. The machine set a single continuous line of type; line breaking and justification were later done by hand. "Type was held in long, narrow boxes from which it was released by a keyboard so that it slid down an inclined channel to a point where it was assembled into a line. The problem of making the type arrive in the right order, although solved by Church, had to be solved over again by Bessemer. He curved the channels to make them of equal length."[204]

The first book typeset on the Young and Delcambre machine was English physician Edward Binns's *The Anatomy of Sleep; or, the Art of Procuring Sound and Refreshing Slumber at Will*, published in London by John Churchill in 1842. This semi-popular work, printed in an edition of 500 copies, was the first book typeset on any type of typesetting machine; it also was one of the first scientific studies of sleep. The verso of the title page stated that the book was "Printed by J. Z. Young by the New Patent Composing Machine, 110 Chancery Lane." Because the edition was only 500 copies, it is likely that the book was printed on a handpress rather than one of the high-speed rotary presses, or printing machines, that were available for larger editions.

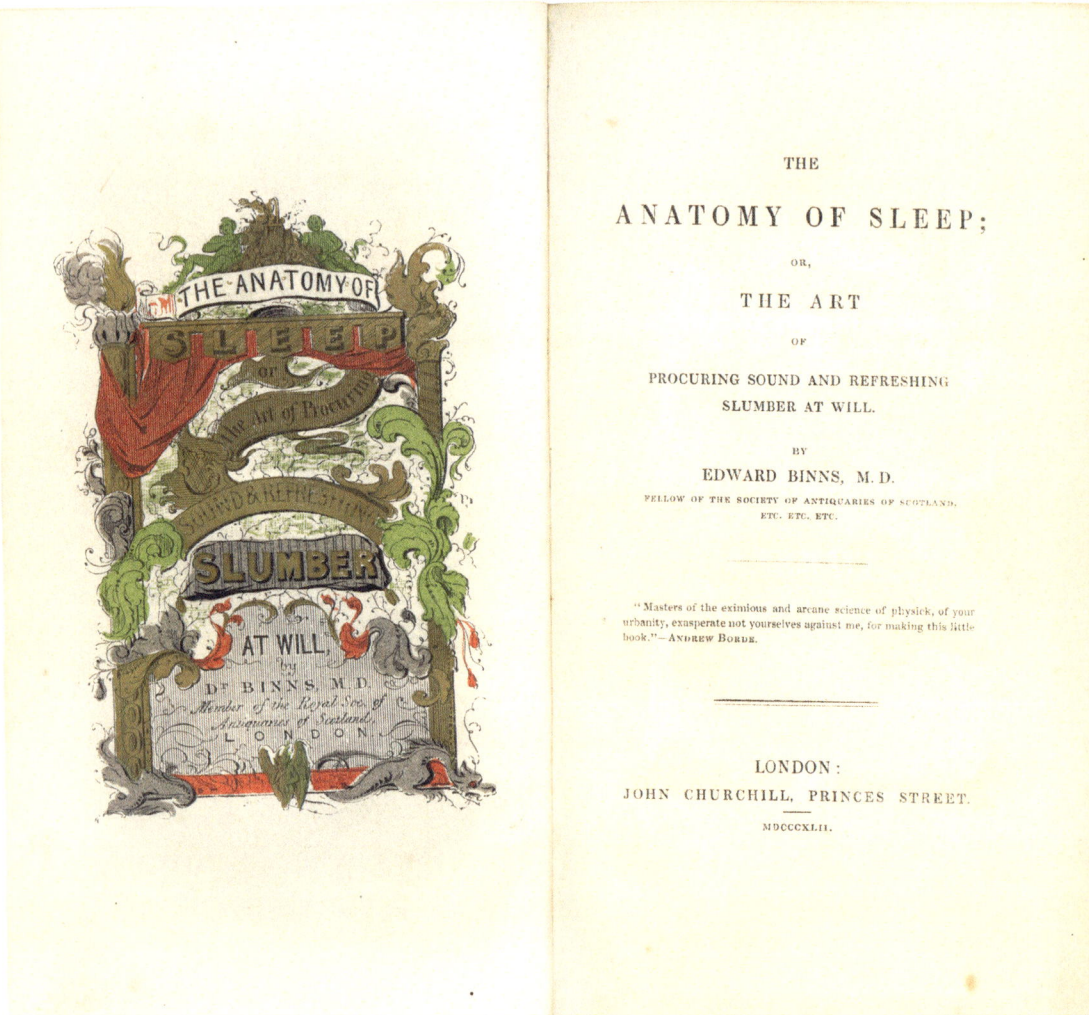

Edward Binns, *The Anatomy of Sleep* (London, 1842)

Binns's book, with its elaborately hand-colored lithographed frontispiece, was the first book typeset by a typesetting machine—the Young and Delcambre Pianotyp, a detail mentioned on the verso of the title page and in Binns's introduction to his book. When it was time to issue a revised edition in 1845, the publisher John Churchill opted for conventional hand typesetting.

In his preface, Binns alluded to the innovative methods employed in typesetting his book:

It would be unjust to the ingenious inventors of the Machine by which this work has been composed, not to say, that we believe it must and will, at no very distant period, supersede in many departments of typography composition in the usual mode. But the use of compositors can never be entirely dispensed with, even supposing the machine to be ten times more perfect than it is. The opposition with which its inventors have had to contend, is what might have been anticipated, but was certainly unexpected by the author. But that in time it will come generally into use, he thinks there cannot be a shadow of doubt. He consequently looks upon the publication of the "Anatomy of Sleep" as an epoch in the history of typography, from which it is possible to conceive a new era in the history of literature may be dated.[205]

Soon after publication of Binns's book, George Biggs applied Young and Delcambre's technology in the publication of *The Family Herald or Useful Information and Amusement for the Million*, the first issue of which appeared in London on 17 December 1842. An image of the Young and Delcambre machine appeared in the title-piece of the periodical. However, the use of the Young and Delcambre machine was opposed by the London Union of Compositors, particularly because women were employed to operate the machine, and the journal ceased publication in this form after twenty-two issues. The Young and

Delcambre machine was also used in 1841 for typesetting another weekly, the Fourierist periodical *The London Phalanx*, where the machine was reviewed with great and seemingly naive enthusiasm in the 28 August issue,[206] and again in the 4 December issue.[207] The 28 August review reprinted a letter from Young and Delcambre to *The Times* in which they claimed, without independent verification, that their machine set type eight times as fast as a human typesetter. On the last page of the 30 April 1842 issue, there is a small notice for Binns's book indicating that it could be purchased from the *Phalanx* office, confirming the relationship between the typesetters of the *Phalanx* and of Binns's book. Beginning in October 1842 with New Series no. 5 (issue no. 62), there is a notice at the end of each issue that the magazine was "Printed at the Composing Machine Establishment, No. 110 Chancery Lane, Fleet Street," which was James Young's business address. This notice was printed in each issue until the magazine ceased publication in May 1843.

In 1845, Binns issued a significantly revised and expanded second edition of *The Anatomy of Sleep* "*With Annotations and Additions by the Right Hon. The Earl Stanhope.*" The second edition, also published by John Churchill, was conventionally typeset by hand, rather than by the Young and Delcambre machine. In the second edition, Binns retained his original preface, noting in footnotes that the first edition had been machine typeset. Binns repeated this mention on the final page of the appendix, acknowledging the unusual feature of the production of the first edition, but not explaining the decision to revert to manual typesetting for the second edition.

However, we may surmise the reasons. Like other early mechanical compositors, the Young and Delcambre machine set a single continuous line of type; line breaking and justification were done later by hand. The inventors also provided a machine for distribution of the type after it was used. To justify the high cost of the machine and still provide a cost saving to printers, the inventors designed the machine to be operated by women, who at this time typically were paid half the wages of men. However, at this stage in the development of machine typesetting, the labor savings in operating the machine and distributing the type did not provide enough economic incentive to replace the manual process, especially since much

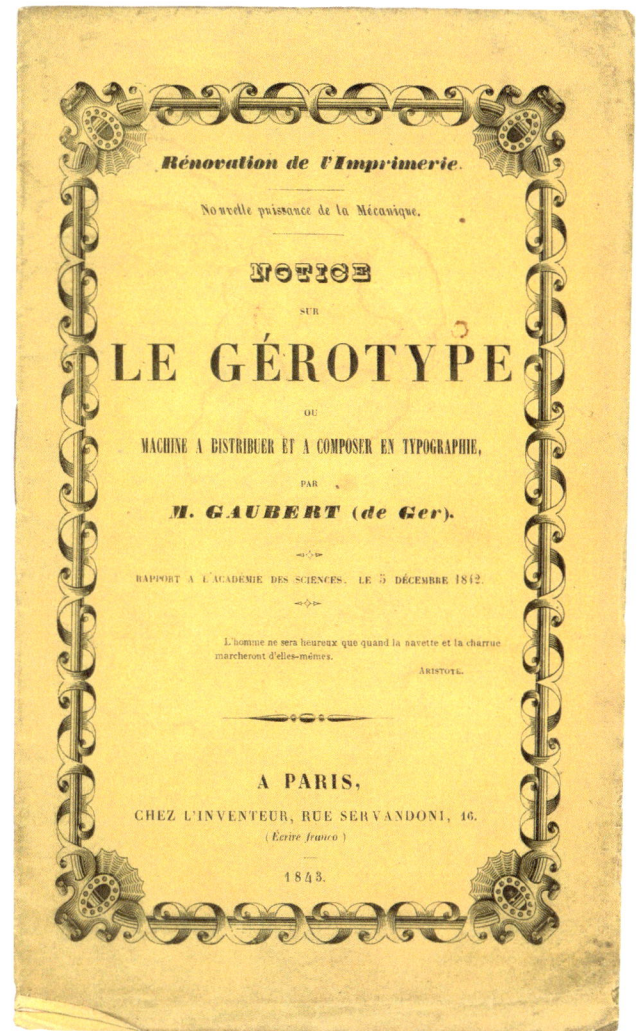

Etienne Robert Gaubert, *Rénovation de l'imprimerie. Nouvelle puissance de la mécanique. Notice sur le Gérotype ou machine à distribuer et à composer en typographie . . .* (Paris, 1843)

Gaubert's small brochure describing his typesetting machinery is one of the few surviving copies of one of the earliest advertisements for any method of mechanical typesetting.

of the manual process, such as type distribution, was typically done by boys, who were paid far less than men in the days of child labor before labor reforms later in the nineteenth century. And John Churchill, the publisher, had to contend with political problems affecting the typesetting of his many other publications since "the use of the Young and Delcambre machine was opposed by the London Union of Compositors, particularly because female labour was employed to operate it."[208]

In 1840, the same year that Young and Delcambre received their patent, Etienne Robert Gaubert, a French mathematician then living in England, received British patent specification no. 8427 for "Certain Improvements in Machinery or Apparatus for Distributing Types or other Typographical Characters into Proper Receptacles, and Placing the Same in Order for Setting up, after being Used in Printing." Gaubert was the first to patent a system of type distribution as well as typesetting. His system, which he named the Gérotype, employed types that were nicked in specific plates unique to each character; these nicks could be detected by his sorting system, which operated by gravity. How much his system was used is unknown; we know of it primarily from his patent, and from various articles that were published about it at the time.[209]

In 1843, Gaubert self-published a fifteen-page pamphlet on the machine titled *Rénovation de l'imprimerie. Nouvelle puissance de la mécanique. Notice sur le Gérotype ou machine à distribuer et à composer en typographie, rapport à l'Académie des sciences le 5 décembre 1842*.[210] In this very rare pamphlet, presumably typeset by his system, Gaubert explained that there were in France ten thousand manual typesetters who set fifty million characters of type per day for ten million readers. He quoted from various published articles about his machine and from the report of the French Academy of Sciences issued on 5 December 1842 that endorsed the machine, but he did not cite a single specific instance of his machine actually being used to typeset a specific publication. He did not even confirm that the pamphlet he published had been typeset using his machine. Nevertheless, Gaubert's machine was viewed ambivalently as both a technological boon to society and potentially a source of unemployment for manual typesetters by the French utopian socialist newspaper *La Phalange*. The 17 February 1843 issue of *La Phalange* published an article, the title of which may be translated as "Monsieur Gaubert's typographic machine: Fate of the workers." The anonymous author, who admired the technological accomplishments of the typesetting machinery, naively believed that mechanical typesetting would soon disemploy manual typesetters, just as machine printing was superseding handpress printing. He wrote [in translation]:

A revolution will take place in the industrial work of printing because mechanics has made a new conquest by invading a terrain that always seemed to belong to intelligent workers....

By the mere fact of the keyboard, printing is already in possession of an admirable improvement. But this was still only half of the major problem to be solved. To function, the keyboard requires first the distribution and classification of the various characters in the specific magazines corresponding to its successive keys. In a word, the following problem had to solved: Given, pell-mell, a bushel of characters of all kinds, with periods and commas, if you pour these characters into a machine, as one pours wheat into a mill, how to do you have them sensed, sorted, classified, and finally distributed, by a machine, into the cells of a hive placed above the keyboard, whose valves, when opened by the keyboard, will produce the composition.

This problem was solved by M. Gaubert in the most admirably ingenious manner, which we will explain. Those familiar with the principle of the Jacquard loom know that this marvelous mechanism is based on a set of needles that feel for patterns on which figures are drawn through holes [in punched cards]. Depending upon whether each needle hits the cardboard or finds a hole, it masks or unmasks a thread, and it is thus, by the touch of all the needles, and the resulting operation of the loom, that the most complicated figures are woven into fabrics. This is a capital principle in mechanics since through it man truly gives machines a mechanical sense of touch.[211]

The author of the article described Gaubert's system of notching types, which allowed the machine to identify each individual character and place it in the correct magazine corresponding to the keyboard keys. He also wrote that Gaubert had also invented a machine that could justify lines of type.

Admiring the new technology but fearing for its consequences, he also wrote:

The revolution in typographic composition, the transformation of the manual labor of this industry into mechanical work, is therefore today a certain fact. Mechanical printing will have preceded mechanical composition by only a few years. It is great progress, a fine conquest of genius, one of the most glorious facts among those that show the divinity of Humanity, its royalty on earth, its infinite power over things. But it is no less true that, in the deplorable society in which we still live, this fact, like most others of the same kind is destined to receive a baptism of tears, misery and despair.

The blindness of the world is truly incredible! Here is a characteristic fact of present-day society. An invention, an instrument of economic creation, an instrument powerfully producing new wealth, is always fatal to the mass of workers engaged in the very work where progress is accomplished. . . .

If three or four hundred printers in France acquire the machines of Messrs. Young, Delcambre, and Gaubert within three or four years, that would mean 8,000 to 10,000 typographic workers, whose position is already very unfavorable, among them many fathers of families, thrown onto the streets, crushed, murdered! Who should we blame? The machine, the invention, the genius, or Society—the absurd, the deplorable society in which a great good, a glorious conquest of the human spirit brings about such a formidable evil? . . . Concerning the typographical question we will try to present some ideas whose application could alleviate the miseries we fear.

The unemployment of manual typesetters that the anonymous author feared would happen because of the widespread introduction of typesetting machines did not in fact occur. Instead, most of the nineteenth-century mechanical typesetting systems found only small niche markets, making it very difficult to determine retrospectively what works, if any, were typeset by each system.

An example of that problem is the Clay and Rosenborg typesetting machine, described in British patent specification no. 8726, "Improvements in Arranging and Setting up Types for Printing." This patent, granted in 1840, was the third patent granted in England for a mechanical typesetting system; the inventors also obtained a supplementary patent (specification no. 9300) in 1842 titled "Composition of Type for Letter-Press Printing." I have never been able to identify any specific printed material typeset on the Clay and Rosenborg machines, though those machines were sometimes publicized along with the Young and Delcambre machines and must have been used to some extent.

Illustrirte Zeitung, vol. 2, no. 45 (Leipzig, 4 May 1844)

The typesetting and type-distribution machines invented by John Clay and Frederick Rosenborg were designed to be operated by women like the Young and Delcambre machines developed at about the same time. The two systems were frequently compared.

These early efforts to mechanize typesetting coincided chronologically with Charles Babbage's pioneering attempts from the 1820s onward to mechanize calculation with his Difference Engines No. 1 and No. 2 and his Analytical Engine.[212] Babbage designed his Difference Engines No. 1 and No. 2 as special-purpose mechanical typesetters that would print their results directly onto stereotype plates. Babbage's motivation was to reduce errors in the manual calculation and manual typesetting of printed mathematical tables, processes that generated numerous errors, especially since mathematical tables were notoriously difficult to proofread. Babbage's goals were not successfully addressed or achieved until World War II and its aftermath, with Howard Aiken's Harvard Mark 1 electromechanical computer and the ENIAC electronic computer developed for the U.S. Army at the University of Pennsylvania.[213]

During the 1850s and 1860s, the Young and Delcambre typesetters and type distributors did produce some identifiable publications. We have little evidence, apart from the Binns book and the two periodicals mentioned above, that Young and Delcambre succeeded in promoting their Pianotyp composing machine in England; however, their system received significant publicity in Europe beginning in the period 1840–1844, often in association with the Clay and Rosenborg machine, with which it was often compared. Among the articles about the machine intended for popular audiences were those in *L'Illustration* (1843) and *Illustrirte Zeitung* (1844). Delcambre, in association with Young, went into business as a typesetter, printer, and publisher in Paris's Montmartre district. Records show that this partnership operated from 1844 to 1849.

Prior to that partnership, Young and Delcambre's Pianotyp was used to set the type for a twenty-three-page pamphlet by Alexandre Dumas, *Un alchimiste au dix-neuvième siècle*, published in Paris by Paul Dupont and Cie. in 1843. This we know because of a notice in the lower part of the typographic border on the upper printed wrapper (which in this instance also served as a title page) reading "Composé à la Machine-Compositeur typographique de MM. Young et Delcambre." This short pamphlet, published by the industrial printer Paul Dupont, has the earmarks of an experimental or promotional publication. Whether Dupont continued with the typesetting machines after this experiment remains uncertain.

In 1847, the Brussels printer F. Verteneuil issued *Oeuvres poétiques de Édouard Smits*, which included a special note "Au lecteur":

Cet ouvrage de M. Ed. Smits est le premier en Belgique dont la composition ait été faite par la machine typographique de M. Isidore Delcambre. Cette Machine, qui intéresse au plus haut point la société tout entière, en imprimant un nouvel essor à la civilisation et, par suite, au bonheur du genre humain, doit exciter un intérêt aussi vif qu'universel.[214]

(This work of Mr. Ed. Smits is the first in Belgium to be typeset by Mr. Isidore Delcambre's typographical machine. This machine, which is of the greatest interest to the whole of society, by giving a new impetus to civilization and, consequently, to the happiness of the human race, must arouse an interest as lively as it is universal.)

Throughout the 1850s, Adrien Delcambre continued his printing and typesetting activity, setting up an industrial printing shop in 1853. It is reasonable to assume that Delcambre had at least some of the books he printed typeset by machinery, providing a way both to exploit and promote his new technology.[215] In 1855, Paul Boizard issued *Paris chez soi: Revue historique, monumentale et pittoresque de Paris ancien et moderne*, an ordinary small quarto printed on ordinary paper; the work's one unusual aspect is the full-page advertisement on the leaf following the title page for the Young and Delcambre typesetting machine that had been used to set the work's type. This is one of a handful of identifiable French publications typeset by the Pianotyp composing machine in Paris around this time. Another book that was both typeset by Delcambre and printed by "Typographie mécanique d'Adrien Delcambre et Cie," was *Le Canada reconquis par la France* by J. G. Barth, published by Ledoyen in Paris in 1855. The advertisement that Delcambre added to this book on its final page is slightly different from that in *Paris chez soi*, with a caption emphasizing the employment of women to operate the Pianotyp. Incidentally, Delcambre clearly identified his firm as printer on the back of the half title and on the rear printed wrapper of this work. In

MACKIE'S STEAM TYPE-COMPOSING MACHINE.

"Mackie's Steam Type-Composing Machine"
From *Every Saturday: An Illustrated Weekly Journal*, vol. 3, no. 105 (18 November 1871)

Alexander Mackie, publisher of the *Warrington Guardian* newspaper, demonstrating the Mackie Steam Composing machine, also known as the "Pickpocket." Mackie is shown holding a role of punched paper tape that drove the machine. *Every Saturday* magazine took more than a casual interest in Mackie's typesetting machine, publishing a long article describing its operation in their 19 August 1871 issue and another article in their 18 November 1871 issue, from which we quote: "Type-setting is still done by the hand, picking up one at a time, although many elaborate keyed instruments have been tried as substitutes. Mr. Mackie's invention, however, seems likely to share a better fate than its predecessors, as it is made fit for every-day hard work in a printing office. It is perfectly anatomical, sets type as fast as eight or ten men, and is governed in its movements on the Jacquard principle, by perforated paper, which may be perforated anywhere, and used again and again for the same size, or any other size of type. The machine has twenty-nine 'pockets' in a circle no larger than the width of a dining-room table, each containing eight kinds of letters. By the action of the perforated paper twenty-four 'pickpockets' are made as they revolve underneath the 'pockets' to pick from one to eight types each [and] every revolution, delivering them in proper order for words and sentences, in which seem endless lines, but which in reality are lines spaced for any width of column or page required."

his ad for his typesetting and distributing machines, he referred to his machine as "Pianotype" so clearly that both this and the briefer version of the name must have been interchangeable.

In 1849, William Martin of St. Pierre-les-Calais, France, who characterized himself as a "mechanist," received British patent no. 12,421 for "Certain Improvements in Machinery for Figuring Fabrics, Parts of which Improvements are applicable to Playing certain Musical Instruments, and also to Printing and other like Purposes." Besides describing improvements to Jacquard punched-card or punched-paper apparatus for weaving and for playing musical instruments (such as a player piano), Martin described in his patent specification how his invention could be applied to typesetting. Martin believed that the punched-card or punched-paper method would speed up the typesetting process; he also believed that it would allow typesetting information to be stored for future use in later editions. In his patent specification, he wrote:

A somewhat similarly constructed apparatus to that just described as applicable for playing musical instruments may also be applied to machinery or apparatus for composing or setting up type for letter-press printing. It is well known that types have been composed or set up by machinery, and that the mechanism of the apparatus generally employed for this purpose consists of a series of levers in connection with the keys of a finger board, arranged in somewhat the same manner as the keys of a pianoforte, organ, or other similar keyed instrument. The types are arranged in vertical columns behind these levers, and by depressing one of the keys with the finger the corresponding lever will be brought into action and a type or letter thereby pushed out from its column into a channel, along which it is carried to the composing stick. My improvements are applicable to any of the machines now in use, and which are worked by means of a key-board or other similar arrangement of levers that are acted on by the fingers of the compositor.

In his patent, Martin specifically described how his invention could be applied to the typesetting machines of Young and Delcambre and of Clay and Rosenborg. The first application of Martin's technology was in the Mackie Steam Composing Machine, also known as the "Pickpocket," invented by Alexander Mackie, publisher of the *Warrington Guardian* newspaper in Warrington, England. Much later, Tolbert Lanston would use punched-paper rolls to drive his Monotype machines.

Among the most interesting type specimens known to have been typeset by an early mechanical typesetting system is found in Mackie's *Italy and France: An Editor's Holiday*.[216] A leaf between the table of contents and the first page of text contains specimens set by the Mackie machine, along with Mackie's claim that the machine could typeset up to 12,000 characters per hour. The whole book was typeset by Mackie's device, and the printing was done on Mackie's newspaper presses. An illustration of Mackie and his device, with Mackie holding a punched-paper roll, was published in *Every Saturday* magazine on 18 November 1871. Another illustration of his typesetting machine was published in J. Luther Ringwalt's *American Encyclopaedia of Printing*.[217]

William H. Mitchel, an American inventor, patented a type composing and distributing machine in 1853; it was the first such machine to be extensively used in America, chiefly for repetitive work. Ten years later, Mitchel traveled across the Atlantic to promote his machines, publishing an illustrated paper on them, "On a Type Composing and Distributing Machine," in *Proceedings of the Institution of Mechanical Engineers*.[218] Mitchel's paper, describing his machine and the advantages it offered in the context of the manual typesetting challenges of the time, appears to be the first paper on the problems of mechanizing and distributing metal types to be published by an inventor of such equipment. The paper offers insight into the problems that inventors faced in mechanizing typesetting and type distribution in the mid-nineteenth century:

Although the process of printing from moveable types has by the power press been accelerated a hundred-fold, the art of composing the types in order to be printed from has not advanced a step since Gutenberg's time. About fifty millions of letters daily are composed and distributed in the United Kingdom; and every one of these is picked from a box by the fingers, and afterwards returned to it in the same manner, without any sort of mechanical aid. Nevertheless during the last twenty years fully twenty attempts have been made by English, French, German Danish, Italian, and American inventors to effect an economy in this direction....

The work of the compositor may be divided into five operations:

Composition, Justification, Making up. Correction, and Distribution....

These five operations represent the work of the compositor in about the following proportions:

Composition ... 55 per cent

Justification ... 17 "

Making up ... 7 "

Correction ... 7 "

Distribution ... 14 "

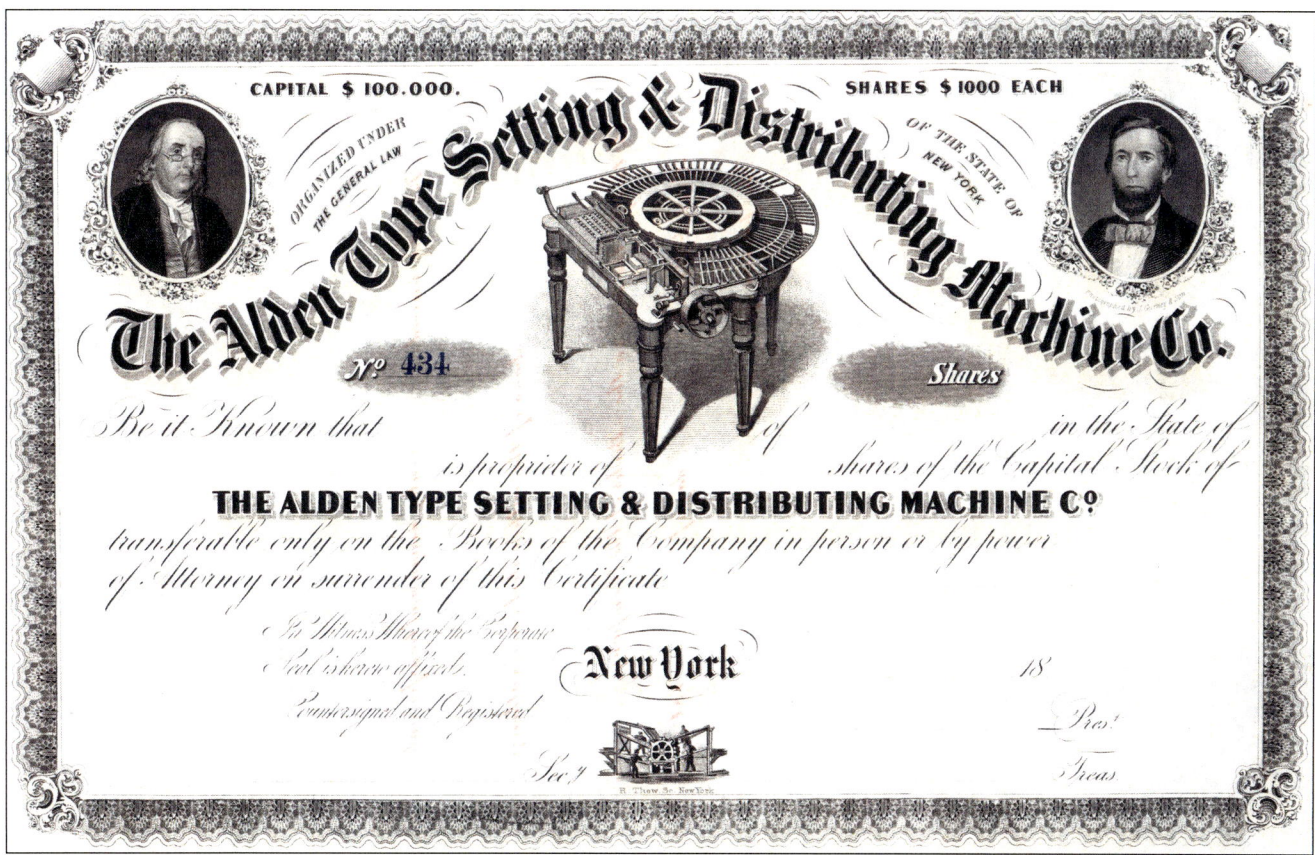

A blank, engraved certificate for shares in the typesetting and type-distributing machine company formed to sell the machine, invented by Timothy Alden, that is illustrated on the certificate. At the bottom center of the certificate we see a small image of a Hoe Type-Revolving Printing Machine. The certificate values the Alden shares at $1,000 each, a huge price at the time.

Of these processes, justification, making up, and correction have never been attempted by machinery. The fifth, distribution, is already very rapidly effected by hand, so that not much scope is left for saving of labour; but inasmuch as all composing machines require the letters to be primarily disposed in lines ready for setting up, it has been found necessary to use machinery to distribute them in that manner.

After a very long explanation of the process of typesetting by his machines, Mitchel noted that the use of typesetting machinery could result in a 43 percent cost saving over manual typesetting, but only if the machines available at this time were used for a basic, repetitive style of typesetting without changes of type size or type style.

Further insight into the challenges of mechanizing typesetting in the mid-nineteenth century comes from an article by the American poet and drama critic William Winter called "Types," published in 1864 in *The Atlantic Monthly*.[219] In this article, Winter noted that while virtually all aspects of book production had been mechanized to a certain extent—including printing, papermaking, and to a lesser extent bookbinding—what had not yet been successfully mechanized on a wide scale was the most complex of the book-production processes, typesetting. According to Winter, the successful mechanization of typesetting was long overdue:

It is only of late that machinery has been successfully employed in the most laborious and expensive process connected with the art of printing,—that, namely, of Composition. In this process, however, iron fingers have proved so much better than fingers of flesh, that it is perfectly safe to predict the speedy discontinuance, by all sensible printers, of composition by hand. . . .

In the interest, therefore, of education no less than health, it becomes imperative that machinery should be substituted for hand-labor in composition. At present, our printing-offices are by no means the sources whence to draw inspirations of order, fitness, and wholesome toil. On the contrary, they are frequently badly lighted and worse ventilated rooms, wherein workmen elbow each other at closely set cases, and grow dyspeptic under the combined pressure of foul air and irritating and long-protracted labor. All this should be changed. With the composing-machine would come an atmosphere of order and cleanliness and activity, making work rapid and agreeable, and lessening the period of its duration. I know that working-men are suspicious of scientific devices. But surely the compositor need not fear that the iron-handed automaton will snatch the bread out of his mouth. To diminish the cost of any article produced—which is the almost immediate result of substituting machinery for hand-labor—is to expand the market for that article."[220]

If anything, Winter was overly optimistic about the efficiency of the typesetting machines, including those of the Americans Alden and Mitchel (see below), that were in use when he published his article. Though Winter does not reference *The Atlantic Monthly* specifically, it is likely that, with the exception, perhaps, of some experimental use of typesetting machinery, nearly all *The Atlantic Monthly* was still typeset by hand.

The drawback of using early mechanical typesetting systems was that their efficiency was only obtained in doing highly structured and repetitive work. Winter was aware that, since 1855, the New York printer and publisher John F. Trow had been employing the William H. Mitchel typesetting machines patented by Mitchel in 1853. Trow initially applied the Mitchel typesetter to set type for his New York City directories—highly patterned and repetitive work. According to Winter, by 1864 Trow was also using the Mitchel typesetters to set the *Continental Monthly* magazine.[221]

Like other exponents of new technology, Winter ignored the unemployment inevitably created by these new machines. He naively argued that because people would always be needed to tend ma-

Scientific American, vol. 84, no. 10 (9 March 1901)

The Paige Compositor, on which Mark Twain lost a fortune. The machine, which was overly ambitious, very expensive, and difficult to operate, was finally brought to market unsuccessfully when cost-effective Linotype and Monotype machines began to dominate hot metal typesetting.

chines, machines could not completely replace labor: "The Sewing Machine has not injured the seamstress. The Power-Press has not injured the pressman. The Type-Setting Machine will not injure the compositor. Skilled labor, which must always be combined with the inventor's appliances for aiding it, so far from dreading harm in such association, may safely anticipate, in the far-reaching economy of science, ampler reward, and better health, an increase of prosperity, and a longer and happier life in which to enjoy it" (p. 617).[222]

Another American typesetting and type-distributing machine that presumably had some users was the Alden-Type invented by Timothy Alden. In 1856, two years before his death, Alden received British patent no. 3089 for "Machinery for Setting and Distributing Printing Types." The elegant stock certificate for the Alden Type Setting and Distributing Machine Co. depicts the machine and includes portraits of Alden and Benjamin Franklin. In its issue of 9 April 1861, the *Wisconsin State Journal* newspaper reprinted an

From Caxton Celebration,1877, exhibition catalogue ... (London, 1877)

These ads for Kastenbein's Composing and Distributing Machines were published at the back of the exhibition catalogue printed for the Caxton Celebration exhibition, where the machines were demonstrated. The manufacturers provided productivity information for the machines. They also claimed that users could be taught to operate their typesetting machines in two days. Kastenbein machines were extensively used at *The Times* of London newspaper during this period.

article from the New York *Home Journal* claiming that the Alden machine could set type as fast as eight men.

Mechanical typesetting and type distributing remained an inventor's challenge till the end of the nineteenth century. Perhaps the most elaborate effort was James W. Paige's Paige Compositor, patented in 1874, the first machine to simultaneously set, justify, and distribute foundry type from a common case using only one operator. The machine was immensely complex, containing 18,000 parts and numerous bearings, cams, and springs. It could average 12,000 ems[223] per hour but had continual need for adjustment based upon trial and error. Mark Twain, who must have been fascinated by the complexity and potential of this mechanism, famously bankrupted himself by sinking $300,000 into the development of this machine, of which fewer than six were built; the only surviving example is in Mark Twain House in Hartford, Connecticut.

The surviving machine was completed in 1887

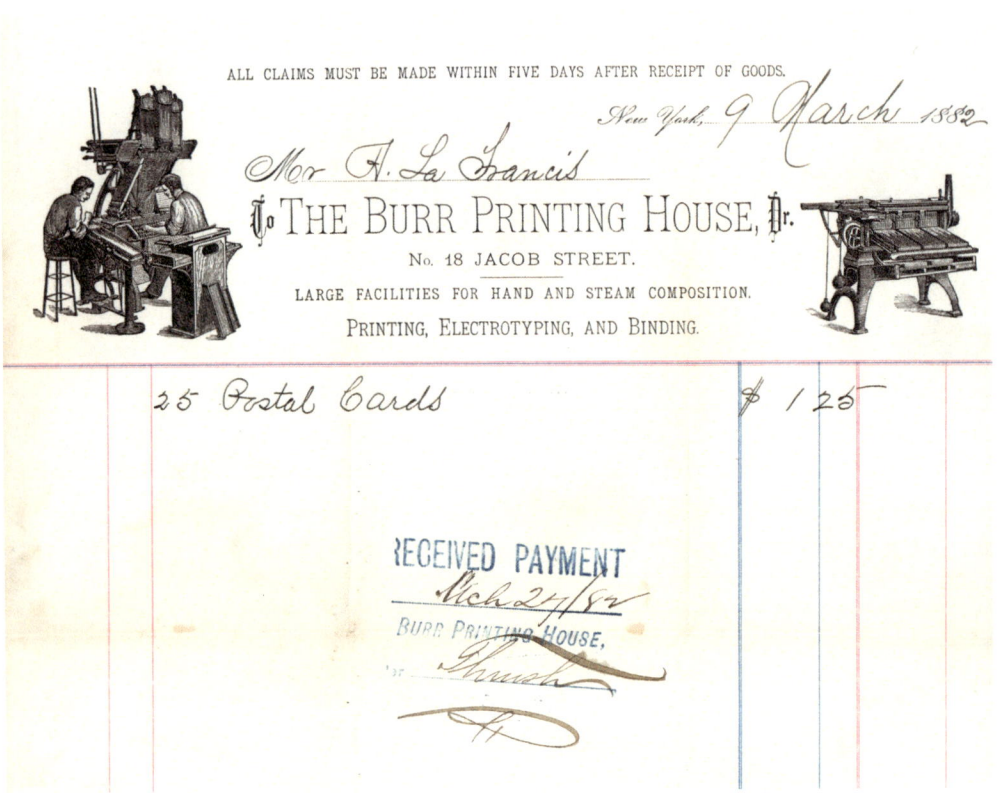

This invoice from the Burr Printing House in New York City, dated 9 March 1882, is the only nineteenth-century invoice from a mechanized typesetter that I found in more than twenty years of collecting. *The New-York Tribune* used Burr typesetters prior to the development of Mergenthaler's Linotype.

under the direction of Charles E. Davis, a mechanical engineer who invented an automatic justifier to be used with the Paige machine. A second machine, completed in 1894, successfully participated that year in a sixty-day trial at the *Chicago Herald*—its only commercial application. The Paige compositor used in the trial was approximately 11 feet in length, 3½ feet wide, and 6 feet high (3.4 × 1 × 1.8 m), and weighed about 5,000 pounds. It was specially designed for newspaper work, using nonpareil type; the distributing, setting, justifying, and leading mechanisms were adjustable to any width of column desired for newspaper or book work.[224] A factor dooming the Paige Compositor was that during its overly long development, the printing industry moved from continuously recycling foundry type, which required the time-consuming process of returning each letter to the case, toward machines such as Linotype and Monotype that melted-down used type and cast new type for each printing project.

In England, during the late 1860s, Charles Kastenbein invented a typesetting and distributing system that was used extensively to typeset *The Times* newspaper in London for at least a decade or two. In 1869, Kastenbein received British patent no. 2031 for his "Improvements in Apparatus or Machinery for Composing and Distributing Type"; at the time, Kastenbein was living in Paris, so the patent was filed under the name of his agent, Charles Denton Abel. In 1872, Kastenbein moved to London; once an English resident, he received a much more detailed patent, no. 2864, with the same title as his patent of 1869. The second patent included six very large, fold-out diagrams showing every aspect of his machines. The manufacturers of Kastenbein's machines demonstrated them at the Caxton Celebration of 1877 and advertised them in the exhibition catalogue sold at the exhibition, claiming that they could set type at the rate of 6,000–8,000 ems per hour and distribute at 4,000–5,000 ems per hour. Apparently, the machines were quite efficient and may have been used to some extent at *The Times* up to at least 1916.[225] In 1879, Kastenbein perfected his machine, increasing typesetting speed considerably.[226] *The Times* eventually replaced the Kastenbein machines with a variety of different hot-metal devices, including Linotype and Monotype.

Scientific American, vol. 60, no. 10 (9 March 1889)

By 1889, the Mergenthaler typesetting machine was known as the Linotype, but the author of the article stated that the machine could just as well have been called the "Tribune" machine, since it had previously been in practical use at the *New-York Tribune* newspaper for two years before it was offered to a wider market.

Between 1883 and 1885, German-American clockmaker and inventor Ottmar Mergenthaler of Baltimore invented the first mechanical typesetting machine, or composing machine, that could set complete lines of type, or slugs. By speeding up typesetting, this machine revolutionized print production, at first in newspaper publication, where speed in producing frequent daily editions was required. The machine eventually became known as the Linotype. A skilled operator of a Linotype could produce up to

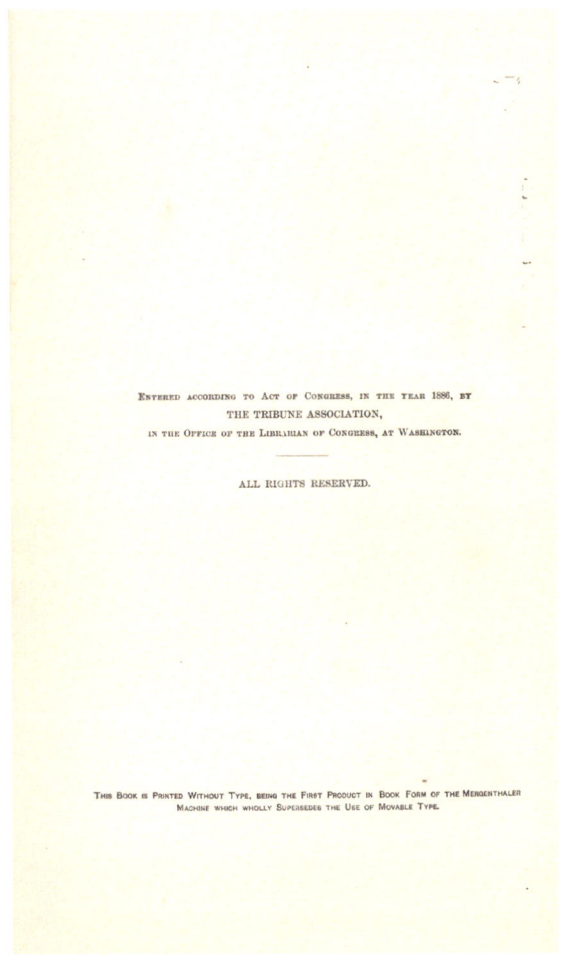

Henry Hall, ed., *The Tribune Book of Open-Air Sports* (New York, 1887)

👆 The notice printed at the foot of the copyright page states, "This Book is Printed Without Type, being the First Product in Book Form of the Mergenthaler Machine which wholly Supersedes the Use of Movable Type." The book was copyrighted in 1886, and by that date, the name Linotype had not yet been given to what was previously called the Mergenthaler machine. The expression "without type" referred to the Linotype's casting of slugs with lines of type, rather than individual letters of type.

👉 👉 This large poster (27½ × 21 in., 70 × 53.5 cm) offered *The Tribune Book of Open-Air Sports* as a premium for subscribers to the newspaper in 1887.

6,000 ems per hour as compared to a hand typesetter, who could set 250 ems per hour.

Mergenthaler developed the first simple prototype in 1883 and 1884. He conceived the idea of assembling a line of dies or female matrices and casting into them molten metal to form a complete slug or line of type. The matrices in these machines were stamped on the edges of upright bars, each bar containing the letters of the entire alphabet. Operation of the keyboard set up stops that allowed these bars to descend to the proper distance when a cast was taken from the aligned matrices. The wedge justifier, the invention of which later became the subject of litigation, was

LOCAL CLUB AGENCY

OF

The New York Tribune

UNDOUBTEDLY THE

BEST WEEKLY NEWSPAPER IN AMERICA

REGARDLESS OF POLITICS,

having a circulation in every State and Territory, and devoted to the welfare of the Whole People. Thoroughly American in spirit, believing in ruling this country for the benefit of the American people. Two pages a week devoted to

HOW TO CARRY ON A FARM.

All the News. "Home Interests" and "Young Folks" for the family. Good Stories. Live Editorials. PROTECTION, TEMPERANCE and PATRIOTIC GOVERNMENT The Tribune's watchword.

WEEKLY, - $1.00 a Year.
Semi-Weekly, $2.00 a Year.

The Tribune Book of Open-Air Sports

IS A

SPLENDID PREMIUM BOOK FOR 1887.

It is a new book, especially prepared by The Tribune. 500 Pages and 150 Illustrations. A book for men as well as boys. Send for circular. Terms, with Weekly, one year, $2.50. With Semi-Weekly, one year, $3.50.

Other premiums are: The fine Waltham Watch; the two Unabridged Dictionaries; Blunt's Rifle Practice; the Waterbury Watch; Wood's Medicine, Hygiene and Surgery, 2 vols., handsomely illustrated, with 200 pages devoted to what to do in Accidents and Emergencies.

SEND FOR SAMPLE COPY AND CIRCULAR

OR

Ask our LOCAL CLUB AGENT for Terms.

THE TRIBUNE, New York.

CHAPTER 11

Brass product label for a Linotype machine
(1½ × 11¾ in., 4 × 30 cm)

Label
Brass, 3¾ × 2¾ in. (9.7 × 7 cm)
Lanston Monotype Machine Co., Philadelphia

An original label for a Monotype machine. This one probably came from a machine produced around 1912, since that is the date of the last patent embossed on the label. The name Monotype refers to the machine's casting of individual letters of type in contrast to the Linotype, which cast slugs with lines of type.

incorporated in the second machine built, in 1885. The impossibility of correcting errors as soon as they were discovered led to the conception of the independent matrix machine, which was built in 1885; this marked the advent of the Linotype as a new factor in the printing world. The first "direct band type casting machine" was in test operation by July 1884, and the first machine with "independent or free matrices" was in operation on 15 July 1885.[227]

Mergenthaler's Blower Linotype composing machine was first used by the *New-York Tribune* newspaper on page four of its issue of 3 July 1886. Whitelaw Reid, editor and publisher of the *New-York Tribune*, had helped to underwrite the development of the machine, and he had the machines installed in opposition to the typographic union. The parts of the 3 July issue composed by the Linotype machine can be distinguished from the hand-set type because of a single wrong-font boldface apostrophe. This appears in only three of the stories, in columns two and three of the page.

Mechanical composing machines resulted in greatly increased production speed and lowered typesetting cost, resulting in longer newspapers. Because of the time involved in hand typesetting, and the constant deadlines to be met, before the Linotype no newspaper consisted of more than eight pages.

In 1887, the *New-York Tribune* newspaper published the first book typeset by Linotype, *The Tribune Book of Open-Air Sports*. Printed on the front and back pastedown endpapers of the well-produced and attractively bound book was a statement that the book could be obtained only with a one-year paid subscription to the *New-York Tribune* weekly, semi-weekly, or daily. Only the title page of this 500-page book was printed from hand-set type. On the verso of the title page is a two-line notice set in small nonpareil capitals and small caps:

Mechanizing Typesetting & Type Distribution

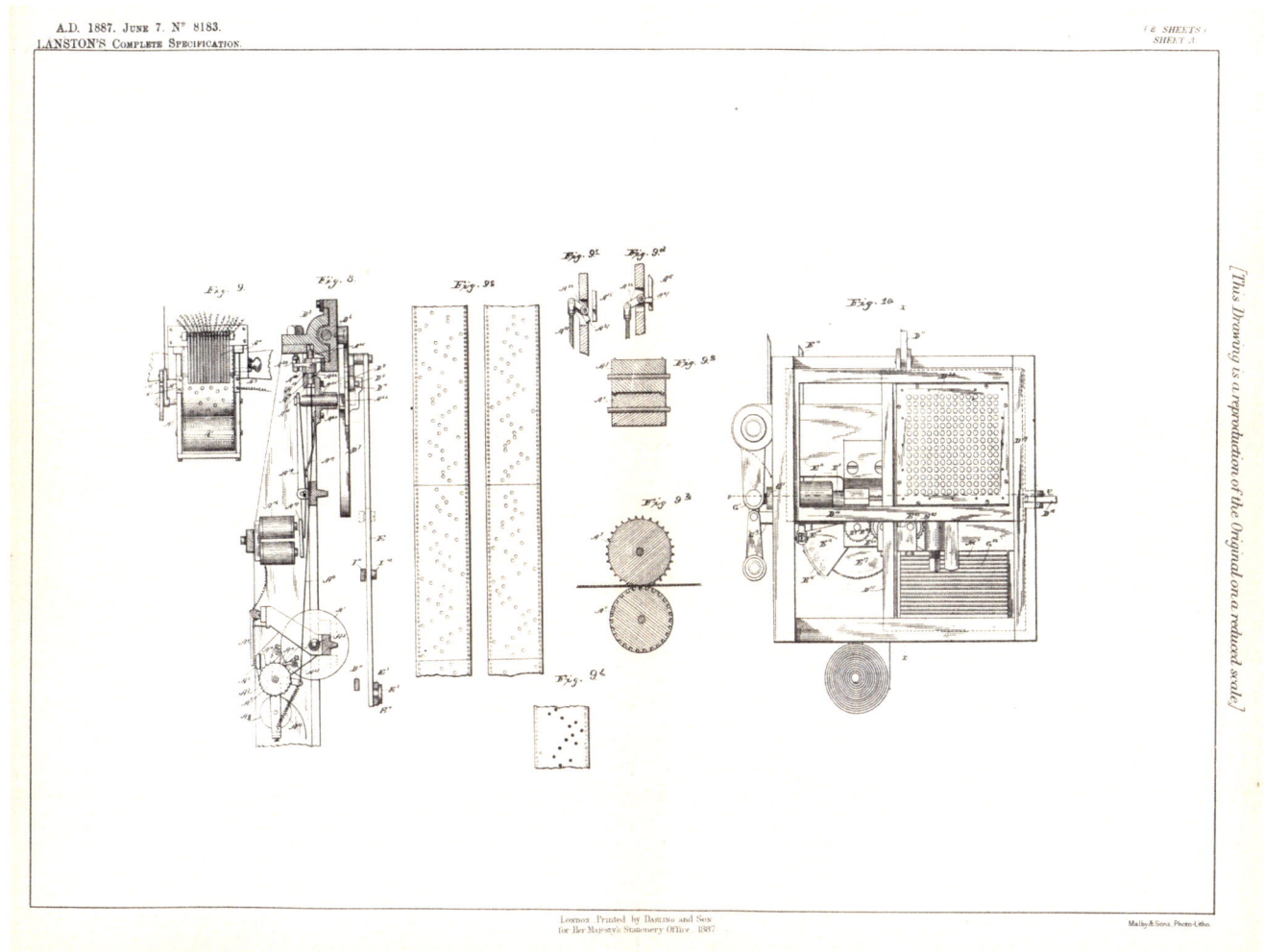

From Tolbert Lanston, patent no. 8183 (1887), in
Improvements in the Art of Printing (London, 1887)

The paper-tape mechanism from the first English patent for the Monotype.

This Book is Printed Without Type, being the First Product in Book Form of the Mergenthaler Machine which wholly Supersedes the Use of Movable Type.

At this time, the Mergenthaler typesetting machine was not yet known as the Linotype.[228] By July 1889, Mergenethaler's invention had been given that name, and the *New-York Daily Tribune* published an article about it in the *Library of Tribune Extras* (vol. 1, no. 7) called "The Linotype."[229] From it, I quote portions:

The New-York Tribune, the first American newspaper to stereotype and the first to print from the Hoe perfecting press, is the first to announce a successful departure from this old and tedious system. For nearly three years, it has been introducing into its office a machine which dispenses with movable type, and whose adaptability to practical newspaper work, in spite of many discouragements at the outset, is at last satisfactorily demonstrated. The machine now does substantially all the composition of this paper. The old-fashioned frames and trays, with their square compartments filled with dingy types, have given place in its composing-room to upright structures of brass and iron, overhung by a network of belts, long lines of shafting, huge metal pipes and revolving wheels. A new industry has, in fact, arisen and among The Tribune typesetters the case and stick are already things of the past. The Tribune's experiments with devices for saving time and labor at the case date back for many years. As long ago as 1866 Mr. [Horace] Greeley, always hospitable to new ideas, made trials of the Alden machine. This like the inventions of Mitchel, Delcambre, Fraser and others, was designed merely to duplicate by machinery the motions of the compositor's hands, and like them, it broke down under the severe test of morning newspaper work. The best of this class of machines were, perhaps, those of the late G. A. Burr, which, up to a few months ago, had been in use in this office for several years. The first was put in in May, 1880, during the lifetime of the inventor and under his supervision, and the number was subsequently increased to three. The results were fairly satisfactory, but there were fatal obstacles to success. In the first place, the original cost was great and the expense of repairs considerable. A specially prepared type was required, which broke easily and wore out very fast. The operator was needed to set the matter, another to justify it, and a third to attend to the distributor (which was a separate piece of machinery). Frequent stoppages were necessary for replenishing the font, and output of each machine, under the most favorable conditions, was only from 4,000 to 5,000 ems per hour so the actual saving was small. Moreover, there were numerous sources of delay, the type 'pied' easily, and typographical blunders were unpleasantly common.[230]

The article, which was very detailed, went on to explain that 33 Linotype machines and their operators had replaced 100 manual typesetters at the newspaper, and though the initial cost of the machines and the cost of the Corliss steam engine and the belt system to power them was substantial, the machines increased "every man's capacity more than two and a half times and reduced the cost of his labor to less than one-third of the usual sum. It should be born in mind that this result was attained without any extra exertion on the part of the operators, or any effort for a spurt of speed."[231]

As much as Linotype machines increased efficiency and saved money, one drawback that was rarely mentioned in early articles was the considerable noise of the machines' operation. Working in a room with thirty-three or more Linotypes in operation would have been extremely uncomfortable for anyone with normal hearing; for that reason, it was sometimes stated that operating a Linotype machine was an excellent career for a deaf person.

Three years after Mergenthaler invented the Linotype, and one year after the Linotype was first usefully applied to production of the *New-York Daily Tribune*, American inventor Tolbert Lanston of Washington, D.C., demonstrated his prototype of the Monotype, a machine that set individual characters of type in justified lines rather than "lines of type" like the Linotype. Lanston's initial typesetting machine consisted of a keyboard producing a perforated record of a job in a paper spool, something like a player piano roll, which controlled an associated machine for fashioning types from cold strips of metal using 196 matrices. The concept of driving a typesetting machine from punched paper, similar to Jacquard cards, had originally been patented by William Martin as early as 1849[232] and had been improved by Alexander Mackie in 1867, but neither of those machines were sufficiently functional to become established in the typesetting community. To produce his new machine, Lanston originally founded the Lanston Type Machine Company in Virginia on 13 November 1886. In his *History of Composing Machines*, Thompson wrote:

The perforated tapes, of which he employed two, caused a strip of type metal to be fed into a compression box and the proper die to be centered above it, a section of the type metal cut off and compressed to form the type, which was then ejected on to the gallery, the entire operation of the typemaking machine being automatic. Justification was provided for on a novel principle. A scale indicated to the operator of the perforating mechanism on completion of a line the amount of space yet unfilled and the percentage which this bore to the filled space, he thereupon striking certain keys to cause perforations to be made at the end of the line. The tape was fed backward through the automatic typemaking machine and these last perforations caused the body of each letter in the line, or, if desired, only the spaces therein, to be increased above the normal such a percentage as to produce a line of justified type. In this machine electromagnets were employed to control the mechanism.[233]

Lanston's U.S. and British patents are dated 7 June 1887. The British patent specification no. 8183, for "Improvements in the Art of Printing," makes sixty-

Robert Thom
Tolbert Lanston and the Monotype (circa 1960)
Oil on canvas, 30 × 24 in. (76 × 61 cm)
From Kimberly-Clark, *Graphic Communications Through the Ages*
Courtesy of RIT Cary Graphic Arts Collection

four claims with respect to a mechanism for line justification and a method of type forming. It consists of twenty-nine pages of text and nine diagrams, of which eight are double-page.

In his patent, Lanston stated:

My invention is a wide departure from the previous methods and proceeds upon a principle, which I believe to be radically new. Instead of producing a line of composition and then justifying it I form my types for a given line in such manner as to cause them when assembled, to form a complete justified line ready for printing direct or for making an impression for stereotype or electrotype purposes without further manipulation.

In attempting to supplant by machinery the ancient process of setting type by hand the advantages to be derived from copying as many of the conditions of such hand set type as possible are manifest. By so doing, the mechanical products will be in harmony with all the other conditions of the art of printing as now practiced, will involve no departure from its usages and will permit the same method of correction of errors, interpolations, shifting of matter &c. as are now in vogue. It is well known that in ordinary composition

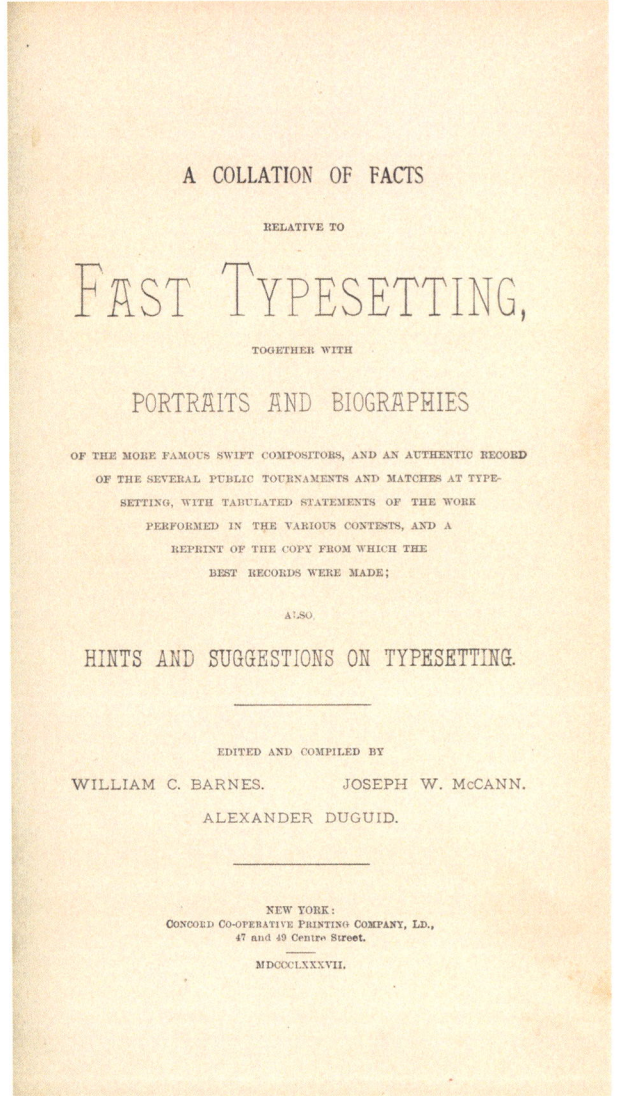

👆 👉 👉 William C. Barnes, Joseph McCann, and Alexander Duguid, *A Collation of Facts Relative to Fast Typesetting* ... (New York, 1887)

This book, issued by three champion manual typesetters, documented the high point of manual typesetting as an occupation before the field was superseded by cost-effective typesetting machines such as Linotype and Monotype. The frontispiece (p. 191) depicts contestants in a Chicago speed typesetting contest. They were nicknamed the "Swifts."

where common type is used it rarely ever exactly fills a line of given length, the rule being that a space of greater or less length is left at the end of the line which must be filled up or absorbed in the process of justification. Now, since it is apparent that in every case this unoccupied space at the end of the line must bear a certain relation to the part of the line filled by the characters or in other words, represent a certain percentage of the combined width of such characters, it follows, that if there be added to the normal width of the body of each of the assembled types a percentage of increase, corresponding to the percentage which said unoccupied space represents to the occupied space, the line composed of types so formed will be rendered self justifying.[234]

To develop his invention, Lanston moved his business to Philadelphia, where he formed the Lanston Monotype Company. As the technology progressed, this company became known as the Lanston Monotype Machine Company. In 1890, Lanston abandoned the concept of stamping letter shapes into cold metal and introduced hot-metal casting techniques. Within a year, he had developed a version of this idea called the Hot-Metal Machine, for which he obtained U.S. patent no. 557994, granted in 1896.

The earliest published technical description of Monotype is *The Lanston Type-Machine. Expert Report of Coleman Sellers, E. D., and Legal Reports of Church & Church and W. W. Gordon*, privately published in New York in 1889. This pamphlet, surviving perhaps in only one copy preserved in the Getty Research Institute, includes examples of actual typesetting and two images of the machine. It was published as a promotional effort to raise money for development of the company, and it includes the earliest published examples of Monotype typesetting.

By 1896, mechanized typesetting by Linotype and Monotype, and other less-documented methods,[235] was gaining acceptance in the printing industry. Even so, many thousands of manual typesetters were still employed around the world, and from time to time, there were speed competitions among manual typesetters.

In 1887, three American competitive manual typesetters, William C. Barnes, Joseph W. McCann, and Alexander Duguid, issued *A Collation of Facts Related to Fast Typesetting, Together with Portraits and Biographies of the More Famous Swift Compositors, and an Authentic*

Record of the Several Public Tournaments and Matches at Type-Setting, with Tabulated Statements of the Work Performed in the Various Contests, and a Reprint of the Copy from Which the Best Records were Made; also Hints and Suggestions on Typesetting. This book documented the high point of manual typesetting as an occupation before the field began to be superseded by cost-effective typesetting machines such as Linotype and Monotype. Considering the difficulty of setting type accurately by hand at high speed, the accomplishments of the so-called "Swifts" were remarkable. In the introduction, the authors noted that, like other competitive activities, the speed of competitive typesetters had increased between 1847 and 1887:

Forty years ago the printer who could set 1,200 ems per hour was deemed a fairly quick hand; at 1,400, he was fast; 1,700 was wonderful, and 2,000 ems per hour was considered among the physical impossibilities. Yet within sixteen years at least seven compositors have in public contests succeeded in surpassing 2,000 ems per hour.[236]

This statement confirms the assertions made by Thomas Curson Hansard Jr. and William Turner Coggeshall, in 1841 and 1861, respectively, regarding the average or typical speed that manual typesetters needed to maintain to earn a good wage: "he must pick letters at the rate of 144,000 per week, 24,000 per day, or 2000 per hour. His rapidity of motion is therefore wonderful, and the exertion is so long continued that the business, although apparently a light one, is in fact, extremely laborious."[237]

L. Graham & Son, *A Brief Description, with Illustrations, of the Printery of L. Graham & Son . . .* (detail). See p. 207.

CHAPTER  12

The Mechanization of Bookbinding: William Burn, William Pickering, Archibald Leighton, and Followers

Like papermaking and printing, at the beginning of the nineteenth century, bookbinding remained a handcraft virtually unchanged from prior centuries. Up until the mid-1820s, bookbinding was largely performed as described and illustrated in Diderot and d'Alembert's *Encyclopédie des sciences*, issued between 1751 and 1772. The *Encyclopédie* includes six plates illustrating the bookbinding process and the few tools used by bookbinders at the time: a bookbinder's sewing frame, used to sew the signatures of a book block; a bookbinder's press; a hammer used to round the spine of a book; a simple press to stamp gold designs, such as coats of arms, onto the covers of books; and various hand tools. Along with papermaking, bookbinding was the primary book-production trade in which women were employed, doing tasks that did not require extensive physical force.

Prior to mechanization, we would hardly expect hand bookbinding to be an occupation involved in trade actions. Compared to industries such as coal mining or textiles, there are few records of strikes or Luddite behavior in the printing trades in England during the second printing revolution. In France, Louis-Nicolas Robert is known to have been motivated to mechanize papermaking at the end of the eighteenth century because of ongoing labor problems in the hand paper mill at Corbeil-Essonnes—problems that were also documented at other paper mills in France.[238] In chapter 7 of this book, I mentioned the destruction or damage of five printing machines by thirty printers who broke into the Imprimerie Royale during the July Revolution in Paris on 29 July 1830. However, these were exceptional instances. One of the few known trade actions in eighteenth-century book production in England involved James Fraser, a well-known London bookbinder, who is depicted in the print illustrated on p. 195. Fraser was one of "The Prosecuting Masters" in a bookbinding dispute in 1786 when five journeymen bookbinders were imprisoned for combination, or forming a union. The print, which commemorated the binders' strike in subtle ways, was published in 1807. It may be the only print that relates to bookbinders' labor disputes published during this period. While the precise reason this print was published in 1807 may be unknown, inevitably labor disputes between bookbinders and their employers occurred. Labor unions were illegal in England until 1824; strikes, which were infrequent, typically resulted in imprisonment, as strikes were illegal in England until 1875.

Details engraved on a paper held by Fraser in the print refer to this early trade-union action when the trade societies called the United Friends, the Brothers, and the City Brothers combined to have their workday reduced by one hour from the existing fourteen-hour schedule in which they worked from 6 a.m. to 8 p.m. Only after these fourteen working hours did the masters pay overtime. Within this grueling schedule, the binders had a half-hour break for breakfast and an hour for dinner. Because striking was ille-

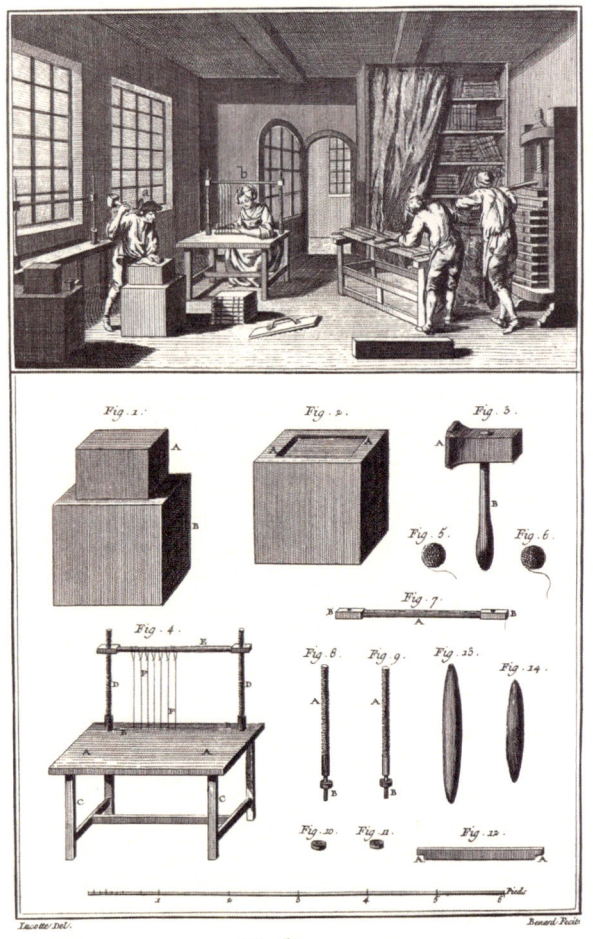
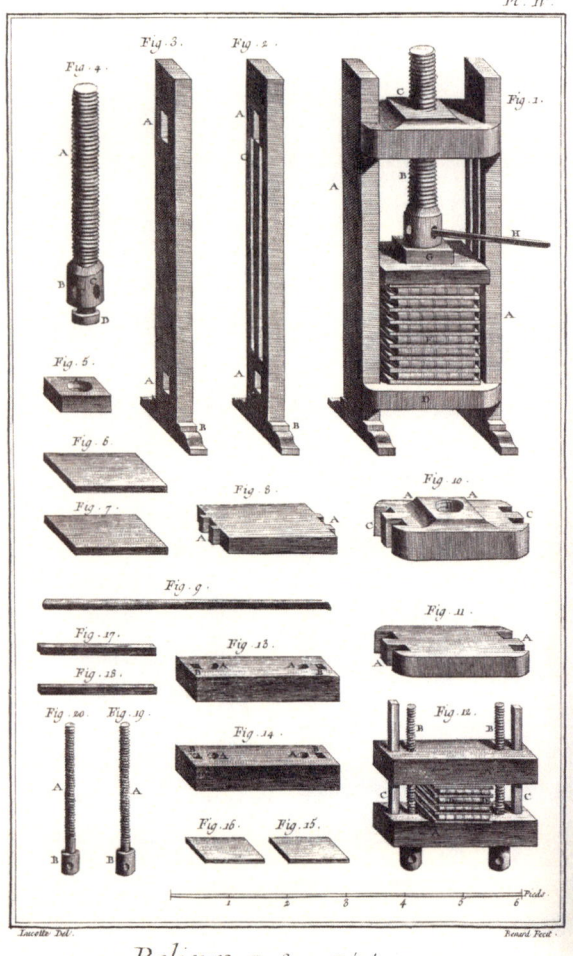

From Diderot et d'Alembert, *Encyclopédie des sciences* (Paris, 1751–1772)

These three images represent the state of the art in bookbinding before mechanization of bookbinding processes began in 1827.

gal, the bookbinders failed in their attempt. The five "martyr" bookbinders received two-year prison sentences; one of them, William Wood, died in prison. Of the others, Thomas Armstrong, William Craig, and Thomas Fairbairn went on to become masters.[239]

Why elements of hand bookbinding began to be mechanized in the late 1820s is uncertain. Probably the need to speed up bookbinding to bind the increasing number of books being published during the second printing revolution was evident to a few inventive men. The first three basic developments in the mechanization of bookbinding—the rolling press, the arming press, and book cloth that could accept gold stamping—all occurred between 1827 and 1832, very shortly after publisher William Pickering introduced publisher's cloth bindings. The rolling press eliminated one of the most menial elements of hand bookbinding, the need to round book spines with a hammer. The arming press provided a means for efficiently stamping designs onto book covers. In principle, this enhanced the appearance of bindings and made production more efficient. Publisher's cloth bindings became a critical element of industrialized book production by providing a relatively inexpensive but permanent binding to replace the temporary wrappers in which many books were sold prior to the

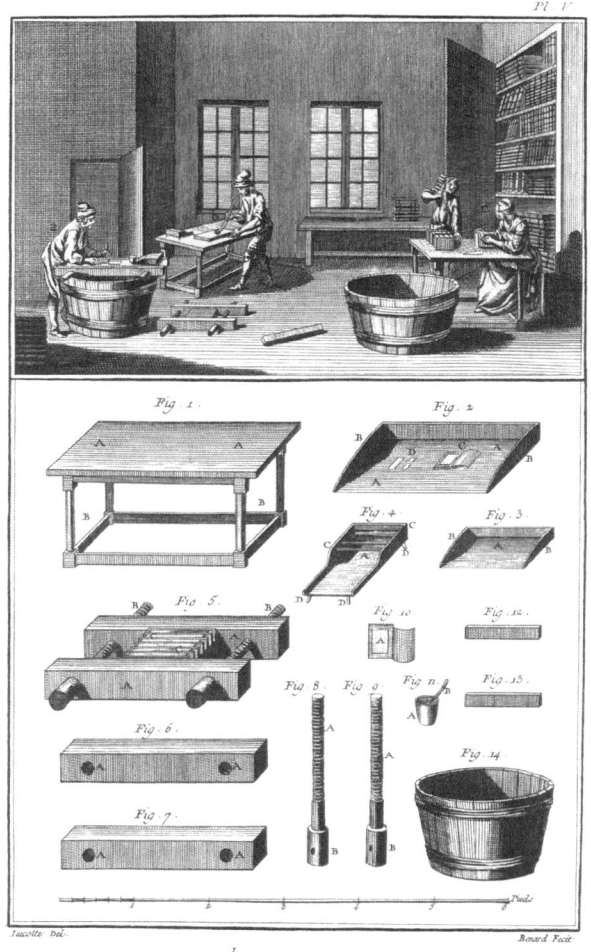

Relieur Doreur.

James Fraser
Print, 16¾ × 13⅓ in. (42.5 × 34 cm) at the platemark

This print commemorates the role of bookbinder James Fraser in a bookbinding dispute in 1786. In the print, details engraved in very small letters on a paper held by Fraser refer to this early trade-union action, when the trade societies the United Friends, the Brothers, and the City Brothers combined to have their workday reduced by one hour from the existing fourteen-hour schedule from 6 a.m. to 8 p.m.

introduction of publisher's cloth. But the origin of publisher's cloth is less precise and somewhat more complex than the invention of the rolling press and the arming press, primarily because we cannot determine precisely when Pickering had publisher's cloth bindings put on the earliest books known to have those bindings.

In his pioneering study of publisher's bindings, *The Evolution of Publishers' Binding Styles 1770–1900*,[240] Michael Sadleir stated, following the opinion of Sir Geoffrey Keynes, that the first example of a publisher's cloth edition binding was Pickering's miniature edition of the works of Virgil (1821), the second volume in his Diamond Classics series. The third volume in the series, Cicero's *De officiis, de senectute et de amicitia* (1821), was offered by Pickering bound in reddish-brown calico cloth. The paper spine labels on Pickering's Diamond Classics were printed with both the volume title and the price (5 shillings, in the case of the Cicero). That these books were issued in publisher's cloth bindings is not in dispute; what is uncertain is the date that Pickering actually applied the cloth bindings to the books.

In 1932, two years after Sadleir published his book on binding styles, London antiquarian bookseller and bibliographer John Carter issued *Binding Vari-*

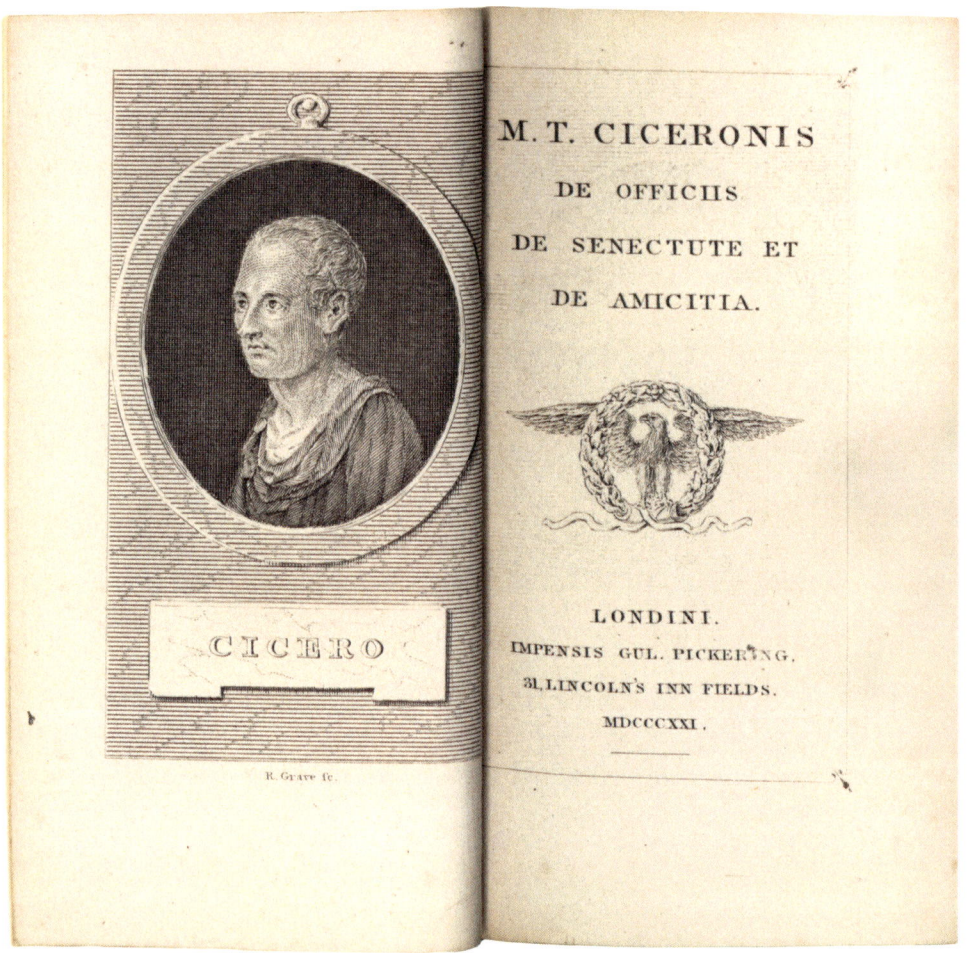

📑 Cicero, *De officiis, de senectute et de amicitia* (London, 1821)

William Pickering's Diamond Classics pocket edition of Cicero was one of the first books offered for sale in a publisher's cloth binding.

👉 👉 *The Plays of Shakespeare*, 9 vols. (London, 1825)

William Pickering's pocket-sized Diamond Classics edition of Shakespeare's plays was probably the first set of books ever offered for sale in publisher's cloth.

ants in English Publishing 1820–1900,[241] in which he argued that it was possible, and even probable, that the publisher's cloth bindings sometimes found on Pickering's *Diamond Classics* were put on a bit later than 1821. Pickering was known to have copies of his books bound up in fairly small quantities as demand warranted, so that the cloth bindings could have been put on later. Carter's argument for this was that, after a very careful search, the earliest reference he could find to Pickering advertising any of his books in cloth bindings was in the "Spring list of 1826."

Three years later, Carter issued another shorter work on the subject, *Publisher's Cloth: An Outline History of Publisher's Bindings in England 1820–1900*.[242] One point Carter mentioned in this work was that the earliest definitive statement that Pickering had been responsible for introducing cloth-edition bindings was made by the publisher and writer Charles Knight

in his 1833 *Penny Magazine* article on printing presses and machinery and bookbinding. Knight wrote, "But within the past seven years the introduction of the cheap and yet neat and substantial binding in cloth, which was first attempted by Mr. Pickering, of Chancery Lane, has created a new branch of business, of equal importance to any of the previously existing branches." Seven years prior to 1833 would date Pickering's introduction of cloth to about 1826.

In the revised edition of his *William Pickering Publisher*, Keynes repeated the indirect evidence for the dating of Pickering's introduction of cloth bindings that he had obtained from John Carter in detail. The story, citing an account recorded in 1855, thirty years after the fact, cannot be depended upon for reliable dating, but it does confirm Pickering's intent to publish half of the first edition of his *Diamond Classics* in cloth bindings:

An anonymous article on "The History of Bookbinding" published in the issue of The Bookbinders' Trade Circular *for March 1855, pp. 9-120, had stated that the use of cloth for binding had been introduced by Pickering in 1823 after noticing the "French red" lining of chintz curtains. He was said to have supplied his binder, Archibald Leighton, of Exeter Street, with the right kind of calico. At first Leighton glazed and stiffened the material with glue; later he found starch to be more suitable for the purpose. Probably the latter part of the story was true, but clearly the date 1823 for the first use of cloth was wrong. This was put right in the next issue of the* Circular *for May 1855, p. 22, where a correspondent, R. E. Lawson, writing from 61 Stanhope Street, contributed an amended version. He agreed that Pickering first introduced cloth into the book trade, but maintained that it was his own suggestion. Lawson was working as a binder for Charles Sully and William Greenfield ("the eminent linguist") and had known Pickering since 1809. His employers had been commissioned to bind the Diamond Classics, and he related the following incident, which presumably took place early in 1820: "Mr. Pickering came one evening—I remember, perfectly well, that the candles were alight—into the shop—I believe No. 2 Upper John Street, Golden Square—and announced to Mr. Sully that he was about to publish the works above named [the Diamond Classics], and wished a quantity done in morocco, and a portion in boards. 'Now,' said he, to Mr. Sully, 'could you suggest some neater mode in which to do the boarded portion than the present one.' I immediately handed to Mr. Pickering, from my side-board, a small oblong quarto of MS music for the guitar, which myself and Mr. Sully were studying under the same master at that time, bound in light blue glazed calico, a remnant of some my mother had been lining her window curtain with, and asked Mr. Pickering 'what he thought of THAT'. 'The VERY THING,' said he, 'and excepting the colour will do admirably.' After a little deliberation, it was decided that they should be done in* couleur du puce, *which was the case, while the old style of 'lettering' was retained in the now rarely-to-be-met with form of the white printed 'label' of the period! The books came in—one thousand copies; five hundred were done in morocco, five hundred in 'cloth' boards; the cloth was purchased at the corner of Wilderness-row, St. John Street, and the whole of the 'CLOTH' copies were covered by myself with glue, Richard Cross, Mr. Sully's apprentice at the time, 'squaring the boards' and 'drawing in'."*[243]

Keynes also pointed out that the use of cloth as a publisher's binding did not add much over the cost of paper wrappers or temporary boards to the cost of books and, from the buyer's point of view, saved a considerable amount of money since it did not require the buyer to rebind flimsy boards or paperbacks. Though the innovation was introduced by Pickering without fanfare in 1820, from our point of view today, the invention of publisher's cloth was one of the most significant developments in the history of commercial bookbinding.

In 2020, Paul W. Nash published *Two Hundred Years of Publisher's Cloth*,[244] which most probably will remain the definitive study of this issue. This long paper, which reproduces many of the earliest publisher's cloth bindings in color for the first time, argues for 1821 as the date that Pickering introduced what we consider the earliest publisher's cloth bindings. Nash presented all the evidence for that date, arguing that the key element was the development of the appropriate type of cloth by the binder, Archibald Leighton, and the promotion of cloth-edition bindings by William Pickering. Despite Nash's expert marshaling of the evidence, the problem remains that we may never know for sure whether or not a publisher's binding was issued exactly at the time that an edition was first sold. Some publishers stored books in sheets and had copies bound as needed over the period of time that the edition was in print; they also sometimes gave parts of editions to different binders, resulting in binding variants. As a result, while we know when Pickering published his books, because he dated them and advertised them on certain dates, exactly when and how he had each copy bound must be inferred, unless the publisher's or the binder's detailed records exist and can be consulted.

In 1827, at almost the same time as the introduction of cloth-edition bindings by William Pickering and Archibald Leighton, London bookbinder William Burn received the Silver Vulcan Medal from the Society of Arts for his invention of the rolling press for bookbinders. The machine consisted of a heavy iron mangle designed to press flat sheets of printed matter before they were folded and sewn; remarkably, it was the first machine to be adopted into the bookbinding trade. Up to this time, this process had been done by workmen hammering the sheets with a fourteen-pound beating hammer—a monotonous, strenuous, and time-consuming job. In evaluating Burn's invention, the Committee of the Society of Arts observed that the rolling press accomplished in one minute what beating with the hammer required twenty minutes to do. In addition, the paper was made smoother, and since the compression was greater, a book that passed through the rolling press was about five-sixths the thickness of a beaten book.[245]

By December 1830, several of the larger bookbinderies in London had installed rolling presses, also known as rolling machines. This threatened the livelihood of the beaters, a semi-skilled class of bookbindery worker who used heavy beating hammers to flatten the surface of the paper. To discuss this problem, the London Society of Bookbinders held a meeting on 7 December 1830, and on 16 December, the society issued *The Memorial of the Journeymen Bookbinders of London and Westminster on the Effects of Machinery, Respectfully Addressed to their Employers*. This document was signed by 498 binders.

In his anonymously published book *The Results of Machinery, Namely, Cheap Production and Increased Employment, Exhibited, Being an Address to the Working-Men of the United Kingdom*, publisher Charles Knight pointedly criticized the position of the bookbinders:

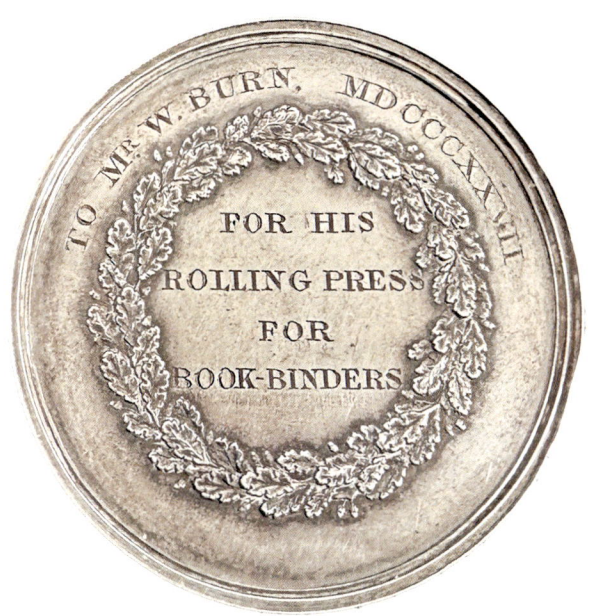

Silver medal
Diameter: 1½ in. (4 cm)

The Silver Vulcan Medal that the Society of Arts awarded in 1827 to William Burn for his invention of the rolling press for bookbinders. The rolling press was the first machine invented to mechanize a manual skill in bookbinding.

CHAPTER 12

Fig. 166. ROLLING.

▸ From Charles Tomlinson, *Cyclopaedia of Useful Arts* (London, 1852), p. 155

A rolling press in operation at a bookbindery.

▸ From *Mechanics Magazine*, no. 472 (25 August 1832)

This article published one of the first illustrations and announcements of the Imperial Arming Press. This device, which was the second machine invented to mechanize bookbinding, was essentially an iron handpress adapted for bookbinders' use. The company also produced an iron handpress for printing called the Imperial Printing Press.

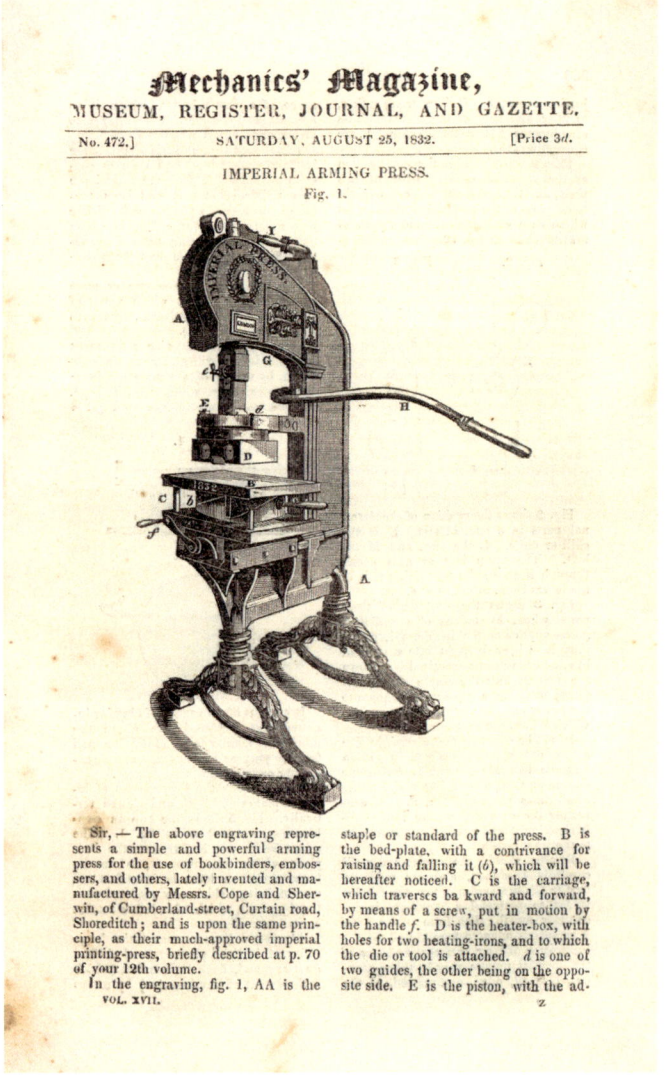

We have a remarkable example of the folly of a particular body of men upon this subject now lying before us. We have a paper, dated sixteenth of December, 1830, issued by nearly five hundred journeymen bookbinders of London and Westminster, calling upon their employers to give up the use of a machine for beating books. Books, before they are bound in leather, were beat with large hammers upon a stone, to make them solid. That work is now done in London by a machine. The workman is relieved from the only portion of his employ which was sheer drudgery—from the only portion of his employ which was so laborious, that it rendered him unfit for the more delicate operations of bookbinding, which is altogether an art. The greatest blessing ever conferred upon bookbinders, as a body, was the invention of this machine.[246]

In response to Knight's criticism, the journeymen bookbinders issued a forty-page rejoinder,[247] but their complaints and arguments proved ineffective in persuading any binderies to discard their rolling presses or beating machines.

Soon after the advent of the rolling press, the Arming Press for bookbinding was invented and manufactured in London by Cope and Sherwin, who also sold the Imperial Printing Press for handpress printing. The Imperial Arming Press was illustrated and described on the cover of the 25 August 1832 issue

of *Mechanics' Magazine*. The device, which was also called a stamping or embossing press, enabled quality stamping onto cloth and leather with excellent control and very high pressure. Cope and Sherwin essentially adapted the concept of the iron printing press to the needs of the binding trade.[248]

The London bookbinder Archibald Leighton, who pioneered the use of cloth publisher's bindings for publisher William Pickering, continued to experiment with cloth as a bookbinding material. In 1832, he developed the first book cloth that could take and retain impressed gilt decoration rapidly and in sufficient quantity to allow for gilt-stamped, cloth-edition bindings. This historical observation was first made by Michael Sadleir, who wrote:

> *Previously, although it had been possible to lay down gold on cloth, each book had to be done by hand and with an expenditure of time and trouble which rendered the process useless for an edition of any size. This date of 1832 is fixed in an interesting manner. In that year John Murray began the publication of a 12mo edition of the works of Byron, bound in dark green cloth with title, etc., in a shield on the spine. Vols. 1 and 2 of that edition were originally issued with dark green paper labels, printed with the title and device in gold; Vols. 3 to the end had the same title and device gold-printed actually on the cloth itself. It was between the issue of Vols. 2 and 3 that Archibald Leighton perfected his process for preparing the surface of the cloth and so introduced the gold-blocking of cloth which has been practised ever since.*[249]

As far as I could determine, Sadleir's observation remains accurate except that the transition from paper label to the gilt label stamped directly on the cloth occurred in the second volume of Murray's set of Byron. The seventeen-volume set, titled *The Works of Lord Byron; with his Letters and Journals and his Life by Thomas Moore, Esq.*, was issued at the rate of one volume per month beginning in January 1832, with publication concluding in 1833. The books were covered in a green, moiré-embossed cloth. The first volume was issued with a green paper label on the spine,

George Gordon Byron, 6th Baron Byron, *The Works of Lord Byron*, 17 vols. (London, 1832)

The second volume of John Murray's edition of *The Works of Lord Byron*, published in February 1832, was the first book bound in publisher's cloth with the gold lettering stamped directly into the cloth. Before Archibald Leighton developed the first book cloth that could take and retain impressed gilt decoration, the title design was stamped on a paper label, as shown in the label mounted on the spine of the first volume.

Charles Babbage, *On the Economy of Machinery and Manufactures* (London, 1832)

The gold stamping on the spine of the cloth binding on the first edition of Charles Babbage's *On the Economy of Machinery and Manufactures*, also published in 1832, resembles the design stamped on the spine of the volumes of the Murray edition of Byron. Babbage's book underwent three editions in 1832, all with the same gilt-stamped cloth bindings, proving that these gilt-stamped bindings were produced in 1832.

with the title and a coronet printed in gold. By the time the second volume was published a month later, the technique of gold blocking on cloth had become a workable proposition, with the result that the lettering and coronet were stamped directly onto the cloth of the remaining sixteen volumes. The historical significance of Murray's edition of Byron in the history of cloth bindings was understood at least as early as 1889, when it was written about in an article on "technical bookbinding" published in *The Bookbinder*.[250]

One correlation that I have not seen mentioned previously by scholars is the coincidence of the availability of the Imperial Arming Press in 1832 and the likelihood that it was used to stamp the bindings of Charles Babbage's *On the Economy of Machinery and Manufactures*,[251] the first three editions of which appeared in 1832 with gold-stamped bindings remarkably similar to those on the Byron set. The stamping of the selling price (6 shillings) beneath the spine title was a Charles Knight innovation that may have been first introduced with this binding. Those three different editions of the book, all of which appeared in publisher's cloth bindings, confirm that these case bindings—especially those on the first and second editions, which were sold out in 1832—were actually put on during the year 1832.

Another early bookbinding innovation was William Hancock's invention of the gutta-percha binding, patented in 1836. These bindings, which bypassed the hand-sewing process, were the forerunner of today's perfect bindings that bind books with flexible plasticized glue instead of sewn signatures. The gutta-percha material lost its flexibility and integrity over the decades, causing the sheets to separate from the binding, but Andrew Ure could not have anticipated that problem when he described the gutta-percha binding process in his *Dictionary of Arts, Manufactures, and Mines* and cited those bindings as "one of the greatest improvements ever made in the art of bookbinding."[252]

Some of the best details of mid-nineteenth century industrial bookbinding, which remained a combination of hand and machine work, appear in the extensive section on bookbinding in Charles Tomlinson's *Cyclopaedia of Useful Arts: Mechanical and Chemical, Manufactures, Mining and Engineering*.[253] Aside from detailed accounts of the binding process, from this we learn that two men trained in cloth-binding technique could complete 100 covers in an hour, and, providing that all the cloth covers were completed in advance, the crew at Remnant and Edmonds bindery in London could bind an edition of 1,000 copies in publisher's cloth in six hours. Within that time, they would perform the gathering and folding of the printed sheets, the sewing of the gatherings, gluing and rounding the backs, cutting the edges, embossing and gilding the covers, and securing the covers to the books. This was, as Tomlinson stated, an example of the power of numbers of skilled workers, and the effect of a refined system of the division of labor.

As with most aspects of book production, more historical research has been conducted on the elements of bookbinding prior to mechanization than on the period after mechanization began. An example of this research is the annotated bibliography of bookbinding manuals up to 1840 by Graham Pollard and Esther Potter.[254] One of the last manuals that Pol-

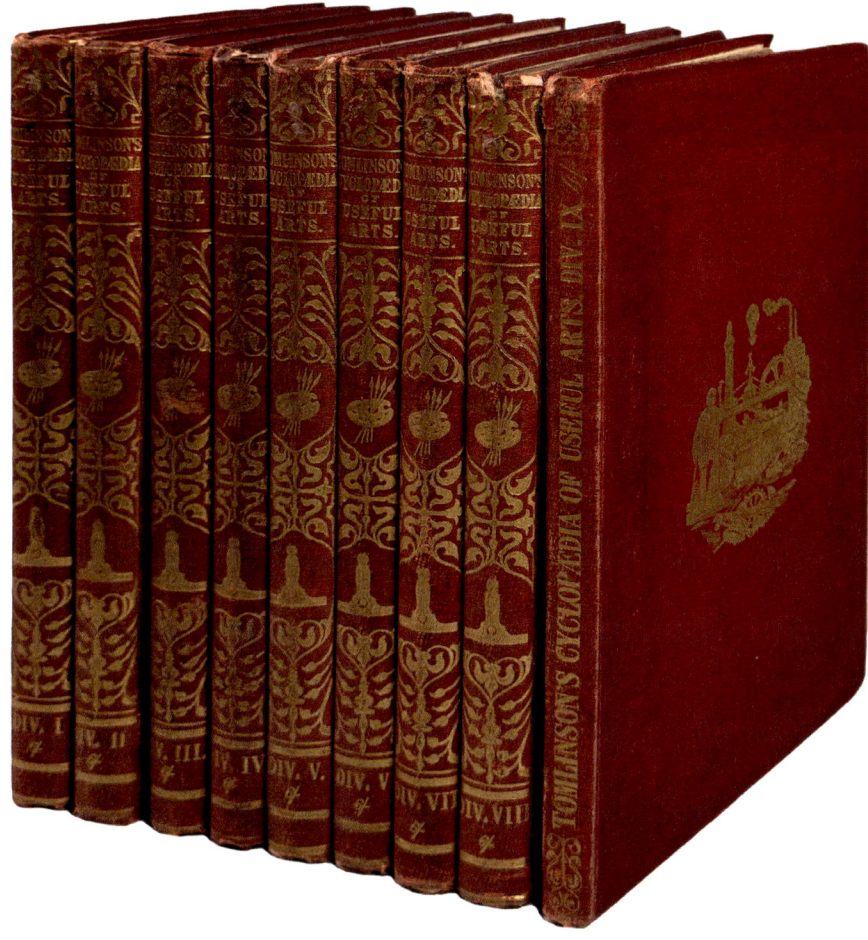

Tomlinson's *Cyclopaedia of Useful Arts* was sold in fascicules, such as the first part illustrated here. It was also sold in publisher's cloth in nine volumes.

lard and Potter included was *Bibliopegia, or, The Art of Bookbinding in all its Branches*, by John Andrews Arnett.[255] In his manual, Arnett featured extensive commentary on the use of the Imperial Arming Press, and he also mentioned, toward the end of his book, the recently invented "Rolling Machine" intended to supersede the necessity of beating the spine of books with a hammer.

In the mid-nineteenth century, various technical manuals were published illustrating the equipment and machines used in bookbinding at the time. One of these was James B. Nicholson's *A Manual of the Art of Bookbinding: Containing Full Instructions in the Different Branches of Forwarding, Gilding, and Finishing. Also, the Art of Marbling Book-Edges and Paper. The Whole Designed for the Practical Workman, the Amateur, and the Book-Collector.*[256] Nicholson's book was intended for the widest audience of professionals, amateurs, and enthusiasts. In it, Nicholson illustrated and described the operation of most basic bookbinding machines used at the time, including an unusually sophisticated hydraulic press invented by Isaac Adams, inventor of the Adams Power Press.

From the standpoint of general historical interest, the best illustrations of mid-nineteenth-century bookbinding equipment and procedures are in the section on bookbinding in Charles Tomlinson's *Cyclopaedia*. Tomlinson began his section with a discussion of the effects of mechanization on bookbinding, stating that the demand for books was greater than at any time in history, and that even though bookbinding "would seem to depend for its production more

Fig. 165. FOLDING.

Fig. 170. SEWING

Fig. 171. STANDING PRESS.

Fig. 172. BOARD-CUTTING MACHINE.

Fig. 174. EMBOSSING PRESS.

Fig. 175. GOLD-BLOCKING PRESS.

Fig. 178. PLOUGHING THE EDGES.

Tomlinson's *Cyclopaedia of Useful Arts* was sold in fascicules, such as the first part illustrated here. It was also sold in publisher's cloth in nine volumes.

Illustrated here are seven bookbinding procedures described and illustrated in Tomlinson's *Cyclopaedia*: folding, sewing (using a sewing frame that had not changed significantly since the eighteenth century), applying a standing press (which had also not changed significantly since the eighteenth century), using a board-cutting machine, using an embossing press, and using a gold blocking press (similar to the Imperial Arming Press). Tomlinson's fig. 174 shows an embossing press that must have been exceptionally heavy and powerful. Fig. 178 shows a bookbinder using a binder's plough to trim the edges of a book block. The device was illustrated in *The Book of Trades* by Jost Amman and Hans Sachs (1568).

upon individual skill than upon large and complicated machinery, yet here, as in so many other cases the manufacturer has, to a great extent, superseded the mechanic."257 He observed that the process of mechanization expanded markets, by lowering costs, and created new jobs, but he also predicted that, as mechanization advanced, certain jobs would be replaced by machine, and he recommended that binders should learn as many skills as possible in case the particular skill which they were accustomed to applying was replaced by machine. He writes:

> *The machinery which has superseded hand labour in some directions has led to a vastly increased amount of hand labour and intelligence in other directions. There is now a greater demand than ever for type-founders and for compositors, for literary men and for artists; for engravers on wood and engravers on metal. And if in the binding of a book, the workshop has expanded into the factory, and if many of the manipulations of the binder are now superseded by machines, there are more folders, more sewers, nay, there are even more finishers than formerly existed; for the great demand for machine-bound books. . . . The same principle is at work throughout the useful arts. A successful machine may supersede certain descriptions of hand labour, and cause for a time much privation and suffering, but it is sure to increase the demand for labour in other branches; a well ascertained fact which ought to be constantly borne in mind by the intelligent workman. He ought to seek every opportunity of acquiring skill in more than one department of his trade, so that should his services in one direction be superseded by a machine, he may be able to apply his skill in another."258*

Just as most historical scholarship on papermaking has focused on hand papermaking, and most scholarship on the history of printing has focused on the handpress period, scholarship on the history of the mechanization of bookbinding is relatively limited.259 One means that we have of tracing developments in this field are the technical manuals that were published. In 1860, German bookbinder Ludwig Brade and German civil engineer Emil Winckler issued *Das Illustrirte Buchbinderbuch* (The illustrated bookbinder's book), one of the *Bibliothek des Wissenswürdigsten* (Library of the most interesting facts) series of volumes published by Otto Spamer in Leipzig. The

L. Graham & Son, *A Brief Description, with Illustrations, of the Printery of L. Graham & Son, Limited, Intended to Convey an Idea of Its Size and Facilities* (New Orleans, 1898)

The building housing L. Graham & Son, Printers and Bookbinders, in New Orleans, 1898—a medium-size printing and bookbinding company. This obscure publication contains excellent images of Graham's equipment and binding facilities, as well as photographs of people using the machinery in their bindery.

first edition of this work, with 80 illustrations on 276 pages, was the most comprehensive nineteenth-century manual on mechanized bookbinding, illustrating all the available machines invented to speed up what remained a largely handcraft process. The first edition was followed by an enlarged second edition of 1868 edited by J. R. Herzog, expanded to 420 pages with 120 text illustrations; the third edition of 1882, expanded with 150 text illustrations, was edited by Robert Metz and published in Halle by Wilhelm Knapp. With the third edition, the publishers issued, at extra cost, an *Atlas zum Illustrirten Buchbinderbuch*

L. Graham & Son, Ltd., 44-46 Baronne St., N. O.

sewing machine. To the left of the operator and connected with the machine before which she sits is a revolving perpendicular shaft to which is attached four radial arms. Each arm presents itself to the operator about six times a minute. As the arm is passing the operator hangs a sheet on it saddle wise. This sheet is then carried to the machine. It does the rest. It takes the separate signatures and perfectly sews its folded leaves together (for as yet the signature is held together only by its folded edges) and forms continuous sewed connection between the different signatures. They emerge from the back of the machine a solid mass of sheets sewed in the backs. The books are separated by an attendant, and after some preparation are passed to the trimming machine.

Here they are trimmed on all three edges to proper size, and thence passed to the forwarding department. At this stage the edges of the books are stained, marbled, sprinkled or gilded, as may be the case.

The next operation is rounding the backs so as to give the concave surface to the front edge of the book. This is done by a beating hammer, and manipulation. The rounded book is then placed between two iron jaws similar to those of a vise, and a swinging roller is brought into contact and passed to and fro over the back of the book until sufficient of it has been forced over the edges of the jaws to make what is called a back. A vertical section of the book thus prepared shows the back in something like an umbrella shape. The use of this will be seen later on. The books are now ready for casing, but we will have to go back to explain this.

While the forwarding has been in progress, the case makers have been at work. As soon as the size and thickness of the book has been ascertained, work on the "cases" or covers can be commenced. The stock is first laid out and cut. Book cloth comes in rolls about 38 inches wide

SEWING MACHINE.

A PAPER CUTTING MACHINE.

L. Graham & Son, *A Brief Description, with Illustrations, of the Printery of L. Graham & Son . . .* (New Orleans, 1898)

- A sewing machine to sew signatures of books.
- A paper-cutting machine.
- An embossing machine.

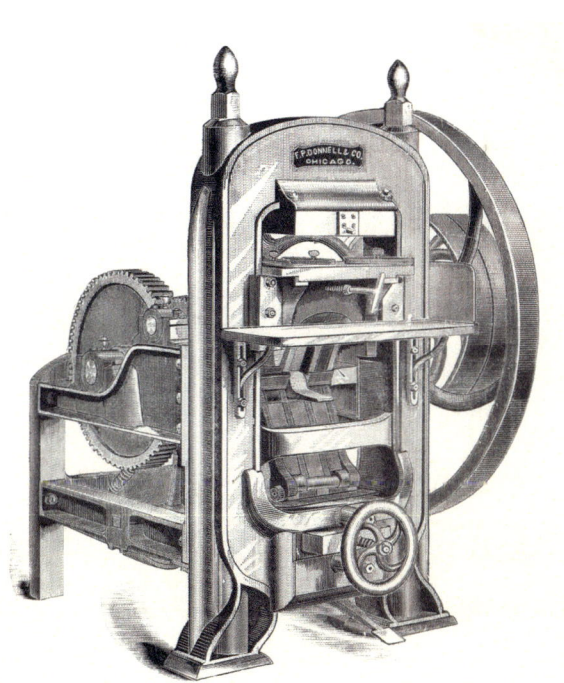

AN EMBOSSING MACHINE.

(Atlas of the illustrated bookbinding book). The work underwent its ninth expanded edition in 1930.

Except for describing machines that inserted wire staples (so-called wire stitches), the third edition of Brade and Winckler's manual did not describe the mechanization of sewing, as no fully satisfactory thread sewing machine for bookbinding was invented until that of the American David Smythe, which did not go into general production until 1886. In the fourth edition (1904), the editors of Brade and Winckler's book illustrated a thread-sewing machine built by Gebruder Brehmer in Leipzig. Several of the other pieces of machinery included steam-powered stamping machines to speed up stamped designs onto covers before the book block was cased in. The 1904 edition also included reproductions of binding examples and actual mounted samples of marbled paper. At the back of the book there were a few advertisements for equipment, supplies, and machines.

Views of the extent of mechanization in bookbinderies by the end of the nineteenth century appear in two very different volumes, from which we have reproduced several images. The first is an obscure illustrated description of a printing and binding operation published in New Orleans, Louisiana, in 1898. It is entitled *A Brief Description, with Illustrations of the Printery of L. Graham & Co., Limited, Intended to Convey an Idea of Its Size and Facilities*. L. Graham and Son took special pride in their bookbindery, which they asserted was the only one of its kind in New Orleans. In their book, they illustrated a book-block sewing machine, a paper-folding machine, a wire-stitching machine, a binder's board cutting machine, a large embossing machine, a large paper-cutting machine, and a book-trimming machine. Two unusual photographs reproduced here show their highly mechanized bindery in operation.

Another glimpse into the state of bindery mechanization at the end of the nineteenth century comes from the French historian and art critic Marius Vachon. In his *Les arts et les industries du papier en France 1871–1894*, published in Paris in 1894, Vachon reproduced many remarkable images, of which several show the state of mechanization of the different elements of paper, print, and book production in France, including Parisian commercial bookbinderies. From these images, we can confirm that, after its slow start in the late 1820s, bookbinding was very extensively mechanized by the end of the nineteenth century.

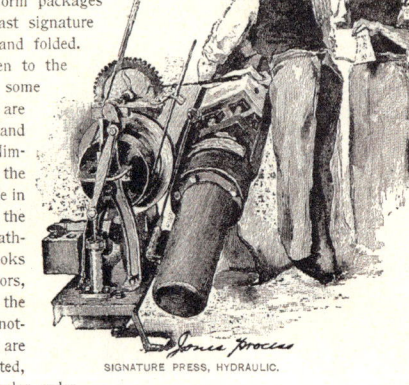

L. Graham & Son, *A Brief Description, with Illustrations, of the Printery of L. Graham & Son . . .* (New Orleans, 1898)

▸ The Jones Hydraulic Signature Press, which applied 50 tons of pressure to remove the indentations caused by letterpress printing.

▸▸ The rear portion of the bindery, showing sewing machines.

▸▸ The front portion of the bindery, showing embossing machines.

FRONT PORTION OF BINDERY, SHOWING EMBOSSING MACHINE.

REAR PORTION OF BINDERY, SHOWING SEWING MACHINES.

From Marius Vachon, *Les Arts et les industries du papier en France 1871–1894* (Paris, 1894)

By the end of the nineteenth century, industrial bookbinding in Paris was extensively mechanized. This image shows people operating machines to gild and print covers at the bindery of the Librairies-Imprimeries Réunies, Paris.

Machines à dorer & à imprimer les couvertures
Ateliers de reliure des Librairies-Imprimeries Réunies

Women operating staple sewing machines at the Maison Lenègre bookbindery. These machines used metal staples rather than thread to sew printed signatures.

Batterie de Machines à coudre au fil métallique
Maison Lenègre

Machines à couper le papier & Presses à imprimer les couvertures
Maison Lenègre

A man operating a machine to cut paper, and other men operating machines to print covers at Maison Lenègre.

Presses pour le tirage en couleur des couvertures
Maison Lenègre

A man operating a machine that printed a cover in color at Maison Lenègre. These may be some of the first bookbindery machines that were much larger than their human operators.

George Leighton, *The International Exhibition—The Nave* (detail). See p. 221.

CHAPTER 13

The Mechanization of Book Production in the United States and Europe, 1851–1904: The Great Exhibition, George Baxter, Manuals and Promotional Books on Mechanized Printing, the Caxton Celebration and Bible

During the second half of the nineteenth century, the development of mechanized book production proceeded on both the industrial scale, with the growth of large printing and bookbinding concerns, such as William Clowes in London (far larger than had been possible before mechanization), and smaller businesses that mechanized on a local, smaller scale. An example of smaller-scale mechanization is recorded in the paper card case advertising the stationery, bookshop, printing office, and bookbindery of Max and Rudolf Zocher founded in Dresden in 1880. From the small images on the box, we can see the machinery used in their bookbindery and the two steam-powered presses powering their printing machines in their printshop, along with two compositors at their type cases. The Zochers' business was probably more typical of the average successful stationery and book-production facility. The opposite extreme was the industrial facility of William Clowes in London, which was able to typeset, print, and bind 10,000 copies of the initial 320-page *Official Catalogue of the Great Exhibition*, with its fifty additional pages of advertising, in only four days. Starting in the 1820s, Clowes had been the first printing firm in the world to apply mechanized printing to book production on an industrial scale.

In a pamphlet titled *Printing: Its Modern Varieties*, London writer George Dodd analyzed the innovation applied in the extremely rapid production of the widely read and used catalogue. The Great Exhibition, which took place between 1 May and 15 October 1851, marked the beginning of the second half of the nineteenth century. It was held in the 990,000-square-foot (92,000-square-meter) Crystal Palace, which had been built for the exhibition in London's Hyde Park. Six million people attended the exhibition, equivalent to one-third of the entire population of Britain at the time. The *Official Catalogue* was one notable example of the technical ingenuity and innovation featured at the Great Exhibition:

> *Few comparisons would present more curious results than that between a printing office in past days and one in 1851. Everything was done by hand, and on the domestic-manufacture system; much is now done by steam, and all on the factory system. Our Clowes, Hansards, and Spottiswoode's in the present time exhibit the factory system in its best aspects; that is combination in some departments, subdivision in others. The well-known rapidity with which Parliamentary Papers are got up has been often noticed; and the recent printing of the* Official Catalogue of the Great Exhibition *was a notable instance of such expedition. We quote a few words from the* Companion to the British Almanac for 1852, *in illustration of this matter:—"The Shilling Catalogue was classified, numbered, made up, and 10,000 copies printed and stitched in covers—in four days. The first complete copy was not produced till 10 o'clock at night on April 30th, and yet 10,000 were at the Crystal Palace before the arrival of Her Majesty on the eventful 1st of May. Two splendid copies, presented*

Color-printed card box, 4 × 2¾ × 1⅓ in. (10 × 7 × 3.5 cm)

This card box was issued by Max and Rudolf Zocher to advertise their business in Dresden in about 1885. It provides some small views of the operations of a mid-size printing and bookbinding operation at the time.

The Zocher print shop, as illustrated on one of the longer sides of the card box (enlarged). Notice the illustration (not to scale) of the steam engine driving the printing machines and the manual typesetting going on at the right of the image.

The Zocher bookbindery, as illustrated on one of the narrow sides of the box (enlarged).

The Zocher book and stationery store on the ground floor, with customers, from one of the longer edges on the Zocher card box (enlarged).

The bookkeeper's office and storeroom from one of the narrow edges of the Zocher card box (enlarged).

The Official Catalogue of Great Exhibition of the Works of Industry of All Nations (London, 1851)

Ten thousand copies of this catalogue were typeset, printed, and bound by the very large industrial printing firm William Clowes Ltd. in four days.

to Her Majesty and the Prince Consort, were bound and gilt in a sumptuous style in six hours.[260]

Reflective of the haste in which the catalogue was produced, a statement on the fifth page of the official catalogue read, "This catalogue necessarily appearing in a state of great imperfection, in consequence of the delay of the reception of Goods at the Building, and the resulting impossibility of completing their allocation in time for the opening on May 1st, will be followed as speedily as possible by a re-issue in a form more accurate and complete."

Relative to printing and book production, the Great Exhibition also stimulated productions from printers in other countries demonstrating their newest and most innovative printing facilities. The Imperial and State Printing Office in Vienna, founded in 1804 by Emperor Franz I, had installed its first two printing machines and a steam engine to drive them in 1836; these were probably the first printing machines installed in Austria. The printing establishment had grown rapidly after that. In 1851, Alois Auer, director of the K. K. Hof- und Staatsdruckerei in Vienna, published *Geschichte der K. K. Hof- und Staatsdruckerei*

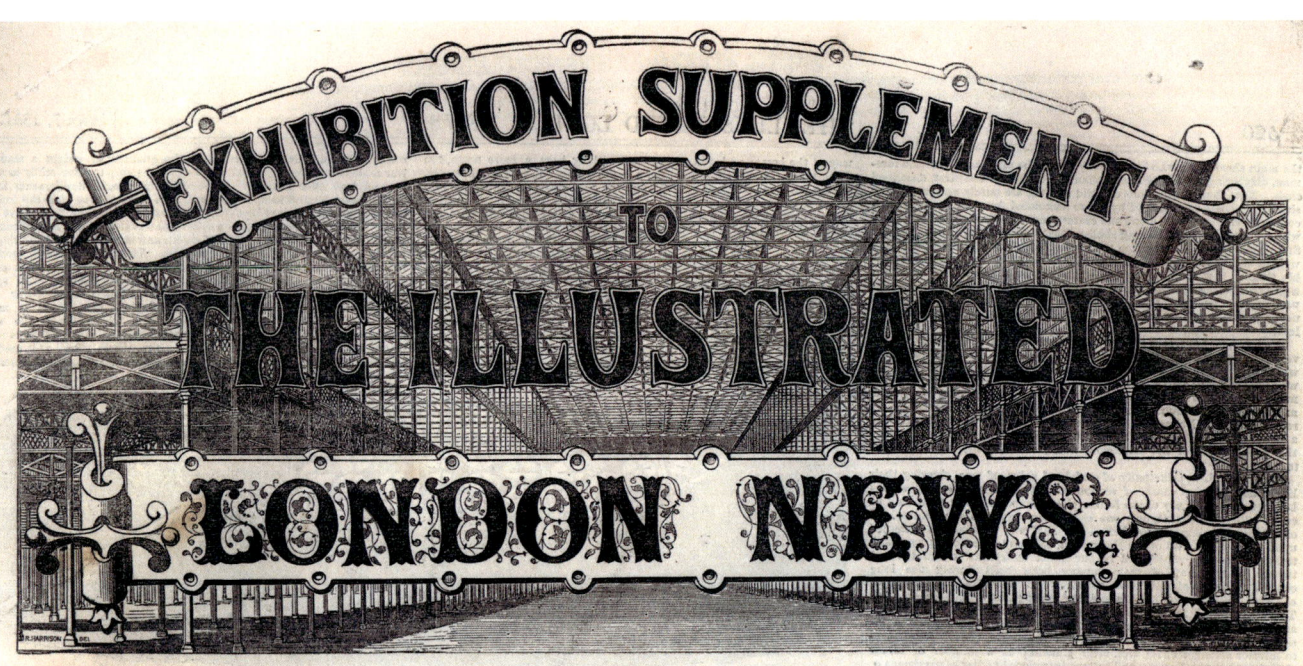

Exhibition Supplement to The Illustrated London News, vol. 18, no. 492 (7 June 1851)

A note at the foot of page 521, in this issue, stated, "THIS SHEET WAS PRINTED IN THE GREAT EXHIBITION."

in Wien, an extensively illustrated multi-lingual history of that state-of-the-art facility. The book included texts in German, English, Italian, and French.

Also for the Great Exhibition, printing machine manufacturers Koenig and Bauer published a sixteen-page pamphlet in German and English titled *Die Ersten Druckmaschinen* (The first printing machines).[261] The purpose of this very detailed bilingual exposition of Koenig's first printing machines was to remind those attending the exhibition of the great historic contributions to printing technology made by Koenig and Bauer—contributions that, by this time, had been overshadowed by developments made in England since 1818 by Edward Cowper, Augustus Applegath, and David Napier. The pamphlet reproduced drawings and details from Koenig's original English patents—the first patents on printing machines.

The coverage of the Great Exhibition by *The Illustrated London News* was remarkable for the depth of its text and for the quantity and quality of its black-and-white woodcut illustrations, with periodic "Exhibition Supplements" covering different aspects of the exhibition. Other than the method of reproducing the many illustrations, the main difference between

From *The Illustrated London News*, vol. 18, no. 492 (7 June 1851)

The Applegath Vertical Printing Machine in operation at the Great Exhibition printed the sheet of the newspaper.

the way this tabloid covered the Great Exhibition and the way it would be covered by a modern tabloid, was the extent of the printed text, set in small type. A difference between the way early and mid-nineteenth-century readers read publications and the way we read them today was the more long-winded style of journalism, fit onto the pages in type that was considerably smaller than what we are expected to read today. The publishers of *The Illustrated London News* also set up an Applegath Vertical Printing Machine at the Crystal Palace so that they could print copies of the *News* during the Great Exhibition, and in their 31 May issue, the *News* published an illustrated article on that machine that included sheets actually printed at the Exhibition. The magazine article described the new printing machine with nearly as much detail as would be required in a patent.

A much more spectacularly illustrated publication on the Great Exhibition was *Baxter's Gems of the Great Exhibition*, printed and published by George Baxter, inventor of the Baxter color printing process. In contrast to the issues of *The Illustrated London News*,

CHAPTER 13

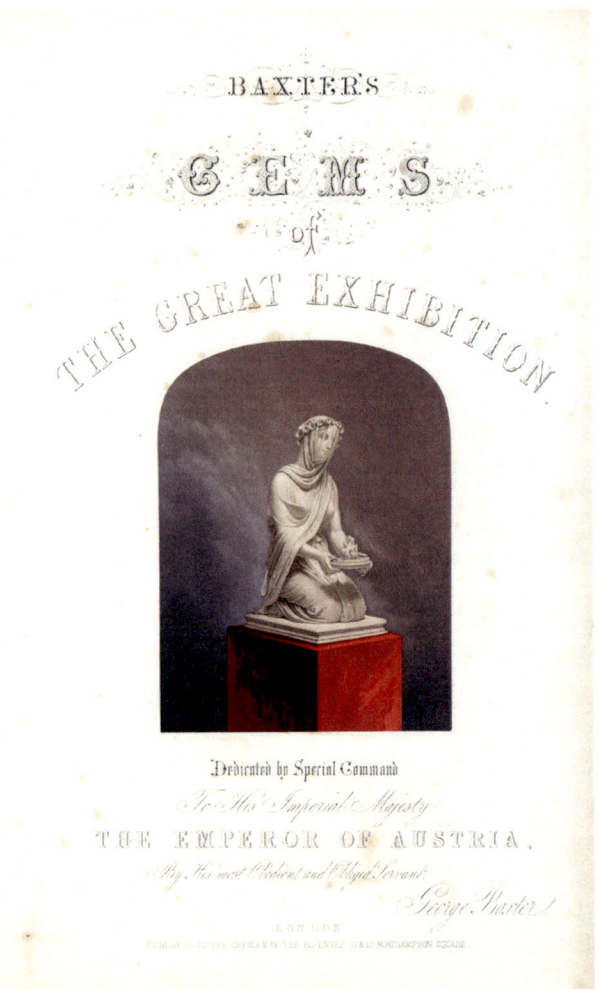

George Baxter, *Baxter's Gems of The Great Exhibition* (London, 1851)

Baxter's color prints, printed on iron handpresses by his patented process, were the most brilliantly color-printed images recording the Great Exhibition.

the color prints produced via Baxter's color printing process were printed on iron handpresses, providing some of the best evidence of the continuity of handpress printing during the first decades of mechanized printing. In 1835, Baxter received patent no. 6916 for "Improvements for Producing Coloured Steel Plate, Copper Plate, and Other Impressions." In this patent, Baxter described a commercial printing process of handpress printing that printed images in color with a level of detail, precision, and brilliance of color superior to any other commercial process of the time, and equal to or better than most hand coloring.

Unlike most patents, Baxter's patent is illustrated with actual examples of his printing process, including a print titled "Cleopatra" that required sixteen or seventeen different impressions on the handpress to complete. The original printed patent, which reproduced Baxter's original color artwork in black and white, could not accurately show the progression of colors involved, but the patent did explain the concepts, which were simple in their planning but technically difficult to execute with high quality. Baxter's process started with an initial printing from a steel key plate, which gave the outline of all the intricate detail and shading that made a fully finished print, but in monochrome. Then he would apply up to twenty different blocks made from wood, copper, or zinc—one for each color. The registration of each block had to be perfect. Baxter's process was elaborate, involving immense challenges in obtaining accurate registration of all the different impressions of colors. From the standpoint of modern labor costs and modern labor-saving technology, it hardly seems practical from a commercial standpoint; however, it was the first commercially viable method of large-scale color printing. Baxter produced prints well through the 1850s, issuing approximately 400 different color prints. Some, including those of Queen Victoria and Prince Albert, might have sold relatively large numbers for the time; C. T. Courtney Lewis claimed that "upwards of 500,000 copies" of the Prince Albert print were sold.[262] Regarding Baxter's overall production numbers, Max Mitzman estimated that "Baxter himself printed over twenty million prints during his career."[263]

The number of ephemeral prints that were sold at the time would have been vastly higher than the number of copies of most books printed, and the survival rate of such prints would have been far lower than that of most books. Such an enormous output of complicated, extremely precise, multi-impression printing without the use of printing machines assumes an average sale of 50,000 copies per print. This seems unlikely at that time, both from the standpoint of actual sales and printing technology. According to Geoffrey Wakeman, Baxter's "key plates, being intaglio, must have been printed on rolling presses, of which Baxter had two. The blocks or electros were printed on hand platen presses; he had 3 Stanhopes, 2 Coggers, and 2 Albions. This would allow work to proceed on the 12 to 20 tint blocks for each key block. Register was obtained by several points on the press, described in the patent."[264] Wakeman probably obtained these press details from the 1860 auction catalogue of Baxter's effects. With respect to these details, we might speculate that Baxter owned, or at least used, more presses when he needed to print the millions of precisely registered impressions to create the

From Edmund Evans, *The Art Album* (London, 1861)

The Gardener's Daughter was one of sixteen images color-printed by Edmund Evans, a licensee of George Baxter's color-printing patent.

circa 500,000 copies of the Prince Albert print, if that many were actually sold.

From georgebaxter.com we also learn about the complexity and care involved in Baxter's process that would tend to contradict historians' claims of such enormous output from Baxter's presses:

Baxter was a perfectionist and personally spent many hours, at least in the early days, engraving all his own steel plates and cutting all the colour blocks. He would only use the best quality materials and mixed all his own oil inks. The paper would be wetted, the key plate applied and the ink left to dry. The paper then had to be dampened again, so that it expanded to exactly the same size as when the key plate was used and the first colour was printed, then again left to dry. This process

Christmas Supplement to *The Illustrated London News* (22 December 1855)

This was the first issue of a newspaper or magazine printed in color. To accomplish this, George Leighton adapted Baxter's color-printing process.

was repeated until all the colour blocks were added. As these presses were all operated by hand this must have been a painstaking process.[265]

Perhaps because of the extreme complexity of his technique, Baxter's patent seems not to have been infringed upon. Eventually, after his patent was renewed in 1849, Baxter licensed the process to other printers, including Joseph Martin Kronheim and Abraham and Robert le Blond, who modified the process and produced commercial color prints more efficiently and presumably in larger editions.

Further experiments in printing wood engravings in colors began about 1850. The two great innovators in this field were Edmund Evans, the printer of Kate Greenaway's and Walter Crane's images, and Baxter's former apprentice George C. Leighton, who successfully adapted Baxter's process commercially. Leighton and his brothers, Steven Leighton and Charles Blair Leighton, established Leighton Brothers in 1849; the firm produced some of the largest color engravings of the Victorian era. Around 1858, George Leighton became the main printer and publisher for *The Illustrated London News*, remaining in that position until 1883.

Leighton used his adaptation of Baxter's process to print the 22 December 1855 "Christmas Supplement" for *The Illustrated London News*—the first issue of a newspaper printed in color.[266] The "Christmas Supplement" consisted of an eight-page insert printed on somewhat thicker paper than the regular issues of the magazine, containing a full-color cover and three additional full-page color images printed from wood blocks by Leighton. Each color print was credited "George C. Leighton Red Lion Square." The remainder of the "Christmas Supplement" was printed in black and white.

Leighton's production of these first color images proved that color printing could be done in high volume and at comparatively low cost without printing machines. "The designs were engraved as woodcuts in the ordinary way, and the impressions from them coloured by etched tone blocks; both blocks and colouring are extremely crude, but the idea caught on with the public and Leighton could not produce the plates fast enough to satisfy the demand."[267] Among the largest color prints Leighton Brothers ever issued was "The International Exhibition – The Nave (Looking West)."[268] The printed surface of this color print, printed from wood blocks, is 18.3 × 25 in. (46.5 × 63.5 cm).

After Baxter, the German printer Joseph Martin Kronheim, working in London, was probably the most prolific producer of Baxter-style color prints. Kronheim printed from zinc or copper plates instead of the wood blocks that Baxter used, often using as many as sixteen plates for a single print. Kronheim's staff operated as many as thirty or forty platen handpresses to produce his prolific output of color prints, involving more than 3,000 designs.[269] When Kronheim discontinued production of Baxter-style color prints is uncertain. The longevity of the Baxter process and its modifications provide an example of how

The Mechanization of Book Production in the United States and Europe, 1851–1904

George Leighton
The International Exhibition—The Nave
Issued as a Supplement to *The Illustrated London News*
(18 October 1862)
Color print; printed surface measures 18⅓ × 25 in.
(46.5 × 63.5 cm)

The registration is remarkably accurate on this large, mass-produced image, printed from a series of wood blocks.

handpress methods persisted while power printing advanced.

Unlike Baxter, who was a stickler for quality, Kronheim was willing to produce both high-quality prints and prints of lower quality for customers who wanted cheaper images. Some of the latter appeared in the 1863 volume of *The Leisure Hour* magazine, published by the Religious Tract Society, a British evangelical Christian organization. The same volume contained some color prints of similar quality produced by Edmund Evans, who was generally known for much higher-quality work.

Manuals and Promotional Books on Mechanized Printing

As industrial mechanized printing establishments developed during this period, several of them published promotional books proudly describing their new facilities. Though the images they contain are undoubtedly idealized, these are the best documentation of the environment in which printing was done at the best facilities as the second half of the nineteenth century advanced. Among the most artistic of these is *Notice et spécimens: Imprimerie – Librairie – Relieure*, published by Alfred Mame et Fils at Tours, France, in 1867. This splendid volume, printed on fine paper in folio format, included outstanding engraved images of Alfred Mame's production facilities, bindery, and bookstore, as well as a large collection of printer's samples providing examples of the kind of work the company printed.

Eleven years later, in 1878, L'Imprimerie Berger-Levrault et Cie. of Nancy, France, published *Notice historique sur le developpement et l'organisation de la maison*,[270] an informative history and development of the company's operations with outstanding illustrations. The work included charts with statistics on the growth of the company's work force between 1855 and 1877, the number of machines they owned, and the number of workers they employed.[271] In the same year, Edmond-Albans Chaix issued *Historique de l'imprimerie et de la librairie centrales des chemins de fer*, a history of the railway timetable printing company founded in 1845 by his father, Napoléon Chaix. The copy illustrated here was inscribed by Chaix to his colleague, the lithographer Joseph-Rose Lemercier, and specially bound for presentation.

As the machine-press industry grew, the first separately published books on the technique and operation of high-speed printing machines appeared in Germany. Germany had initially lagged far behind England and America in the Industrial Revolution in general and in the adoption of printing machinery; however, after a slow beginning, Germany was gradually becoming competitive in the field. The first book on the subject, published in 1853, was *Der Maschinenmeister an der Schnellpresse. Ein Handbuch für Buchdruckereibesitzer, Factore, Maschinemeister und Mechaniker* (The master of the high-speed press. A handbook for printing shop owners, factors, masters of the machine, and mechanics], by Andreas Albert, the factory master at Koenig and Bauer.[272] This was followed eight years later by *Die Schnellpresse, ihre Mechanik und Vorrichtung zum Druck aller typographischen Arbeiten* (The high-speed press, its mechanics and technique for printing all typographic works), written and published by C. F. Wittig and C. F. Fischer of Leipzig; revised editions of this book appeared in 1866 and 1878. A third book on high-speed machine printing, *Die Schnellpresse, ihre Construction, Zussammenstellung und Behandlung: Praktischer Leitfaden für Buchdrucker und Maschinenbauer* (The high-speed press, its construction, assembly and handling: A practical guide for printers and mechanical engineers),[273] was published in Leipzig by Andreas Eisenmann in 1865.

The most comprehensive German manual on printing published in the nineteenth century was Alexander Waldow's *Die Buchdruckerkunst in ihrem technischen und kaufmannischen Betriebe* (The art of printing in its technical and commercial operations), published in two volumes (plus atlas) between 1874 and 1877. The second volume of text and the atlas constitute the most comprehensive nineteenth-century works on printing by machinery issued in any language. That text volume contained numerous unusual plates, including several in color; the atlas was mainly a collection of images of machinery, including many unusual printing machines not shown in English-language works. Waldow issued these and many other works from his own highly respected publishing company, which specialized in publishing books for book printers. He was also the publisher of the periodical *Archiv für Buchdruckerkunst und verwandte Geschäftszweige* (Archive for the art of printing and related businesses).

The first separate work on machine printing in French, Adolphe-Lucien Monet's *Le conducteur de machines typographique. Études sur les différents systèmes de machines mise en trains* (The typographic machine operator. Studies on the different systems of setting up and operating), was issued in 1872.[274] The first edition was a small octavo of 401 pages with only minimal illustrations. Monet continued to develop his text and, in 1878, published a second edition under the title *Les machines et appareils typographiques en France et à*

IMPRIMERIE — ATELIER DE PRESSES MÉCANIQUES

Alfred Mame et Fils, *Notice et spécimens: Imprimerie-Librairie-Relieur* (Tours, 1867)

The machine room; no handpresses are visible. Mame made a special effort to provide good lighting, with a skylight as well as gaslights above each machine.

GALERIE DES MACHINES.

L'Imprimerie Berger-Levrault et Cie., *Notice historique sur le développement et l'organisation de la maison* (Nancy, 1878)

In this image of men working in a machine room, an iron handpress is visible in the lower left corner.

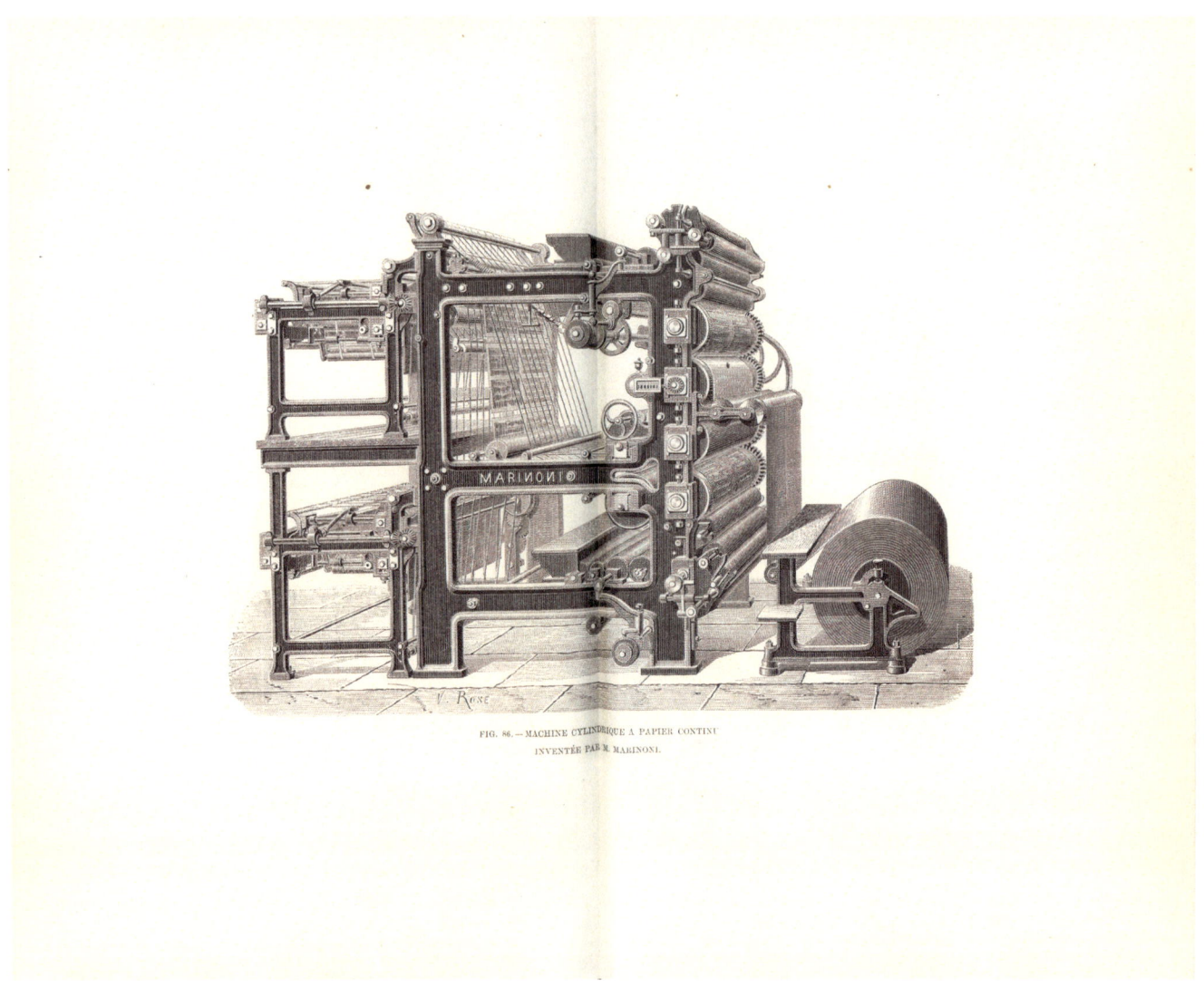

From A.-L. Monet, *Les machines et appareils typographiques en France et à l'étranger* (Paris, 1878)

This web press, invented by Hippolyte Marinoni about 1880, was among the most widely used of Marinoni's high-speed printing machines. Monet's book was the first French book on printing machines and their operation. It underwent three increasingly comprehensive editions between 1872 and 1898. Monet described and illustrated only French machines, primarily by Marinoni, Alauzet, and Jules Derrier.

l'étranger (Printing machines and apparatus in France and foreign countries), an extensively revised and well-illustrated work in large-octavo format covering foreign machines as well as French productions. This edition was translated into Spanish in 1879. A further revised and expanded third edition appeared in 1898, titled *Machines typographiques et procédés d'impression. Guide pratique du conducteur—Traité complet* (Printing machines and processes of printing. Practical guide for the conductor—Complete treatise). By reviewing and comparing the three editions of Monet's work, we can follow the technological development of mechanized printing through the last quarter of the nineteenth century.[275]

The first printing manuals published in the United States appear to have been intended for the owners of small to medium-sized printing shops, who would not necessarily have used high-speed presses in their operations. The first edition of Thomas F. Adams's *Typographia* (1837), a widely used American printing manual, contains no discussion of "machines," but the revised second edition of 1844, published under the title *Typographia, or the Printer's Instructor*, includes a brief discussion of "machine printing."[276] This section was expanded in several of the following eleven editions; the 1857 edition discusses Hoe's rotary press and particularly recommends the Adams Power Press, invented by Isaac Adams, calling it "the best power press for book work."[277] By 1864, the date of the last edition of *Typographia*, Adams's book had become outdated; it was replaced by Thomas MacKellar's *The American Printer*, which first appeared in 1866 (MacKellar's work was a general guide to all the operations of a printing shop; I have not seen any nineteenth-century American books designating themselves specifically as manuals on the operation of printing machines). Regarding the Adams Power Press, MacKellar stated that "most books . . . are printed on the bed-and-platen power-press invented by Isaac Adams, of Boston,—the only machine-press yet discovered that is capable of producing fine work and exact register. It will give from six to eight thousand impressions per day."[278] While MacKellar discussed basic handpress printing, the great emphasis throughout nearly all the text was on mechanized printing, and MacKellar provided fine images of some of the latest machine presses. From this we may conclude that, by 1866, a high percentage of book printing in America was being done by machine; however, it should not be forgotten that many American book printers continued to use iron handpresses, especially of the Stanhope, Columbian, and Albion types, throughout the nineteenth century. Mackellar's book went through eighteen editions, the last appearing in 1893.

The first manual published in English exclusively on machine printing, Jackson Gaskill's *The Printing-Machine Manager's Complete Practical Handbook; or, The Art of Machine Managing Fully Explained*, was published in 1877. Gaskill's book contained numerous text illustrations of available machine presses, as well as an eleven-plate color insert showing the eleven press passes required for printing a six-color plate. Gaskill divided his book into chapters on machine presses used for special purposes, including "Book-Work Machines," "Book and Magazine Machines," "Book, Magazine, and News Machines," "Magazine and News Machines," and "Book and General Commercial Work." He ended his book with extracts from the rules of the London Society of Machine Managers, established in 1873. Another notable feature of Gaskell's work was its twenty-three pages of advertisements at the back for a wide variety of presses and related equipment and supplies.

Two years later, Frederick J. F. Wilson produced the second English-language manual on machine printing, *Typographic Printing Machines and Machine Printing. A Practical Guide to the Selection of Bookwork, Two-Colour Jobbing, and Rotary Machines*.[279] Without acknowledging the existence of Gaskill's book, or the prior works in French and German, Wilson began his preface by stating, "At the time this Handbook was originally planned and announced, no modern practical work on Printing-Machines or Machine-Printing existed." Wilson's book represents a significant advance over Gaskill's brief handbook, from the standpoint of both text and illustrations, and it must have experienced a much wider circulation than that of Gaskill, as it had gone through four editions by 1884. One of the unusual features of the fourth edition is an illustrated discussion of various types of steam engines appropriate to power printing machines. Wilson's last word on the topic was a much-expanded book issued in 1888, written with Douglas Grey; it was titled *A Practical Treatise upon Modern Printing Machinery and Letterpress Printing*.[280] This may be considered the most comprehensive nineteenth-century book in English on mechanized printing.

In the United States, the largest printing operation during the nineteenth century was the Government Printing Office (GPO) in Washington, D.C. In 1881, R. W. Kerr, an employee of the GPO, published *History of the Government Printing Office, (at Washington, D.C.) with a Brief Record of the Public Printing for a Century, 1789-1881*.[281] This octavo volume included some excellent illustrations of the government book-production facilities.

In 1894 and 1896, two remarkably similar deluxe volumes commemorating type foundries were published in Germany and America. In 1894, J. G. Schelter and Giesecke of Leipzig issued a deluxe folio volume on their seventy-fifth anniversary titled *Fün-*

Jackson Gaskill, *The Printing-Machine Manager's Complete Practical Handbook* (London, 1877)

Gaskill's pocket-sized book was the first English manual exclusively on printing by machine. It underwent only one edition.

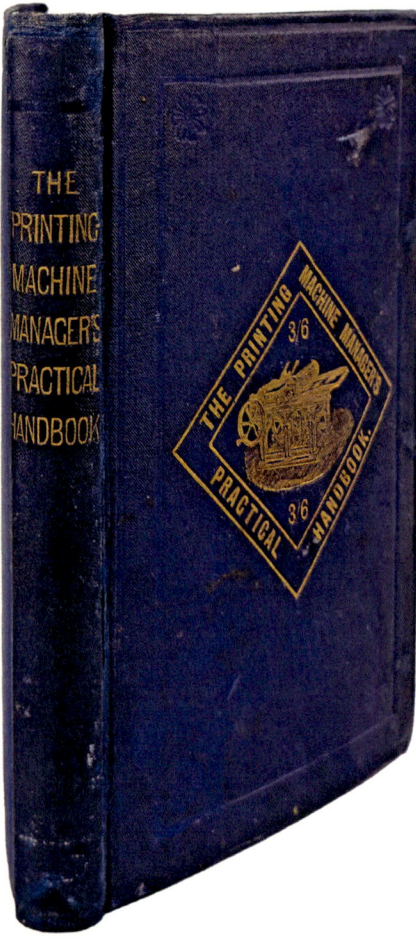

fundsiebzig Jahre des Hauses J. G. Schelter & Giesecke in Leizpig (Seventy-five years of the J. G. Schelter and Giesecke Company in Leipzig).[282] This book was beautifully bound and illustrated with watercolors integrated in the text and many full-page, photogravure illustrations.

Undoubtedly inspired by Schelter and Giesecke's volume, the MacKellar, Smiths, and Jordan type foundry in Philadelphia issued almost an exact copy of it in 1896 titled *1796–1896. One Hundred Years*.[283] The American publication was the same folio size as the German, with a similar printed cloth binding, a similar illuminated title page, and similar page layouts and similar photographs of their plant. The only thing Mackellar, Smiths, and Jordan did not do was credit Schelter and Giesecke's book as the model for their publication.

Although this book primarily concentrates on developments in England, France, Germany, and the United States, I would also like to mention some works from two other European countries. In the Netherlands, the *Vereeniging ter Bevordering van de Belangen des Boehkandels* (Association for the promotion of the interests of booksellers) held an International Exhibition for Book Trade and Related Trades in the Palace of Industry in Amsterdam in July and August 1892. This resulted in a landmark catalogue[284] containing samples from many printers, and a commemorative chromolithographed title page depicting the progress in book production from the formation of the society in 1817 to the date of the exhibition in 1892. The modern presses illustrated on the binding and the chromolithographed title page were steam-powered lithographic presses. I also found a copy of the poster published to advertise this exhibition—an enlarged copy of the chromolithographed title page for the catalogue.

In 1894, G. Fritz, the vice-director of Austria's state printing house, published *Die K. K. Hof- und Staatsdruckerei und deren technischen Einrichtungen (The imperial and royal court and state printing office and its technical facilities)*,[285] a modest pamphlet on the printing house's new technical facility illustrated with photographs and drawings. Built on several floors constructed of reinforced concrete and steel, the facility was powered by two huge horizontal steam engines: The first operated all the printing machinery through an elaborate system of steel cables driving belt drives on four floors of the building; the other generated power only for lighting, since electric lighting was only gradually being introduced to individual buildings in Vienna at this time. Fritz, who was particularly proud of the wiring system distributing electrical power throughout the building, included a schematic of the control panel in the pamphlet.

Ten years later, in 1904, the K. K. Hof- und Staatsdruckerei issued *Zur Feier des einhundertjährigen Bestandes der K.K. Hof- und Staatsdruckerei*,[286] a far more remarkable publication commemorating the hundredth anniversary of the institution. In my opinion, this is the finest and most elegantly designed publication on a printing establishment issued up to this date. *Zur Feier des einhundertjährigen Bestandes der K.K. Hof- und Staatsdruckerei* was influenced in its overall design by the Vienna Secession or possibly Jugenstil (Art Nouveau); its black-and-white text and illustra-

From R. W. Kerr, *History of the Government Printing Office (at Washington, D.C.) with a Brief Record of the Public Printing for a Century, 1789–1881* (Lancaster, Pennsylvania, 1881)

In 1881, the Government Printing Office of the United States published an elaborately illustrated volume on their production facilities.

☞ Printing Department—Main Press Room

PRINTING DEPARTMENT—Main Press Room.

☞ Printing Department—Folding Room

PRINTING DEPARTMENT—Folding Room, North Capitol street wing.

☞ Binding Department—Forwarding Room

BINDING DEPARTMENT—Forwarding Room.

CHAPTER 13

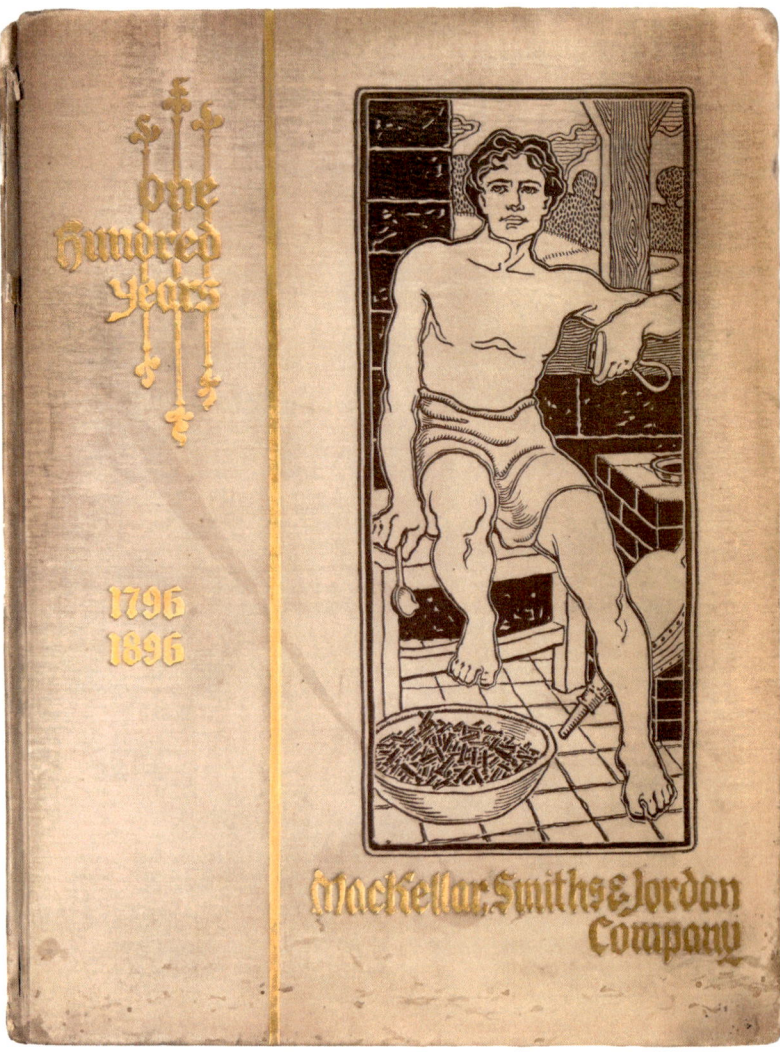

Schelter & Giesecke, *Fünfundsiebzig Jahre des Hauses J. G. Schelter & Giesecke in Leizpig* (Leipzig, 1891) and MacKellar Smiths and Jordan Foundry, *1796–1896: One Hundred Years* (Philadelphia, 1896)

The two deluxe, small-folio volumes published by type foundries in Germany and America in 1894 and 1896 closely resemble one another in format and overall design and content.

tions were also influenced by that style. The book depicts the role of the facility as both a commercial and an artistic production center, and the emphasis in all the images is on individual people or the small groups involved in all the aspects of production, by hand or by machine. In keeping with the emphasis on individual people on the staff of the institution, the company also published a twelve-page brochure in similar typographic style. This brochure listed all 1,684 employees of the K. K. Hof- und Staatsdruckerei since its foundation in 1804 by name, with the date of their first employment. They were arranged by function, from typesetter, type caster, lithographer, specialist for photography or galavanoplasty, woodcutters, printers, both for letterpress and lithography, etc., including some unusual professions such as *Manipulanten* (manipulation). Women workers were separately listed.

Vereeniging ter Bervordering van de Belangen des Boekhandels, *Internationale tentoonstelling, Juli–Augustus 1892* (Amsterdam, 1892)

Chromolithographed title page

A larger chromolithographed poster for the Amsterdam exhibition of 1892 was duplicated in reduced format as the title page for the printed catalogue. At the bottom left of the image, we see an eighteenth-century-style handpress and, on the right, the kind of press that was probably used to print the catalogue.

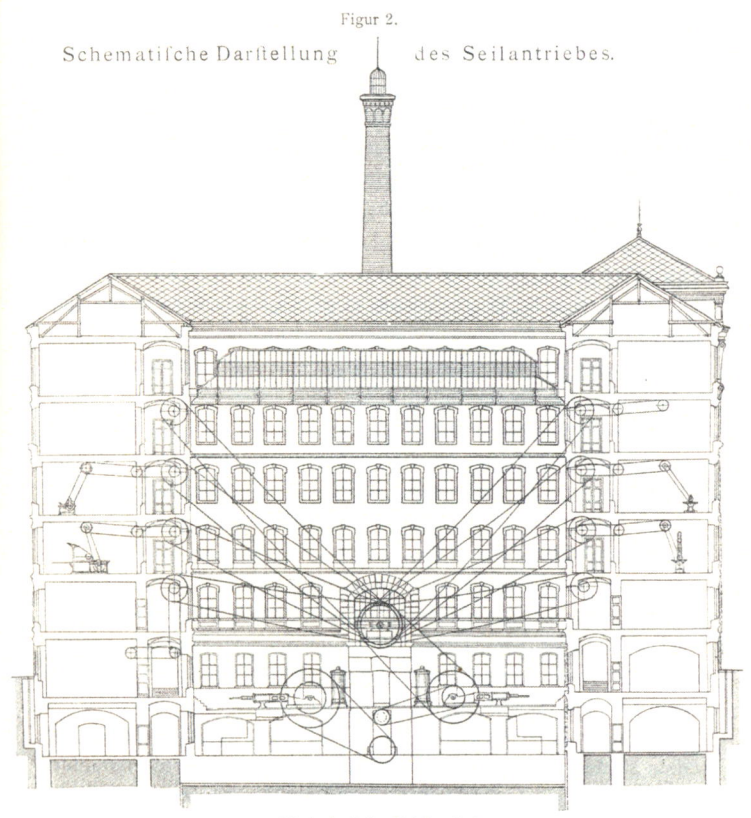

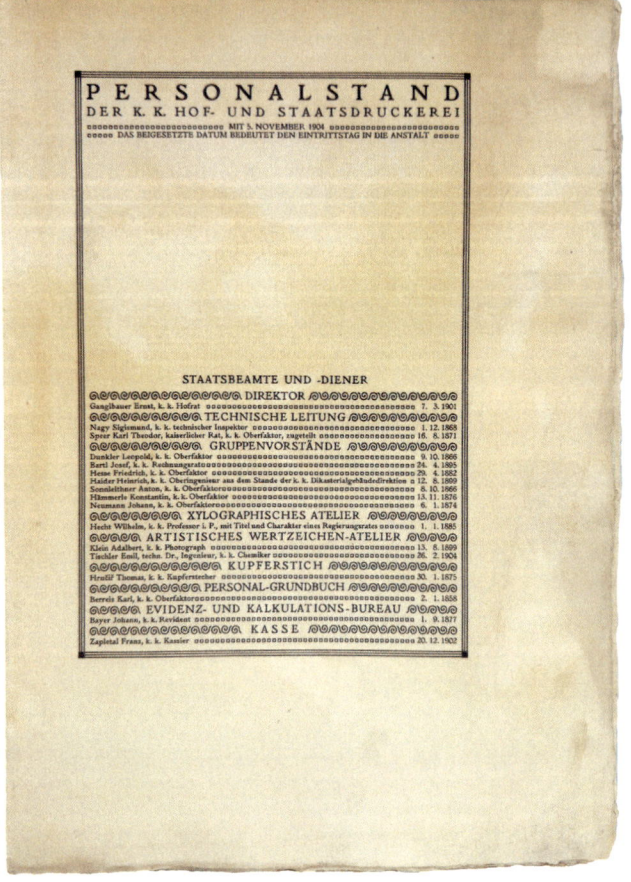

▎ G. Fritz, *Die K. K. Hof- und Staatsdruckerei und deren technischen Einrichtungen* (Vienna, 1894)

A schematic of the elaborate system of steel cables powered by a steam engine, driving belts running machinery on the four floors of the K. K. Staatsdruckerei in Vienna. A second steam engine on the same level generated electric power to light the building, but not enough power to run the machinery.

▎ From K. K. Staatsdruckerei, *Zur Feier des einhundertjährigen Bestandes der KK Hof- und Staatsdruckerei* (Vienna, 1904)

The Imperial and Royal Court and State Printing Office in Vienna printed this elegant Vienna Secession-style commemorative book on its hundredth anniversary. The pages were printed on handmade paper with a carpet-like watermark covering most of the pages. The elaborate watermark can be viewed if the pages are held up to the light. The thin tissue guards protecting some illustrations have a similar subtle printed pattern. The text for the book and an inserted list of employees who had worked at the printing office over the century were typeset by hand without any paragraph indentation.

The bookbindery.

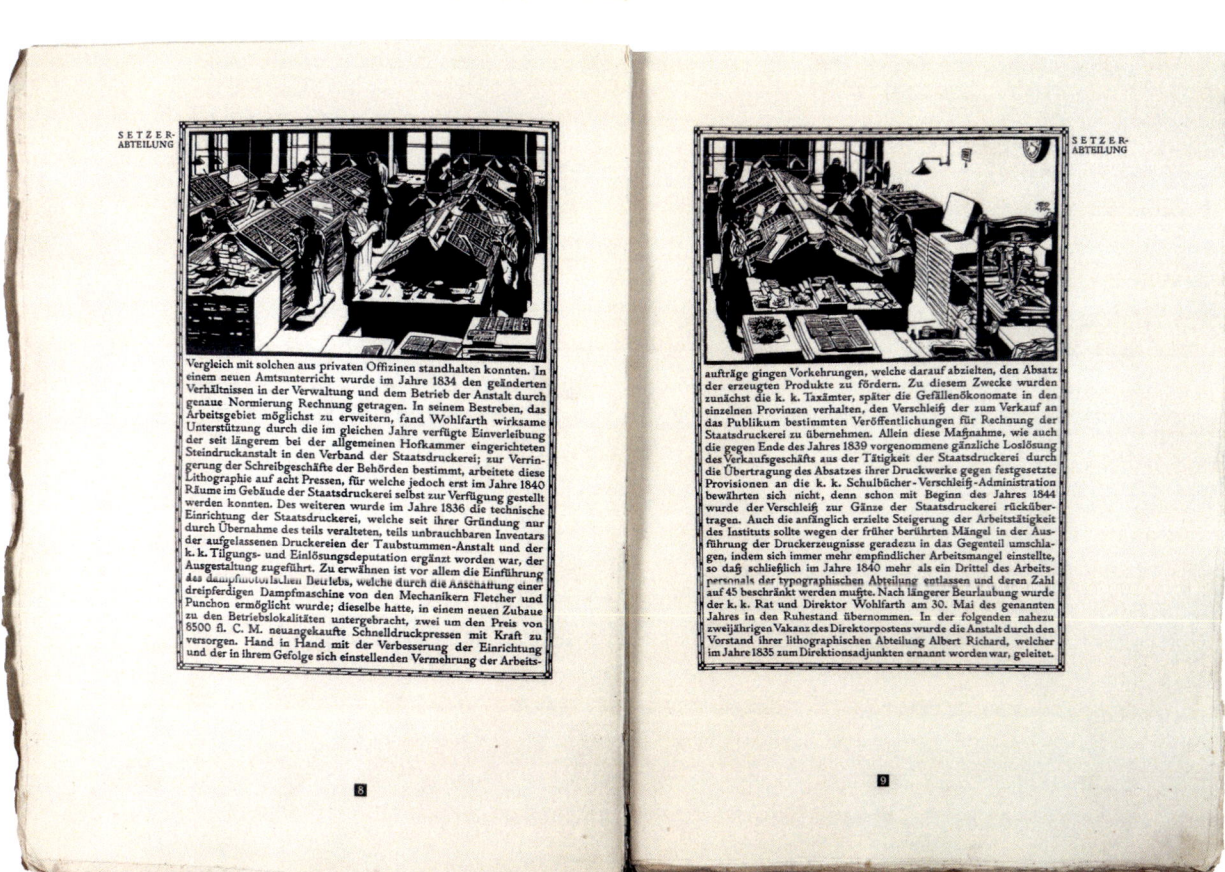

The typesetting department was entirely manual.

CHAPTER 13

The Caxton Celebration of 1877 and the Caxton Memorial Bible

For insight into the way that printing and book production—both their histories and technological advances—were perceived toward the end of the nineteenth century, we may consider the Caxton Quadricentennial Celebration that took place in London in 1877. Held during the era in which print was the primary medium for distributing and storing information, before the invention of any of today's electronic media, the Caxton Celebration was also probably the largest exhibition on the history of printing that ever took place.

In the summer of 1877, four hundred years after printer William Caxton published *The Dictes or Sayengis of the Philosophres* (The sayings of the philosophers), the first book printed in England, the Caxton Celebration opened at the South Kensington Museum (now the Victoria and Albert Museum) in London. The exhibition was organized by typefounder and politician Sir Charles Reed, working with a committee consisting of industrial printer William Clowes, printer and Caxton bibliographer William Blades, and several members of the printing firm Eyre and Spottiswoode. Two hundred or more people participated in some way as patrons or members of committees, representing a "who's who" of the printing industry in England and Europe at the time, along with leading scientists, scholars, librarians, and collectors. A few Americans, such as printing machine designer and builder Richard March Hoe, were also involved in committees. The exhibition was open for two months, from 30 June to 1 September 1877. According to David McKitterick, the exhibition "attracted a reported 23,684 visitors"[287]—an impressive number considering the population size and literacy levels of the time. For an exhibition that lasted only two months, the celebration generated a surprisingly large number of publications—probably more than were ever produced in association with a single exhibition of books and printing.

Planning for the exhibition started many months before it opened, and publicity was extensive. The illustrated newspaper *The Pictorial World*, in their issue of 24 February 1877, reported on a preliminary meeting of planners, including Sir Charles Reed and William Spottiswoode, held in the Jerusalem Chamber of Westminster Abbey, the location of Caxton's first printing office 400 years earlier, and published an engraving showing twelve mostly bearded men sitting around a table, including a secretary taking notes. Publicity for the show seems to have included marketing to children, or at least to parents who read to children:

> *Maclise's celebrated painting of Caxton showing the first specimen of his printing to Edward IV had been painted twenty years earlier, and when it was engraved in 1858 it was in the possession of John Forster (d. 1876). In 1877 it was on loan from its new owner, Lord Lytton, to the South Kensington Museum. The central part of the engraving was reissued in April 1877, as a contribution to the festivities. Reproductions were available at reduced prices to readers of* Young Folks *and* Young Folk's Weekly Budget. *As the original copies of the steel engraving had cost 4 guineas and upwards, the special offer price of a shilling plus vouchers from the magazines was a considerable bargain.*[288]

It is hard to imagine how comparable incentives related to the history of printing would have a wide appeal to children, or their parents, in the third decade of the twenty-first century.

In their issue of 30 June 1877, the opening day of the exhibition, the British illustrated weekly newspaper *The Graphic* published a double-page image captioned "The Caxton celebration. William Caxton showing specimens of his printing to King Edward IV and his queen." In their issue of 1 July 1877, *The Illustrated London News* published a collection of images related to the exhibition called "Caxtoniana"; the same newspaper, in their issue of 7 July, published an article on the opening of exhibition and a large image captioned, "Mr. Gladstone at the Caxton Memorial Exhibition, South Kensington, on Saturday last." The image showed Prime Minister William Gladstone watching printing done on a "Gutenberg-style" handpress. *The Illustrated London News* described the opening ceremony of the exhibition as follows:

> *The opening ceremony was brief and simple. The leading part was borne by the Right Hon. W. E. Gladstone. He was met by Sir Charles Reed, chairman of the*

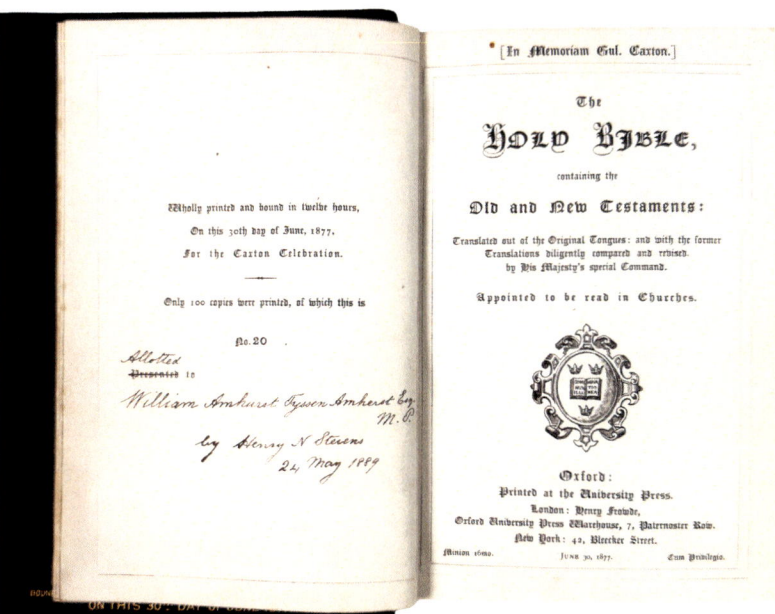

The Holy Bible (Oxford, 30 June 1877)

Printing, binding, and production of the elaborate pull-off cases for 100 copies of the Caxton Memorial Bible were accomplished within only twelve consecutive hours on 30 June 1877. The little pamphlet describing the Bible was printed on thin India paper and bound in silk so it would fit alongside the Bible in the slipcase.

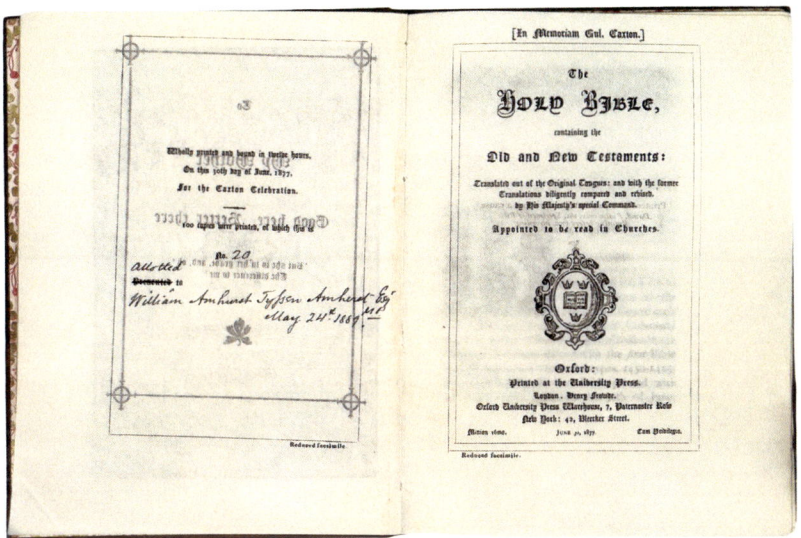

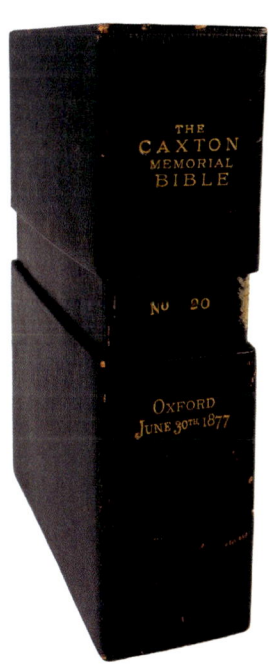

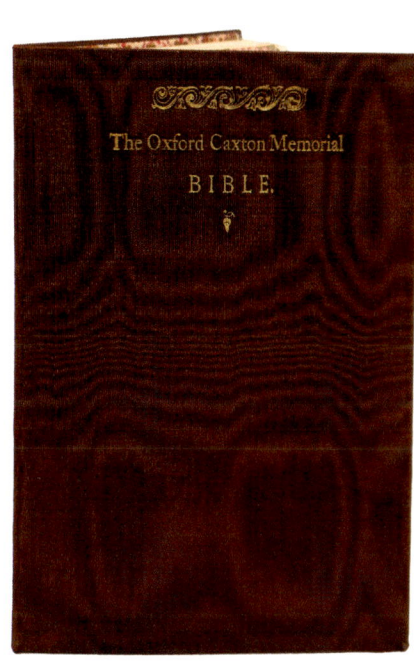

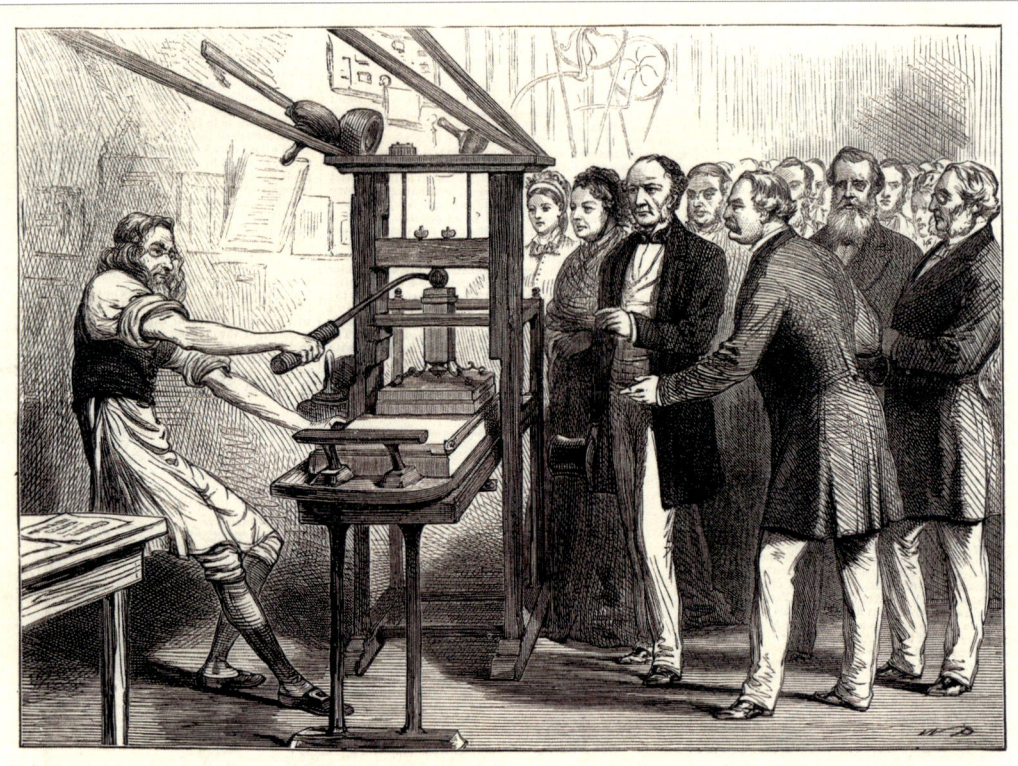

The Illustrated London News, vol. 71, no. 1982 (7 July 1877), Supplement

Prime Minister William Gladstone observing a handpress being pulled at the Caxton Memorial Exhibition.

committee; Mr. W. Blades, the biographer of Caxton; and the other gentlemen we have named, with the Archbishop of York. A large assembly of ladies and gentlemen filled the rooms assigned for this ceremony, as well as the adjacent galleries. After a special dedicatory prayer offered by the Archbishop, Sir Charles Reed read a short statement of the occasion and the objects of the Exhibition. Mr. Hodson, secretary to the Printers' Pension Corporation, handed to Mr. Gladstone a copy of the Exhibition Catalogue. The right hon. gentleman then declared the Exhibition to be opened. This formal declaration was immediately hailed by a flourish of trumpets from the band of the Royal Horse Guards Blue. Mr. Gladstone was conducted through the exhibition, which he examined with attentive interest. Our Illustration shows him looking at the working of an old press. There was a luncheon provided by the Conservatory of the Horticultural Society's Gardens. The chair was occupied by Mr. Gladstone, at whose right hand sat his Majesty the Emperor of Brazil, but the Emperor left the table before the toasts were proposed. His Majesty's health was, of course, duly honoured next to that of our Queen and Royal family. In his principal speech, giving the memory of William Caxton for the chief toast, Mr. Gladstone commented upon the invention of printing, with his usual copiousness of thought and knowledge, and expressed his admiration of the results now attained. The other speakers were the Bishop of Bath and Wells; Dr. Joseph Parker; Mr. Hall, of the Oxford University Press; M. Chaix, of Paris; Herr Fröbel, of Stuttgart; Sir C. Reed, and Mr. G. Spottiswoode. Subscriptions and donations to the Printers' Pension Corporation fund were announced, amounting to £2000, besides which there will be the receipts from the Exhibition.[289]

On 21 July, the American illustrated magazine *Harper's Weekly* printed a version of the image of Caxton and Edward IV originally published in *The Graphic*. A few days later, *The Tablet: The International Catholic News Weekly* took an interest in the exhibition, reviewing it in its 28 July 1877 issue. An Irish novelist, Catherine Mary MacSorley, commemorated the anniversary by publishing a historical novel for young people about Caxton titled *The Earl-Printer. A Tale of the Time of Caxton.*[290]

As a record of the exhibition, a catalogue was issued titled *Caxton Celebration, 1877, Catalogue of the Loan Collection of Antiquities, Curiosities, and Appliances Connected with the Art of Printing*. The catalogue, edited by George Bullen, keeper of the printed books at the British Museum, underwent several revisions during the exhibition. In its final form, this 472-page book listed (sometimes with descriptive bibliographical notes) a total of 4,734 items exhibited, a record of what was probably the largest exhibition of rare books, prints, and printing equipment ever held. It encompassed works from the Gutenberg Bible and the Mainz Psalter up to 1877, including about 190 books printed by William Caxton, classics illustrating the spread of printing, landmarks of book illustration, examples of music printing, books on papermaking, notable achievements in color printing, examples of historic, unusual, or new technologies in printing, as well as printing presses and typesetting and typefounding equipment. The catalogue contained no images; presumably, it was a sufficient challenge to publish a bibliographical record of such an enormous exhibition, crediting the numerous lenders to the show.

Probably the largest section of the exhibition was devoted to Bibles. These were separately recorded in a book by the supervisor that portion of the exhibition, the American bibliographer and antiquarian bookseller Henry Stevens, who lived in London. After the exhibition, in 1878, Stevens published a separate catalogue of the Bibles exhibited.[291]

More significantly from the historical viewpoint, at the instigation of Henry Stevens, Henry Frowde of Oxford University Press undertook the publication of a Bible that would demonstrate the advances in printing technology since its introduction in England by Caxton. By Stevens's account, this was a last-minute idea, undertaken by the press only a few days before the opening of the exhibition. The Bible was printed on printing machines from standing type at Oxford, by Oxford University Press, and bound in full black morocco leather, in matching full black morocco pull-off cases, by Oxford University Press in London in an edition of 100 numbered copies, with the printing and binding occurring in a twelve-hour period on the opening day of the exhibition. Printing began at 2:00 a.m. and the first bound copies were delivered at the opening of the exhibition at 2:00 p.m. the same day. Copy no. 2 was presented to Prime Minister William Gladstone when he opened the show, copy no. 1 having been reserved for Queen Victoria. In March 1878, Stevens published a small-format, thirty-page book titled *The History of the Caxton Memorial Bible Printed and Bound in Twelve Consecutive Hours on June 30, 1877*. In this book, Stevens told the story of this remarkable achievement, in which copies of the 1,052-page volume in sextodecimo format were printed from standing type on paper that had been specially made for the edition by Oxford University Press only a few days before printing. The printed sheets were artificially dried, and the copies were hand bound in morocco leather and enclosed in special slipcases by 101 binders assigned to the task. Stevens calculated that, had type composition been necessary, it would have taken "2,000 compositors and 200 readers to set up and properly read the Bible in these same twelve hours."[292]

In comparison with the speed of papermaking, printing, and binding in Caxton's time, the speed of production of the Caxton Memorial Bible represented an enormous advance, so great that it would be difficult to quantify, and it is evident that the producers of the Caxton Memorial Bible were very proud of these advances. What would have taken perhaps a year or more in Caxton's time—or indeed prior to 1800—was accomplished in less than a day.

If we compare the speed of completion of these 100 Caxton Memorial Bibles with twenty-first-century printing technology, it is probable that the papermaking and printing could be done as rapidly or perhaps even faster than what was achieved on 30 June 1877. However, I doubt that 101 hand binders capable of binding the volumes and enclosing them in the elaborate pull-off slipcase could be found and organized to do the task without exceptionally elaborate and time-consuming preparation—maybe years of training. To achieve anywhere near this speed of production of such an elaborate binding and slipcase today, the work would have to be done by machine, and today's machines could not equal the quality of the binding or the slipcase on the Caxton Memorial Bible. What we have gained through mechanization we have lost through the diminishment of handcrafts.

R. Hoe & Co.'s web press in operation at the Philadelphia Centennial Exposition,
9 December 1976 (detail). See p. 248.

CHAPTER

The Mechanization of Newspaper Production in the United States and Europe, 1826–1900: The Hoe Family and Hippolyte Marinoni

The first American newspaper to acquire a printing machine did so eleven years after Koenig and Bauer's steam-powered printing machine began printing *The Times* of London newspaper. In 1825, the *New-York American* newspaper imported one of David Napier's Double Imperial printing machines from England. As a measure of how far behind the United States lagged British printing technology, in 1825 there was no equivalent printing machine produced in America, so importing a machine was necessary. Napier's Double Imperial, or double-cylinder machine, was specifically designed for newspaper production; its manufacturer, the Scottish engineer David Napier, had been building printing machines in London since 1822. Due to the scarcity of small steam engines in America at the time, the *New-York American*'s press was operated by human power turning a crank.

The first newspaper in the United States to be printed on an American-made printing machine was Nathan Hale's *Boston Daily Advertiser*, which began being printed on one of Daniel Treadwell's Power Presses in 1829. Because the circulation of the *Boston Daily Advertiser* was comparatively small, the Treadwell press was apparently adequate for a time. Treadwell had designed the Treadwell Power Press for the printing of books, rather than newspapers, which had to be printed faster and usually required more copies than books. Treadwell's machines became well-established in Boston, and by 1835, about fifty Treadwell Power Presses were in operation from Boston to Washington, mostly for the printing of books.[293]

On 4 January 1826, the editors of the *New-York American* ran an article on the front page of the newspaper headlined "The Napier Printing Machine," covering the newspaper's acquisition of their new machine and the details of the first month of its operation. The author of the article noted that having a man turn the flywheel to power the machine was cheaper than using a horse directed by a boy, and more reliable! The editors' decision to devote a front-page article to their new printing machine reflected the great novelty of the machine for an American newspaper at the time and the interest that publishers assumed that their readers would have in its operation.

At this time, the circulation of the *Daily Advertiser* (another newspaper printed by the *New-York American*) was only 2,700 copies, and we may assume that the *New-York American*'s circulation was about the same. The publishers acquired the printing machine to be able to boost their circulation beyond the 4,000 copies per day that could be printed by handpresses, since the Napier machine that they acquired was capable of printing 2,400 sheets or even 2,880 sheets per hour. The newspaper's description of its new machine gives us a valuable impression of the operation of an early double-cylinder printing machine:

The machine weighs about two tons, and is composed of a great variety of parts, but so compact and beautiful in all its movements, that it occupies no greater space

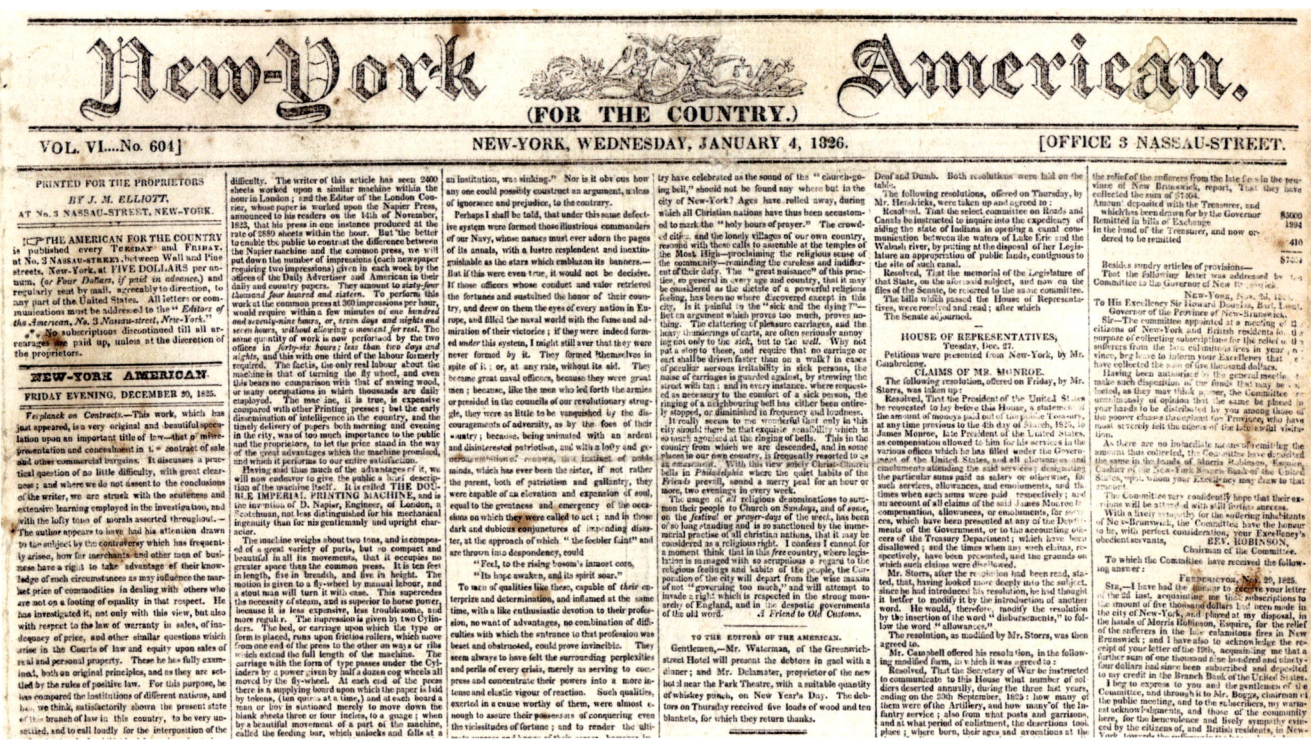

New-York American, vol. 6, no. 604 (4 January 1826)

The front page of this issue reported on the newspaper's newly acquired Napier printing machine, imported from London. This double-cylinder perfecting machine was the first printing machine installed by an American newspaper.

than the common press. It is ten feet in length, five in breadth, and five in height [3 × 1.5 × 1.5 m]. The motion is given to a fly-wheel by manual labour, and a stout man will turn it with ease. This supersedes the necessity of steam, and is superior to horse power, because it is less expensive, less trouble, and more regular. The impression is given by two Cylinders. The bed, or carriage upon which the type or form is placed, runs upon friction rollers, which move from one end of the press to the other on ways or ribs which extend the full length of the machine. The carriage with the form of type passes under the Cylinders by the fly-wheel. At each end of the press there is a supplying board upon which the paper is laid by tokens (ten quires at a time,) and at each board a man or boy is stationed merely to move down the blank sheets three or four inches [7.5 or 10 cm], to a gauge; when by a beautiful movement of a part of the machine, called the feeding bar, which unlocks and falls at a given moment, the sheet is held to the cylinder until it is taken by tapes and cords, running upon twenty-two brass reels (eleven to each cylinder,) and carried by them close round the cylinder to the form, over which the cylinder passes and gives the impression. Then the sheet is lifted by a cord in the centre, and one on each margin of the paper at the sides, clear of the type, and carried by another cord, called the delivering cord, alternately at each end of the machine, to two little boys, who are stationed underneath the supplying boards to receive the sheets and lay them even. These cords can be shifted at any moment so as to suit newspapers of any size. The cylinders are made to rise and fall by slides, with springs at the bottom of each side, to ease the impression.—The moment the type are passing under the cylinder, it is brought down by means of levers, which are continually changing their position by the action of a universal joint, having a wheel on one end, and playing upon a rack under the form, to which the bed or carriage with the types is fastened; and this pinion wired, being alternately on the top and bottom of the rack, produces the change in the levers, and acts upon the slides, which causes the rising and falling of the cylinders. The moment a cylinder has printed a sheet, it raises and is held in its

🔖 *The Sun*, no. 1 (3 September 1833)

The first issue of *The Sun* newspaper, priced at one penny.

🔖 *Morning Herald*, vol. 1, no. 1 (New York, 6 May 1835)

The first issue of what was later named the *New York Herald* newspaper, a penny paper competing with *The Sun*.

position sufficiently high to let the type pass under it, to go to the other, which has got its sheet closely fastened in its surface by the tapes, and is waiting for the type to perform its operation; and then immediately rises, to let the bed return to the other end; and so, rising and falling alternately, throws out the sheets the moment they are printed, as fast as the paper can be supplied.[294]

Not mentioned in the above article is the fact that the *New-York American*'s new Napier machine was set up and maintained by Robert Hoe I, founder of the famous R. Hoe & Co., which would go on to dominate the manufacture of printing equipment in America in the nineteenth century. Hoe, an Englishman, emigrated to the United States in 1803 and two years later partnered with two brothers, Matthew and Peter Smith, to establish Smith, Hoe & Company, manufacturers of printers' equipment in New York City. The company prospered and was responsible for several innovations, including the introduction of the cast-iron frame to replace the wooden frame in handpresses. After the deaths of his partners, Hoe renamed the firm R. Hoe & Co., and in 1829, he began working on his own version of the Napier cylinder press, using the intimate knowledge of the press's engineering that he had acquired while setting up the *New-York American*'s machine. The Hoe cylinder press ended up supplanting Napier presses for newspapers in the United States.

Robert Hoe's machines supported the nascent development of American mass media. In September 1833, inspired by the enormous success of Britain's *Penny Magazine*, Benjamin H. Daly launched *The Sun*, the first successful penny daily newspaper in the United States. Within four months of its launch, *The Sun* had 4,000 subscribers—equivalent to the largest circulations of traditional newspapers in the United

From "New York Printing Machine, Press and Saw Works, R. Hoe & Co.," *Graham's Magazine*, Philadelphia, vol. XL, no. 6 (June 1852), p. 566

The R. Hoe & Co. factory buildings on Broome Street, New York City, in 1852.

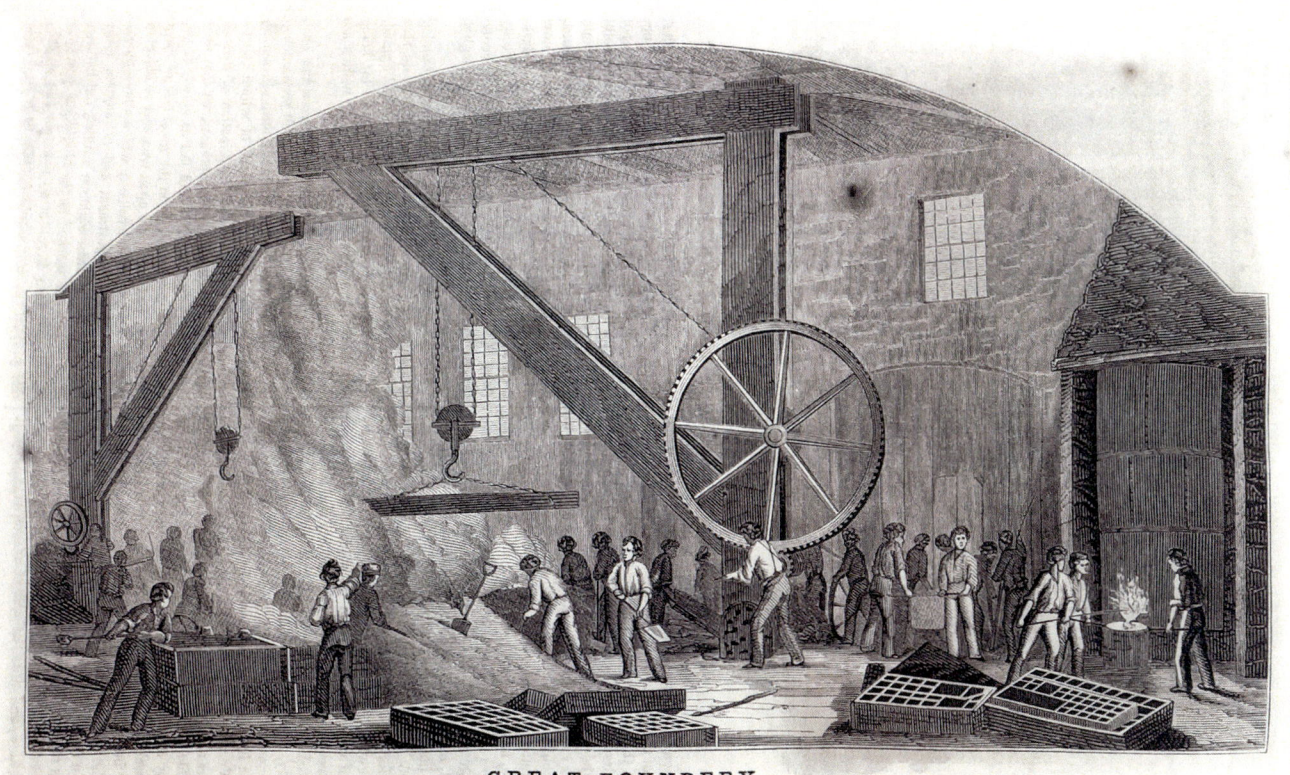

GREAT FOUNDERY.

From *Graham's Magazine*, vol. 41 (1852), p. 572

The "Great Foundery" [sic] at the R. Hoe & Co. factory. It was necessary for Hoe to produce all or most of the iron parts for their printing machines, including the giant gear in the center of the image.

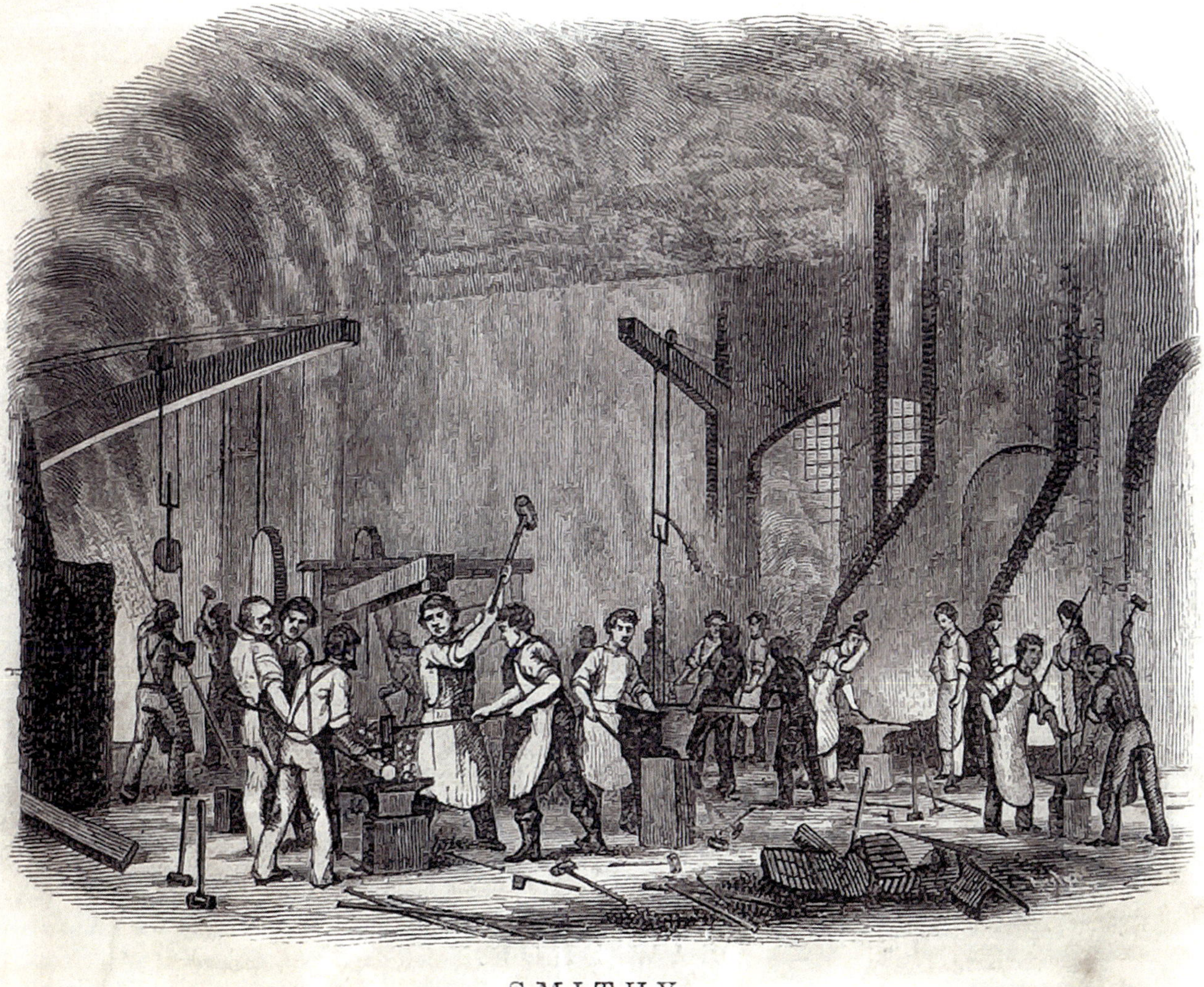

SMITHY.

From *Graham's Magazine*, vol. 41 (1852), p. 573

The "Smithy" blacksmith department at R. Hoe & Co.

States at the time—and by late 1834, it had the largest circulation in the United States, over 10,000 copies. Competing with *The Sun* was *The New York Herald*, published by James Gordon Bennett Sr., which issued its first number under the banner of *Morning Herald* on 6 May 1835. By 1845, *The Herald* had surpassed the circulation of *The Sun*, becoming the most popular and most profitable daily newspaper in the United States, with a circulation of 84,000 copies.[295] Both *The Sun* and *The Herald* printed their editions on R. Hoe & Co. printing machines.

After Robert Hoe's death in 1833, control of R. Hoe & Co. passed to his son, Richard March Hoe, who had joined the firm at the age of fifteen. Between 1843 and 1847, Hoe invented a rotary drum printing press that dramatically increased the speed of printing; it became known as the lightning press. Instead of the traditional flatbed, Hoe's press attached the type to a central cylinder, around which revolved four to ten impression cylinders. Hoe perfected the press through a series of patented advances, receiving the key patent (U.S. no. 5,199) on 24 July 1847.

Richard March Hoe's invention was introduced to the printing industry on 17 March 1847, in the offices of the *Public Ledger* in Philadelphia. When the press was installed in the *Public Ledger*, an issue of the *Philadelphia Dollar Newspaper* was printed on it as a trial; the press was then used to print the *Public Ledger*, beginning with the issue of 22 March 1847. The machine's ability to print 8,000 copies per hour revolutionized newspaper printing.

In June and July 1852, *Graham's Magazine* of Philadelphia published an extensive illustrated article titled "New York Printing Machine, Press, and Saw Works.

From *Graham's Magazine*, vol. 41 (1852), p. 10

The pattern room at R. Hoe & Co. The workman in the foreground is working out the design of a huge gear probably used in Hoe's Type-Revolving Printing Machines.

PATTERN ROOM.

LARGE FAST PRESS BUILDING.

LARGE FAST PRINTING PRESS.

🔖 From *Graham's Magazine*, vol. 41 (1852), p. 8

Construction of one of the Type-Revolving Printing Machines. The largest model, which may have been under construction in this image, measured 33 feet long and was 14 feet 8 inches high and 6 feet wide (10 × 4.5 × 1.8 m); the central cylinder was "about four and a half feet [1.4 m] in diameter."

🔖 *Graham's Magazine*, vol. 41 (1852), p. 11

One of the larger Hoe Type-Revolving Printing Machines on-site at a newspaper. This one would have required at least six people for operation. As usual, the connection to the steam engine powering the machine was not shown.

R. Hoe and Co."[296] The article illustrated Hoe's buildings and described the various features of its manufacturing facilities for saw blades, printing machines, and printing presses. The account of Hoe's printing press and printing machine works appeared in the July 1852 issue of the magazine; the images published in this portion of the article are the only series of illustrations of the elements of a printing-machine factory from the mid-nineteenth century in operation that I have been able to find. We have reproduced images depicting their buildings, the forge, the production of handpresses, and printing machines, including the eight-cylinder rotary printing machine at the *Sun* office in New York. This machine measured thirty-three feet long and was fourteen feet eight inches high and six feet wide [10 × 4.5 × 1.8 m]; the central cylinder was "about four and a half feet [1.4 m] in diameter."

The anonymous author of the article believed that the new printing technology would increase jobs rather than eliminate them, writing:

It is a pregnant fact, and one singularly corroborative of the soundness of the writer's view . . . relative to the effect of machinery increasing rather than diminishing the number of hands employed or likely to be employed in the business of printing, in consequence of the daily augmenting demand for printed matter arising from its cheapness and perfection—that, since the introduction of the fast-printing machines the call for handpresses has greatly increased.[297]

It is probable that Hoe's Four-Cylinder Type-Revolving Press made it possible for Harper and Brothers to increase the circulation of their monthly to 50,000 copies and of *Harper's Weekly* to 200,000. The article's author pointed out that *Harper's* required higher-quality printing for their magazine than was required by newspapers, which were the typical customers for these presses, so *Harper's* operated their Hoe lightning press at half its maximum speed—5,000 impressions per hour.

A footnote in Steven Tucker's *History of R. Hoe & Company, 1834–1885* provides more detail on the early success of Hoe's presses:

In 1856 the New York Correspondent of the Manchester Guardian *reported the success of this press: 'The largest presses ever built are the eight cylinders, which will throw off 20,000 sheets per hour, 333 copies per minute! These presses cost 25,000 dollars each. There are but three in existence. The first pair were built for the* Philadelphia Ledger, *a paper which circulates 80,000 daily, or more than any other daily journal in the United States. The proprietors were forced to build these fast presses in order to meet the enormous demand for their paper. Soon afterwards the New York* Sun *ordered one, which it uses in connection with a four cylinder one, and by which means it can strike off 30,000 copies every hour. The* Herald, *of this city, uses two four and one six cylinder presses, which enables it to print hourly 40,000 papers. The* Times *and* Tribune *have each a four and six cylinder; the* Commercial Advertiser *and* Post *a four cylinder; the* Boston Journal *one six cylinder;* Traveller, Times, *and* Transcript, *also of Boston, one four cylinder each; the* Baltimore Sun *two four cylinders, and the* Cincinnati Commercial *one. The Messrs. Hoe are also building a four cylinder for the* Boston Herald, *and another for the* Philadelphia Sunday Despatch. *The four cylinder press will run off 10,000 sheets an hour, and costs 12,500 dollars; the six cylinder, 15,000 sheets, and costs 18,000 dollars; and the eight cylinder, 20,000 sheets, and costs 25,000 dollars.' (Quoted in* Typographical Circular, *New Ser., XXVIII (1856), p. 233.) The cost of the eight cylinder press made for the* New York Sun *in 1849 was about $20,000.00.*[298]

The 1867 R. Hoe & Co. catalogue, with its beautifully printed chromolithographed title page, rated the speed of the various versions of the Hoe rotary press as follows:

The speed of these machines is limited only by the ability of the feeders to supply the sheets. The Four cylinder machine is run at a rate of over ten thousand per hour, the Six cylinder machine fifteen thousand an hour, the Eight cylinder machine twenty thousand, and Ten cylinder machine twenty-five thousand. This system combines the greatest speed in printing, durability of machinery, and economy of labor.[299]

The catalogue also stated that the largest version of the four-cylinder machine had to be driven by a 3-horsepower steam engine, which was also available from R. Hoe & Co.

The next major advance in high-speed, high-vol-

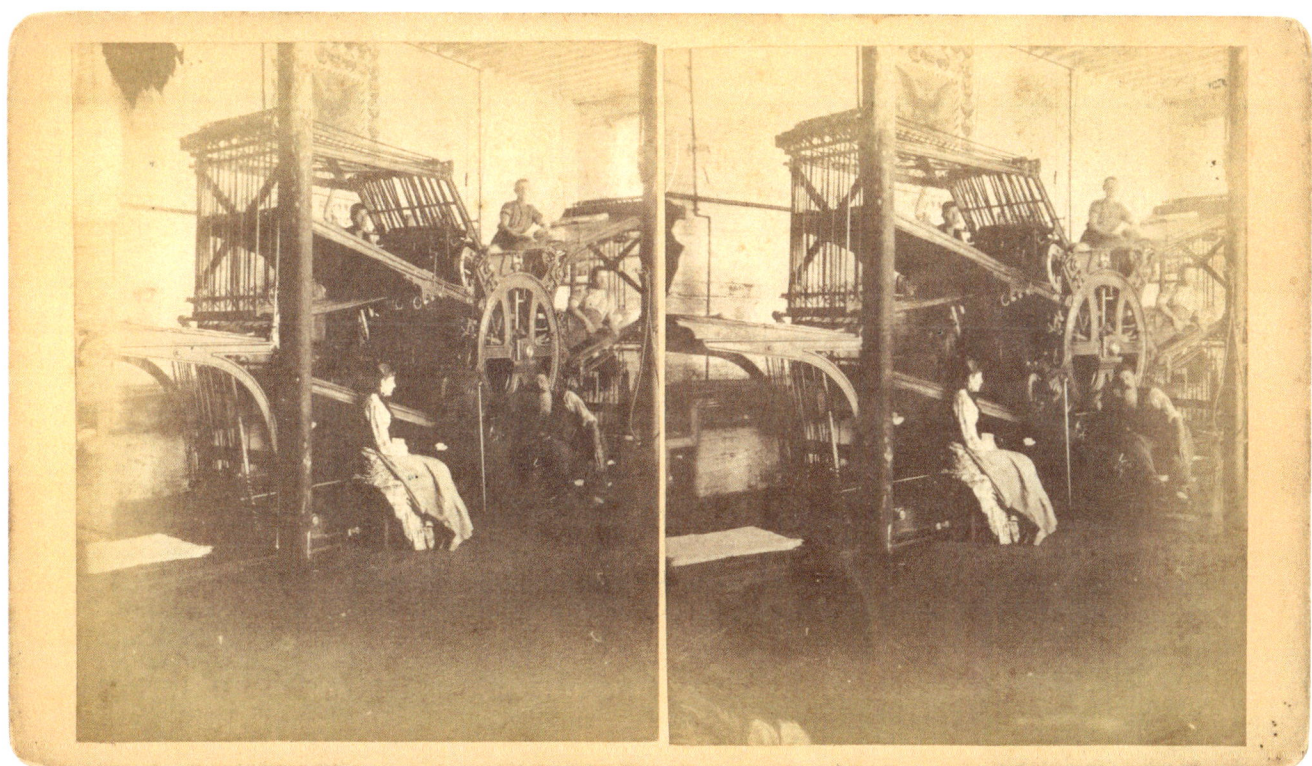

🎩🎩 R. Hoe & Co., *Printing Machines* (New York, 1867)

A Hoe Four-Cylinder Type-Revolving Printing Machine, operated by four men. The catalogue specified the size of all the components of the machine, the weight of the largest component, the weight of the machine boxed, and the space required for its installation. The machine, which could be built in eight sizes, weighed between 18,500 and 29,000 pounds and required a 2- or 3-horsepower steam engine for operation.

🎩 Stereo image: 3½ × 6⅓ in. (90 × 160 mm). Mount: 4 × 7 in. (100 × 178 mm)
Photographer or publisher not identified

A stereo photograph of Hoe's Four-Cylinder Type-Revolving Printing Machine, probably the same model as shown in the illustration from Hoe's 1867 catalogue. This may be one of the only surviving period photographs of any large printing machine from the mid-nineteenth century. The woman sitting on the machine, and wearing a full skirt, would have probably been employed to feed paper into the machine.

Type Revolving Book Perfecting Press.

As the name indicates, this is a rotary machine the forms being secured on the surface of two large horizontal cylinders. This construction, as it does away with the reciprocating motion, admits of a greater speed in printing than any other, and the distribution of the ink also is more perfect, there being room for from six to twelve inking rollers to each form. The machine is equally well adapted to letterpress, stereotype, and wood-cut work, and will print from 1500 to 2000 perfected sheets per hour, the only limit to its speed being the capability of the feeder to supply the paper. As it dispenses with the registering apparatus, and is furnished with a self-acting sheet flyer only one attendant is required for a press of the largest size.

DIMENSIONS AND PRICES.

				$
No. 1,	Bed 28 × 41 inches,	Matter 24 × 37 inches,	. . . .	
No. 2,	" 31 × 46 "	" 27 × 42 "	. . . .	
No. 3,	" 34 × 50 "	" 30 × 46 "	. . . .	
No. 4,	" 37 × 54 "	" 33 × 50 "	. . . .	
No. 5,	" 40 × 60 "	" 36 × 56 "	. . . .	

Boxing and shipping, or carting and putting-up, extra.

Grand, Broome, Sheriff, Columbia, and Gold Sts., New-York; and Dorset St., Salisbury Square, London, England.

From *Harper's Weekly*, vol. 20, no. 1041 (9 December 1876)

Portrait of Richard March Hoe, styled "Colonel" from his service in the U.S. National Guard.

🔑 R. Hoe & Co., *Printing Machines* (New York, 1867)

Color-printed title page for the 1867 catalogue.

🔑 R. Hoe & Co. *Manufactures of Type-Revolving, Perfecting, Single and Double Cylinder and Adams' Printing Machines* . . . (New York, 1873)

A Type-Revolving Perfecting Press designed for printing books. The catalogue entry emphasizes that the speed of the machine was limited only by the speed at which paper could be fed into it by hand. In this large format catalogue, Hoe did not yet include a web press, which printed from a roll of paper and was not constrained by the limitations of humans feeding sheets of paper into the machine.

ume printing for newspapers was the web press invented by William Bullock of Pittsburgh, which marked a substantial improvement upon Richard March Hoe's rotary press. Bullock's press was the first press built especially for curved stereotype plates, and the first to be automatically fed by a continuous roll of paper (web), eliminating the laborious hand-feeding system of earlier mechanized presses. It also printed on both sides of the paper at once, cutting the paper before or after printing, and could print up to 12,000 sheets per hour. Bullock filed his first patent for the machine in London, receiving patent specification no. 955 on 4 April 1862; this preceded Bullock's U.S. patent (no. 38,200) by about one year.[300]

In a bizarre accident, Bullock was killed by his own invention. On 3 April 1867, while making adjustments to one of his new web presses being installed

HOE'S "WEB PRINTING MACHINE," EXHIBITED IN MACHINERY HALL.—[Photographed by the Centennial Photographic Company.]

From *Harper's Weekly*, vol. 20, no. 1041
(9 December 1876)

A picture of Hoe's Web Printing Machine in operation at the Philadelphia Centennial Exposition. The magazine article stated that Hoe's machine could print both sides of 15,000 copies per hour of an eight-page or four-page newspaper. Lifting the extremely heavy rolls of paper to the height evidently required by this press design, as shown in the image, would have required a special mechanism.

for the *Philadelphia Public Ledger* newspaper, Bullock tried to kick a driving belt onto a pulley when the machine was running, and his leg was crushed when it became caught in the machine. After a few days, he developed gangrene and died on 12 April during an operation to amputate the leg.

In 1873, R. Hoe & Co. issued what may be the largest-format catalogue of printing machines and equipment published during the nineteenth century: *R. Hoe & Co. Manufacturers of Type-Revolving, Perfecting, Single and Double Cylinder and Adams' Printing Machines, Washington and Smith Hand-Presses, Self-Inking Machines, Etc. Every Article Connected with the art of Letter-Press, Copper-Plate, and Lithographic Printing and Bookbinding, Stereotyping and Electrotyping. . . .*[301] The catalogue was notable for its inclusion of bookbindery equipment, a department not covered in catalogues

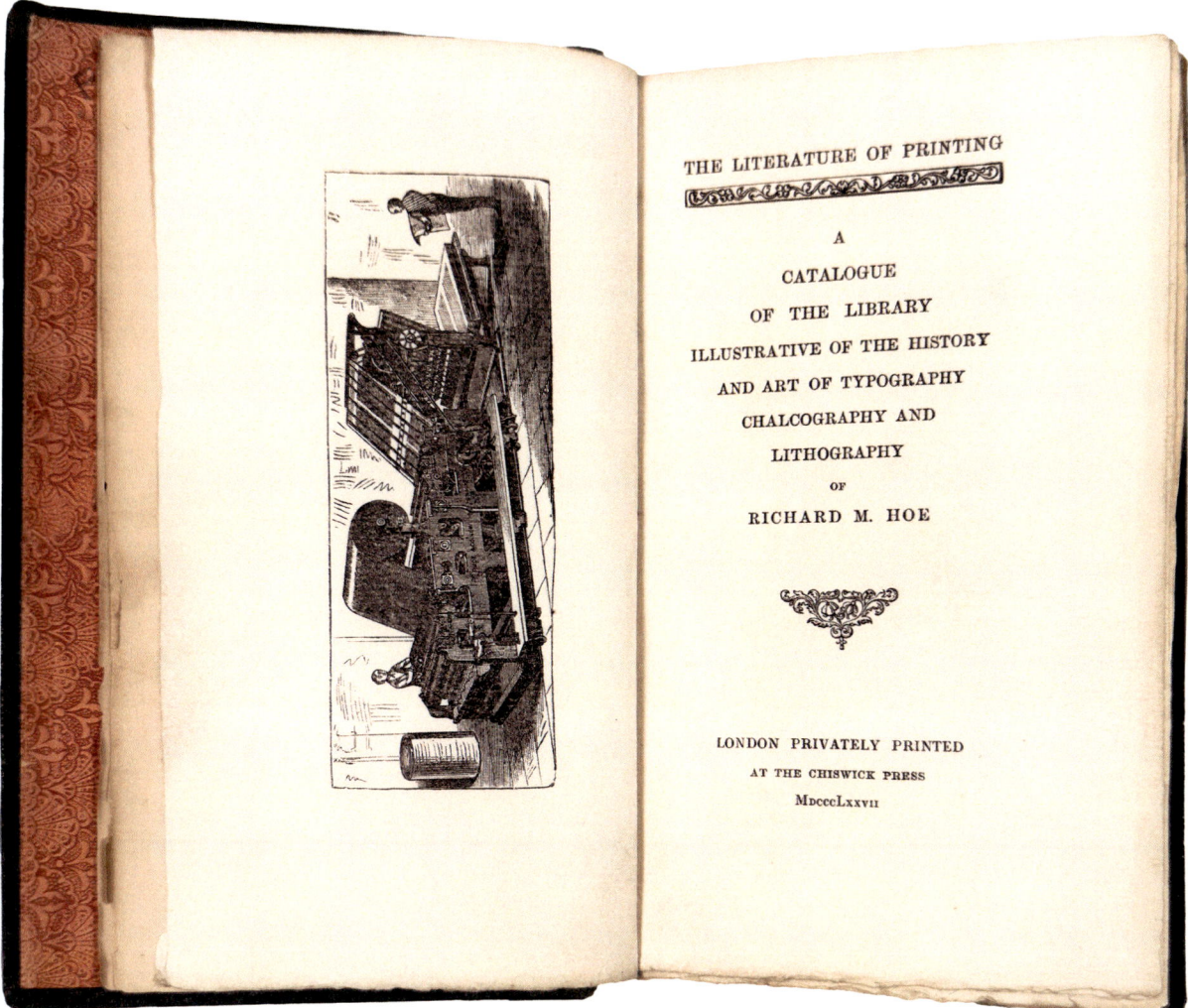

[Edward M. Bigmore,] *The Literature of Printing. A Catalogue of the Library . . . of Richard M. Hoe* (London, 1877)

Hoe published an image of his latest, fastest web press as the frontispiece to the catalogue of his historical library, which was finely printed by handpress on handmade paper. For readers understanding the contrast, the juxtaposition of the newest and fastest with the old, slow, and handmade would have been noticeable.

previously issued by Hoe that I have seen. The catalogue also included handpresses, which Hoe continued to sell late into the nineteenth century. No prices are listed; instead, spaces for prices were left blank, presumably for a salesman to fill in based upon the interests of a particular customer and whatever the company was charging for a given product at the time (presumably Hoe & Co. wanted to be able to use this large-format, clothbound catalogue for several years without having to change prices). The catalogue's introduction highlights the company's new products. Hoe & Co. also continued to offer printing machines that could be operated by hand cranks, since even by 1873, steam engines remained in short supply in parts of America.

On 9 December 1876, *Harper's Weekly* published an article about Colonel Richard March Hoe & his web printing machine, which was then being exhibited at the U.S. Centennial Exposition in Philadelphia.[302] The article described and illustrated the Hoe web perfecting press that Hoe invented and patented in 1870 and 1871. The press used a continuous roll of paper 5 miles long, which was passed through the machine at the rate of 800 feet (240 meters) a minute. As the roll emerged, it passed over a knife that cut pages apart; they were next run through an apparatus that folded the pages. These completely printed and folded newspapers were delivered "as quickly as the eye could follow them." Hoe's web press, which produced 18,000 newspapers an hour, was first used by the *New-York Tribune*. The large format catalogue Hoe published in 1873 did not include the web press; instead, on an introductory page dated January 1873, it stated, "We have also a Perfecting Web or Endless Sheet Press in

La Presse, no. 1 (1 July 1836)

The first official issue of Émile de Girardin's *La Presse*. Prior to 1 July, Girardin printed a single-page pilot issue on 15 June. The ad for Panthéon Littéraire, which may be the first display advertisement ever printed in a newspaper, appeared on the last page of the issue of 9 August 1836. The image is from Gallica, the digital library of the Bibliothèque nationale de France.

progress, nearly completed, to print with great speed and economy."

In 1877, Richard Hoe published *The Literature of Printing: A Catalogue of the Library Illustrative of the History and Art of Typography Chalcography and Lithography*. A New Yorker, Hoe had this catalogue privately printed on handmade paper at the Chiswick Press in London in a small but unspecified number of copies. Its only illustration was a frontispiece showing Hoe's high-speed web perfecting press—the latest and greatest in printing technology. Hoe's decision to have the catalogue of his historical library printed on a handpress is an excellent example of the way that traditional handcraft printing technology was retained throughout the second printing revolution.

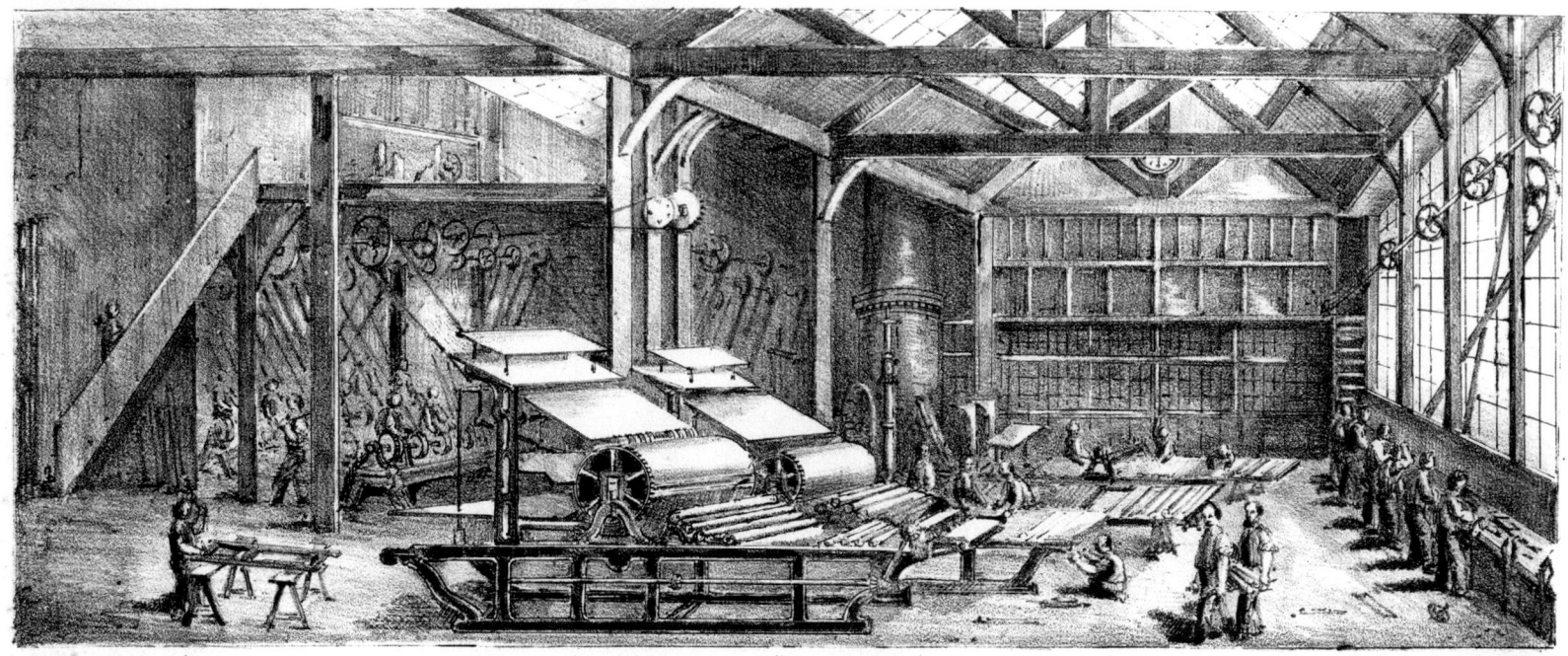

ATELIERS de la MANUFACTURE de PRESSES TYPOGRAPHIQUES de M. ALAUZET MÉCANICIEN, RUE BRÉA, 7. PARIS.

Théodore Dupuy
From *Album des célébrités industrielles contemporaines*, vol. 2
(Paris, 1865)
Lithograph

Lithograph of the printing machine factory of Dupuy's colleague and competitor P. Alauzet in Paris. The text for this advertisement for Alauzet's business claims that Alauzet's presses could print 4,000–6,000 copies per hour, and that the machines were being used to print French magazines like *L'Illustration*, *Le Magasin pittoresque*, *L'Univers illustré*, and others.

Marinoni, Machines à imprimer; Machines à vapeur [Paris, c. 1875]

An image of Hippolyte Marinoni's printing machine and steam-engine factory in Paris from an 1870s brochure printed on pink paper. At this time, Marinoni claimed to have sold 6,000 machines.

According to the *Catalogue of the William Blades Library*,[303] *The Literature of Printing* was compiled by bibliographer and antiquarian bookseller Edward C. Bigmore, coauthor, with the printer and publisher Charles W. H. Wyman, of *A Bibliography of Printing* (1880–1886). That Hoe, an American at the center of the American printing industry, chose to have the catalogue of his private library on the history and technique of printing published in London leads me to believe that he was motivated by the Caxton Quadricentennial Celebration, which occurred in London in that year; in the Caxton Celebration's catalogue, Hoe's name appears on the list of the General Committee for the celebration.[304] Because of this, I think we might reasonably speculate that Hoe wanted to show some of his fellow printing-history enthusiasts in England the treasures that he had gathered in America on the history of the subjects. Whatever Hoe's motivation, his catalogue was the first American bibliography on the history of printing and typography.

My copy of Hoe's catalogue has Hoe's signed inscription to the American minister, journalist, and politician Samuel J. Barrows, who had worked for Richard Hoe as a very young man. In skimming through the contents of the catalogue, I was intrigued to learn that Hoe owned a copy of Freylinghausen's *An Abstract of the Whole Doctrine of the Christian Religion. . . .* (1804), which the catalogue described as the "first stereotyped book." This statement was not strictly correct, but it was the first book printed by the Stanhope stereotype process; it was also the first book printed on machine-made paper, and the first book printed on the completely iron Stanhope Press— details that would have been unknown to Bigmore and Hoe.

On 15 June 1836, French journalist and politician Emile de Girardin inaugurated penny-press journalism in France with the pilot number of his new popular daily conservative newspaper *La Presse*. Girardin was not a stranger to early mass media. Just prior to Charles Knight's launch of *The Penny Magazine* in 1832, in October 1831 Girardin had launched the *Journal des connaissances utiles*, a monthly for which he claimed that 100,000 copies per month were printed. Printing

Vue des Ateliers Marinoni
Construction de Machines à imprimer

Maris Vachon, *Les Arts et les industries du papier en France 1871–1894* (Paris, 1894)

The interior of Marinoni's press factory toward the end of the nineteenth century. A wide variety of machines appear to be under construction on the factory floor. One of Marinoni's famous rotary presses is under construction in the left foreground.

100,000 copies of a monthly was far less of a production challenge than printing a high-circulation daily newspaper. The first issue of *La Presse* was published on 1 July 1836. Its subscription price of was only forty francs a year, half the cost of traditional newspapers in France. *La Presse* was the first newspaper in Europe to lower its price, to extend its readership, and to increase its profitability through advertising. It is highly likely that initially Girardin had *La Presse* printed on Applegath and Cowper machines imported from England, since at the time Koenig and Bauer were selling machines primarily in Germany, and the early French manufacturers of printing machines, Pierre Alauzet and Hippolyte Marinoni, did not start manufacturing printing machines until 1846 (Alauzet) and 1847 (Marinoni).[305] Marinoni, in collaboration with Alexandre Yves Gaveaux, designed and built his first Presse à Reaction for Girardin in 1847 and 1848.[306]

In a small brochure published after the Exposition Universelle of 1878, presumably in the early 1880s,

Hippolyte Marinoni illustrated the front of his Paris factory, where he produced printing machines and steam engines of his own design. He indicated that his company had sold 6,000 printing machines by that date. In that brochure, he illustrated the original machine that he had first designed for Girardin, calling it Machine à Journaux; this machine was still for sale twenty years after its introduction. Marinoni also illustrated his latest high-speed web press, the Nouvelle Machine Rotative, developed between 1875 and 1878; the machine printed 20,000 copies per hour. He stated that he had already sold over 100 of those web presses, chiefly in Europe, but also two in the United States. Monet, in his *Les machines et appareils typographiques en France et à l'étranger*[307] devoted pages 141-146 to that machine and reproduced images of it on plates 84 and 85. It was that particular machine for which Marinoni became especially known, to the extent that he reproduced an image of it on bearer bonds issued by his newspaper, *Le Petit Journal*, in 1896. An image of the same machine was included in the graphic design of bearer bonds issued by Librairies-Imprimeries Réunies as recently as 1964.

Moise Polydore Millaud published the first issue of *Le Petit Journal* in Paris on 1 February 1863 in an edition of 83,000 copies. Within two years, *Le Petit Journal* was printing 259,000 copies, making it the largest daily in Paris. By 1870, it had reached 340,000 copies, twice the figure for the other major dailies put together. This progress was made possible by printing machines designed by Marinoni in 1866 and installed at the *Journal* in 1872.

Despite their apparent success, the Millaud family found themselves in financial difficulties and, in 1873, sold their interests in *Le Petit Journal* to a group headed by Emile de Girardin. In 1882, Hippolyte Marinoni took control of *Le Petit Journal*, succeeding Girardin. In 1884, Marinoni introduced the *Supplément Illustré*, a weekly Sunday supplement that was the first to feature color illustrations. This became so popular that, in 1889, Marinoni developed a color rotary press that could print 20,000 sheets per hour. By 1895, one million copies of the supplement were being printed every week and *Le Petit Journal* had a press run of two million copies, 80 percent of which went to the provinces, making it, for a time, France's predominant newspaper and making Marinoni the high-profile owner of one of the early mass-media empires.

By this time, Marinoni had developed a huge business supplying rotary presses for newspapers. In a catalogue of his printing machines issued in 1886, Marinoni noted that he had sold 8,500 machines up to that date, and stated that he had sold his presses to the following countries: Germany (3), England (80), Australia (2), Austria (7), Belgium (9), Brazil (4), Canada (4), Chile (1), Spain (9), France (97), Hungary (1), (Italy (12), Netherlands (2), Argentina (5), and Russia (3). Of the early entrepreneurs of mass media, Marinoni, who began his career as a mechanician, was an exception in owning both the print media and the means of their production. To a certain extent, he can be considered a precursor of twenty-first-century internet entrepreneurs and corporations such as Apple that also produce digital entertainment.

[Hippolyte Marinoni,] *Le Petit Journal* (Paris, 1902)

A calendar for the year 1902, printed in color on Marinoni presses. The calendar depicts people buying and reading copies of the newspaper from vendors wearing *Le Petit Journal* hats with visors. Vendors display folded copies mounted on rods, one of which a vendor in the middle of the image is holding up to buyers riding on the roof of the horse-drawn bus. The calendar claimed that five million people per day read *Le Petit Journal* and that each week they printed a million copies of the color supplement.

Bronze medal, diameter: 2¾ in. (7 cm)

A medal issued by R. Hoe & Co. in 1900, honoring Gutenberg and the Hoe Octuple Press, which printed 192,000 four-page newspapers per hour.

CODA

William Morris, Theodore Low De Vinne, and Robert Hoe III Reflect upon Nineteenth-Century Developments in Book Production

During the second printing revolution that occurred during the nineteenth century, printing output increased nearly a thousandfold, from Joseph Moxon's seventeenth-century "token" of 200–250 copies per hour to the 200,000-plus copies per hour achieved by high-speed rotary presses. In 1890, toward the end of this second printing revolution, printer Theodore Low De Vinne, who was introduced in the prologue to this book, reflected on the remarkable developments in book production that he had personally witnessed in an essay titled "The Printing of 'The Century'."[308] In this illustrated essay, De Vinne eloquently described and illustrated advances in printing machines, paper, ink, the reproduction of illustrations, and advances in bindery machinery that had improved magazine production over the previous twenty years, to the point where production quality equaled or in certain cases exceeded what could be accomplished on a handpress.

In his initial paragraph, De Vinne referred to Charles Knight's groundbreaking *Penny Magazine*, first published in 1832, which was the first extensively illustrated high-circulation magazine, and one of the publications that launched the era of mass media.[309]

As much as the reproduction of images had improved over the nineteenth century, elements of the social organization of book production had hardly changed, according to De Vinne's account. Most printing processes were still done by men, and bind-

John Milton, *Aeropagitica. A Speech of Mr. John Milton for the Liberty of Unlicensed Printing, to the Parliament of England* (New York: The Grolier Club, 1890)

One of 325 copies, designed and printed on Holland paper by Theodore Low De Vinne. This is De Vinne's personal copy, with his "*Aere Perennius*" engraved bookplate, in a special binding by Zaehnsdorf, signed and dated 1899 on the front, full-leather, pastedown endpaper. The binding was protected by a silk-lined, straight-grained, morocco-covered dust jacket.

From *Les Amis des livres. Portraits* (Paris, 1899), [plate 36]

👉 Robert Hoe III, one of the original organizers and first president of The Grolier Club after it was organized in January 1884. The image is from Gallica, the digital library of the Bibliothèque nationale de France.

👉 Robert Hoe III's *A Short History of the Printing Press* (New York, 1902)

ing operations were primarily done by women. At this point in time, all typesetting at *The Century Illustrated Monthly Magazine* was still done by hand. De Vinne made a point of saying that women typesetters at *The Century Magazine* were paid at the same rate as men, suggesting that this practice remained an exception rather than the rule at this time.

De Vinne concluded his essay with the following paragraph:

Twenty years is but a short interval in the chronology of an art that is more than four hundred years old, but a good deal has been done for the improvement of printing between the years 1870 and 1890. Cylinder presses have supplanted hand and platen presses in printing woodcuts and large editions of fine books. Dry paper has taken the place of damp paper. In many large printing houses the appliances for dampening have been abolished, or set aside to be used only for rough and hand-made papers. Smooth-surface papers of moderate price have been introduced that take a sharper impression and show cleaner grays and more vigorous blacks than can be had from impressions on the luxurious India and Japan papers. Easy working and durable black inks are as common now as they were scarce twenty years ago. Electrotype plates are made of smooth surface, and are curved with unharmed lines, to fit the cylinders of rotary printing machines on which they produce presswork that fully meets the most exacting requirements. Last, but not least, the final pressing of the printed work, which makes a solid and shapely magazine, is done more quickly and more thoroughly by pressing in the fold than was ever done when the work was pressed in sheets. Some of these items may seem of trifling importance to the reader. Singly, they may be; collectively, they are not. Whoever compares the first number of this magazine with the latest, must

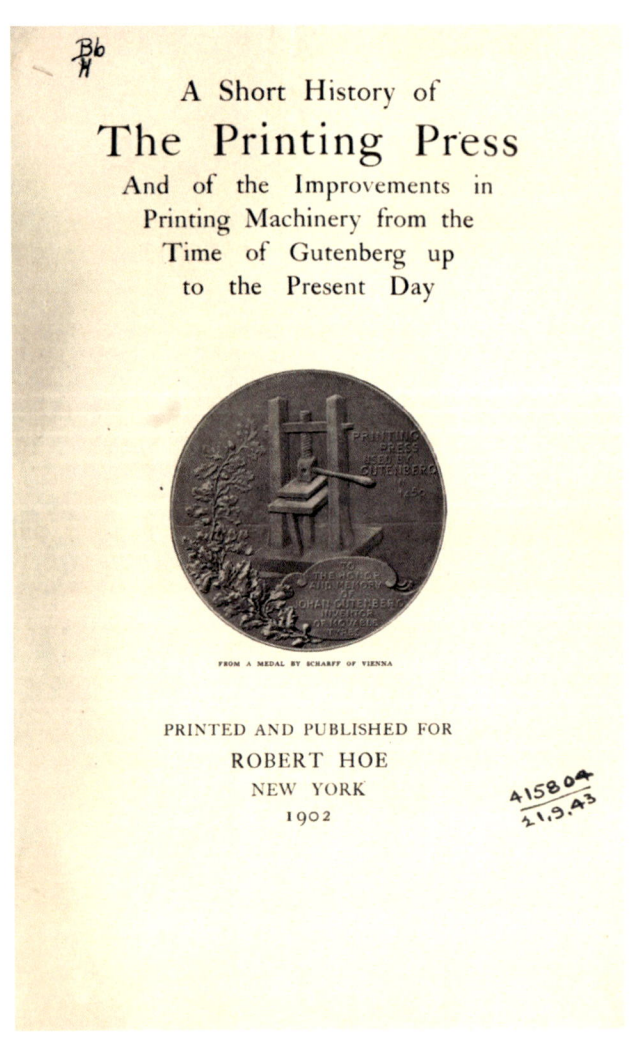

admit that decided improvements have been made in magazine printing. In the literary workshop of which John Milton dreamed, "the pens and heads, sitting by studious lamps, musing, searching, revolving new notions and ideas," were those only who thought and wrote. Now, the thinkers have mechanical helpers. In machine shops and paper mills, in printing houses and electrotype foundries, are other studious men equally busy in mechanical devices that aid the writers in realizing this dream of the "Areopagitica."

Like De Vinne, Robert Hoe III, the first president of The Grolier Club, had a keen sense of the history of printing and printing technology. A great book collector, Hoe was the owner of R. Hoe & Co., the company founded by his grandfather, Robert Hoe I, and built up by his father, Richard March Hoe. R. Hoe & Co. began producing printing machines in the 1820s and, by the mid-nineteenth century, bolstered by Richard March Hoe's extraordinary inventions, had become the leading American developer and manufacturer of printing machines of all sizes, specializing in the largest and fastest machines for newspaper production. In 1902, Robert Hoe III published *A Short History of the Printing Press and of the Improvements in Printing Machinery from the Time of Gutenberg Up to the Present Day*,[310] an illustrated review of advances in printing technology from the time of Gutenberg to the end of the nineteenth century; because so many advances in printing presses had occurred during the nineteenth century, from the invention of the Stanhope iron press onward, Hoe focused primarily on that period. On the title page of his *Short History*, Hoe reproduced a bronze medal (illustrated here), issued by R. Hoe & Co. in 1902 to honor the 500th anniversary of the birth of Johannes Gutenberg and to commemorate the Octuple Press manufactured by R. Hoe & Co., which printed 192,000 four-page newspapers per hour.

From the medal issued by R. Hoe & Co., we see the enormous size and complexity of the machine that printed 192,000 four-page newspapers per hour. By the end of the second printing revolution, high-speed printing machines of many varieties dwarfed men in their size and capabilities. This was also true with respect to the enormous Fourdrinier machines in paper mills, and even some of the larger machines in mechanized bookbinderies. To obtain faster and cheaper production, people in the book-production trades became tenders or expert operators of the machines—specialized technicians and mechanics more than individual users of smaller, human-sized tools, such as handpresses or typesetters' composing sticks, into which they placed individual pieces of type with their hands, putting their individual human stamps on the products they produced. Perhaps exceptions were individual operators of complex typesetting machines, such as Linotypes and Monotypes, who could sometimes recognize the subtle, individual results of their work on the printed page, or typographers who designed the typefaces used by the machines, or book designers, graphic designers, illustrators, and photographers, who could still see and appreciate the results of their art. Connoisseur printers like Theodore Low De Vinne appreciated the subtle gradations of machine-made paper and high-quality web press printing. Just as handpress printing was viewed as an art, there was an art to obtaining the best results from printing machines, but the manufacturing aspects of industrial printing, such as machine papermaking, high-speed printing, and machine bookbinding, had become impersonal. Manufacturing processes that had started at the beginning of the nineteenth century as handcrafts in which workers could see the products that they had made with their hands, using basic tools, had become impersonal, efficient, assembly-line productions.

Throughout all these revolutionary mechanistic advances that led to mass media, some workers who had seen the transition from handcraft to industrialization remembered the satisfaction that expert craftsmen had obtained by making things by hand or with simpler tools or machines. In 1892, English textile designer, poet, artist, and writer William Morris issued his beautiful, fine-press edition of John Ruskin's *The Nature of Gothic* from his Kelmscott Press in Hammersmith in West London. Ruskin's essay, which first appeared in the second volume of *The Stones of Venice* (1851–1853), argued that Gothic ornament, handcrafted by individual workers, was an expression of the artisan's joy in free, creative work; it became a manifesto for the Arts and Crafts movement, an international trend in art and design that rejected industrialization and mass production in favor of traditional handcraftsmanship and the use of natural materials. Morris printed his limited edition on a new Hopkinson

Robert Thom
William Morris (c. 1960)
Oil on canvas, 30 × 40 in. (76 × 102 cm)
From Kimberly-Clark, *Graphic Communications Through the Ages*
Courtesy of RIT Cary Graphic Arts Collection

and Cope Improved Albion handpress (no. 6551) built in 1891, using handmade paper and his own Golden Type, inspired by the fifteenth-century typefaces of Nicolas Jensen; hand bound in vellum, it presents a striking and dramatic contrast to the mass-produced publications from mechanized presses.

That Morris was able to produce this edition on an iron handpress of recent manufacture exemplifies the persistence of handpress printing and handcraftsmanship in books throughout the second printing revolution. This echoes what happened during the first printing revolution that occurred in the mid-fifteenth century. In mechanizing the process of producing texts for reading, Gutenberg faced resistance to innovation. His invention of printing from movable type was an enormous advance over manuscript copying, but in order to make books printed from movable type acceptable to readers steeped in the medieval manuscript tradition, fifteenth-century printed books retained the traditional medieval codex form, page format, and typefaces based on letterforms that had evolved through centuries of medieval manuscript copying. Since not all readers accepted this radically new way of producing books, manuscript copying persisted to a diminishing degree well into the sixteenth century. In the nineteenth century, the second printing revolution resulted in the development of mass media and new kinds of publications,

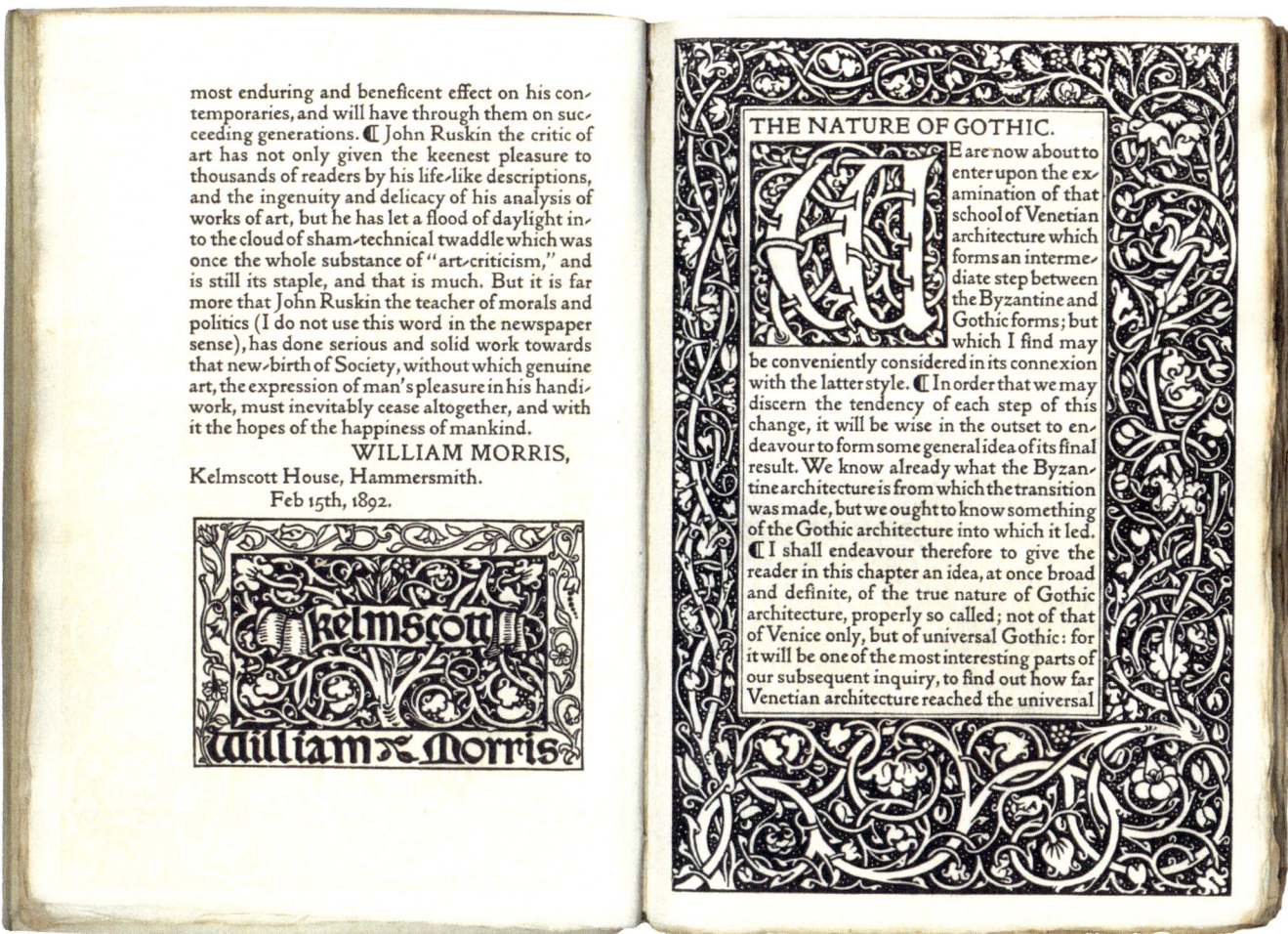

but it never completely supplanted the traditional handpress printing technologies invented in Gutenberg's time. Perhaps because handpress printing had been the standard for 350 years, handpress printing continued to be practiced throughout the nineteenth century and remains an art practiced today. Throughout the rapid media changes of the twentieth century and the present twenty-first century world of digital presses and electronic publishing, handpress printing, handmade papermaking, and hand bookbinding still persist, resulting in handcrafted volumes, also innovative, limited only by their creators' imagination.

John Ruskin, *The Nature of Gothic* (Hammersmith: Kelmscott Press, 1892)

Ruskin criticized the dehumanizing effects of industrialization, arguing for a more meaningful and fulfilling relationship between workers and their work. The book had a profound effect upon William Morris, who issued it, along with his own introduction, as one of the first publications of the Kelmscott Press. In reaction to mass-produced machine-printed books, Morris printed his fine limited editions set in type of his own design, on handmade paper, using an iron handpress, and had them bound by hand. Morris was readily able to acquire a new iron handpress when he founded the Kelmscott Press in 1891 because iron handpresses were still being manufactured at the end of the second printing revolution.

Notes

1. De Vinne's essay, originally published in *The New York Evening Post* newspaper, was reprinted in *The Nineteenth Century: A Review of Progress During the Past One Hundred Years in the Chief Departments of Human Activity* (New York: G. P. Putnam's Sons, 1901), pp. 387-400.

2. De Vinne, "Printing in the Nineteenth Century," p. 399.

3. De Vinne, "The Printing of 'The Century'," *The Century Illustrated Monthly Magazine*, vol. 41, November 1890, pp. 87-99. Circulation of the magazine peaked at 200,000 copies in 1890. For more on De Vinne, see Irene Tichenor and Michael Koenig, *The Dean of American Printers: Theodore Low De Vinne and the Art Preservative of All Arts* (2014), the catalogue of an exhibition held at The Grolier Club in 2014.

4. Richard D. Altick, *The English Common Reader*, 2nd ed. (Columbus: Ohio State University Press, 1998), p. 75: "There had never been anything like it in the history of English books. In the first six weeks (March 3-April 18, 1795) 300,000 copies of the various tracts were sold at wholesale; by July of the same year, the number had more than doubled; and by March 1796, the total number sold reached the staggering figure of 2,000,000. Two printing houses, Samuel Hazard's at Bath and John Marshall's in London, were kept working at capacity to supply the English demand alone."

5. Altick, *English Common Reader*, p. 47.

6. Edward C. Bigmore and Charles W. H. Wyman, *A Bibliography of Printing* (London: Quaritch,1880-1886), Vol. II, p. 54. Bigmore and Wyman describe a second edition of 1693-1701, and a third edition of 1703, a portion of which they say is the "fourth edition." It is not unusual for surviving sets of this very rare work to combine parts from different editions. Prior to Moxon, the first accurate illustration of a printing press appeared in *Petri Scriverii Laurecrans voor Laurens Coster van Haerlem: Eerste vinder vande boeck-druckery* (Tot Haerlem: Adriaen Rooman, 1628), p. 117.

7. William Moxon, *Mechanick Exercises on the Whole Art of Printing*, edited by Herbert Davis and Henry Carter. 2nd ed. (London: Oxford University Press, 1962), pp. 262, 484-486. James Moran, *Printing Presses: History and Development from the Fifteenth Century to Modern Times* (London: Faber, 1973), p. 32.

8. Bernardino Ramazzini, *De morbis artificum Bernardini Ramazzini diatriba. Diseases of Workers*. The Latin text of 1713, revised, with translation and notes by Wilmer Cave Wright (Chicago: University of Chicago Press, 1940), pp. 414-419.

9. Franklin's original letter has not survived. Its text is published in Volume 37 of the Franklin Papers (franklinpapers.org).

10. One of the first authors to research the history of stereotyping was Armand-Gaston Camus, the author of *Histoire et procédés du polytypage et de la stéréotypie* (Paris: Baudouin, An X [1801]). Camus was a French revolutionist who founded the Archives nationales de France.

11. According to the standard deflator, £50 in 1789 would be worth roughly £7,500 today; however, the actual value then was much higher since coal miners' wages were far lower in those days relative to the cost of living than they are today.

12. Arkwright received patent no. 931 in 1769 for "A new Machinery never before found out, practiced, or used, for the Making of Weft or Yarn from Cotton, Flax, and Wool, which would be of great Utility to a great many Manufacturers in this His Kingdom of England, as well as to His Subjects in general, by Employing a great number of Poor People in Working the said Machinery, and by Making the said Weft or Yarn much Superior Quality to any ever heretofore Manufactured or Made."

13. William Felkin, *A History of Machine-Wrought Hosiery and Lace Manufactures* (London: Longmans, Green and Co., 1867) p. 227.

14. Hargreaves's patent no. 962 for "A method of Making a Wheel or Engine of an entire New Construction (and never before made Use of), in order for Spinning, Drawing and Twisting of Cotton, and to be managed by One Person only, and that the Wheel or Engine will Spin Draw, and Twist Sixteen or more Threads at One Time by a Turn or Motion of One Hand a Draw of the other."

15. Arkwright's patent no. 1111, issued in 1775, for "Certain Instruments or Machines which would be of publick Utility in Preparing Silk, Cotton, Flax and Wool, for Spinning, and constructed on easy and simple Principles very different from any that had ever been contrived."

16. *The trial of a cause, instituted by Richard Pepper Arden, Esq; his Majesty's Attorney General, by Writ of Scire Fascias, to repeal a patent granted on the Sixteenth of December 1775, to Mr. Richard Arkwright, for an invention of certain instruments and machines for preparing silk, cotton, flax, and wool for spinning . . .* (London: Hughes and Walsh, 1785). A copy in my library is of particular interest since its final leaf bears the inscription of "Sykes and Taylor, Birkacre nearly Chorley." This was a cotton-spinning factory operated on a site owned by Richard Arkwright, or near a site owned by Arkwright, until 1788. In October 1779, Arkwright's mill at Birkacre was destroyed by Luddite activity.

17. Louis Dunham, in *The Industrial Revolution in France, 1815-1848* (1955), wrote, "It is difficult to say when the first steam engine was used in France as a source of power for running machinery. It is known that one was brought from England for use in pumping late in the eighteenth century, and another was set up near Saint-Quentin in 1803, although we do not know for what it was used. A third was set up near Mulhouse in Alsace by Dollfus-Mieg in 1812 and was used to run a cotton-spinning mill. We have reports of at least one

steam engine in use in a cotton mill at Lille in 1817, of another in a woolen mill at Louviers in the same year, and of two in woolen mills at Elbeuf in either 1819 or 1821. We are told that steam was introduced into the cotton-spinning mills of Rouen between 1817 and 1820." (p. 113).

18. 25th Congress, 3rd Session, House of Representatives, Treasury Dept., doc. no. 21. Steam-Engines. My copy is unusual for having drawings in ink of a steamboat and flying "eagle" on the upper cover and a steam engine on the lower cover. These were drawn by S. Lincoln to decorate this copy for presentation to David Thomas in 1840.

19. A facsimile of Haas's pamphlet was published by the Gutenberg Museum in Bern in 1955.

20. Moran, *Printing Presses*, p. 41.

21. "Premier mémoire sur l'impression en lettres, suivi de la description d'une nouvelle presse," *Mémoires de mathématiques et de physique des Scavans Etrangers*, vol. × (1785), pp. 613-650.

22. British patent no. 1431 for "A Method of Making Plates for the Purpose of Printing by or with Plates instead of the Moveable Types commonly used, and for Vending and Disposing of the said Printing Plates and the Books or other Publications therewith Printed, whereby a much greater degree of Accuracy, Correctness, and Elegance will be introduced in the publication of the Works both of the Ancient and Modern Authors than had hitherto been attained."

23. David McKittrick, *A History of Cambridge University Press. Volume 2: Scholarship and Commerce 1698-1872* (Cambridge: Cambridge University Press, 1998), pp. 272-273.

24. Simon Eliot, *The History of Oxford University Press. Vol. II: 1780-1896* (Oxford: Oxford University Press, 2013), p. 383.

25. Moran, *Printing Presses*, pp. 49-57. Horace Hart, *Charles Earl Stanhope and the Oxford University Press*, reprinted from *Collectanea* 111 (1896) of the Oxford Historical Society with notes by James Mosley (London: Printing Historical Society, 1966).

26. Dard Hunter, *Papermaking: The History and Technique of an Ancient Craft* (New York: Dover Publications, 1978), p. 341. Labor problems were particularly pervasive in the French hand papermaking industry throughout the eighteenth century, as Leonard Rosenband has documented in his history of the Montgolfier paper mill at Vidalon-le-Haut. Besides the predictable disputes between workers and their masters, poaching of skilled workers from one paper mill by the owner of another was common, causing interference with production. To address these issues, the French government passed many laws in the eighteenth century regulating the papermaking process—dealing with the industry's labor issues, standardizing the paper sizes produced in the mills, and setting the start of a papermaker's workday at no earlier than three a.m. See Leonard N. Rosenband, *Papermaking in Eighteenth-Century France: Management, Labor and Revolution at the Montgolfier Mill, 1761-1805* (Baltimore, Maryland: Johns Hopkins Press, 2000).

27. Colin Cohen and Geoffrey Wakeman, *The Art of Making Paper Taken from the Universal Magazine* (Loughborough: The Plough Press, 1978).

28. These portions were reprinted with an introduction by Colin Cohen and Geoffrey Wakeman and issued in an edition limited to 200 copies by the Plough Press, Loughborough in Leicestershire, England, 1978. The complete work was first translated into English by Richard MacIntyre Atkinson more than 200 years after its original publication, in a splendid full-size edition limited to 405 copies (Kilmurry [Ireland]: The Ashling Press, 1976). This new English translation included all the plates, printed on blue handmade paper made by Ashling Papermakers.

29. Denis Diderot and Jean le Rond d'Alembert, *Recueil de planches sur les sciences, les arts libéraux et les arts mécaniques, avec leur explication. Imprimerie-relieure* (Paris: Inter-Livres, 2001).

30. Rosenband, *op. cit.* p. 4.

31. Rosenband, *op. cit.* p. 122.

32. Leonard N. Rosenband, "Comparing Combination Acts: French and English Papermaking in the Age of Revolution," *Social History*, vol. 29, no. 2 (May 2004), pp. 165-185. The citation is from pp. 180-182.

33. Joel Munsell, *Chronology of the Origin and Progress of Paper and Paper-Making*, 5th ed. (Albany, New York: J. Munsell, 1876), p. 62.

34. R. H. Clapperton, *The Paper-Making Machine. Its Invention, Evolution and Development* (Oxford: Pergamon Press, 1967), pp. 12-13.

35. Jean Abbot Dusautoy, *Paper-Maker's Ready Reckoner, or Calculations to Shew the Prime Cost of Writing or Printing Paper* (Romsey: Printed by J. S. Hollis, and sold by the Author, 1805).

36. The Consortium of Online Public Access Catalogues locates copies at the British Library, Edinburgh University, and Glasgow University. OCLC (was the Online Computer Library Center) adds a copy in the Netherlands; no copies in American libraries are recorded.

37. Dickinson received British patent no. 3191 for "Certain Improvements on my former Patent Machinery for Cutting and Placing Paper, and also certain Machinery for the Manufacture of Paper by a new Method."

38. Clapperton, *The Paper-Making Machine*, pp. 65-77.

39. Coleman, *The British Paper Industry 1495-1860* (Oxford: Clarendon Press, 1958), p. 91.

40. In 1827, there were four papermaking machines in France. In 1833, the number exceeded twelve, according to Albert Prouteaux, *Practical Guide for the Manufacture of Paper and Boards, with additions by L. S. Le Normand*. Translated from the French with notes by Horatio Paine (Philadelphia: Henry Carey Baird, 1866), p. 19. For the most part, these machines were built in England.

41. Louis Andre, *Machines à papier: Innovation et transformations de l'industrie papetière en France 1798-1860.* (Paris: Éditions de l'École des Hautes Études en Sciences Sociales, 1996), p. 89.

42. Dard Hunter, *Papermaking: The History and Technique of an Ancient Craft*, 2nd ed. (New York: Knopf, 1957), p. 353.

43. John Anastasius Freylinghausen, *An Abstract of the Doctrine of the Christian Religion. With Observations. From a Manuscript in Her Majesty's Possession . . .* (London: John Wilson, 1804).

44. [Richard Gough,] "Review of new publications," *The Gentleman's Magazine*, vol. 75 (1805), p. 250.

45. Joseph Wilson, *A History of Mountains* (London: Nicol, 1807-1810). Coincidentally, Wilson's book was printed in London by Thomas Bensley, who, on 31 March 1807, would become the chief financier of the development of Friedrich Koenig and Andreas Bauer's first steam-powered press, which would become fully operational in 1814.

46. British Library shelfmark: Maps M.T.6f.1.(1). The cataloguing of that copy reads, "[A composite design, in aquatint, showing comparative heights of mountains.] R.A. Riddell pinxit. Engraved by J. Merigot of Paris. With a scale to measure the height of the mountains depicted]. London: R.A. Riddell, 1806. 1295 × 908 mm."

47. Wilson, *A History of Mountains*, vol. 1, p. xvii.

48. House of Commons, *Report from the Select Committee on Fourdrinier's Patent; with the Minutes of Evidence, and Appendix.* London: House of Commons, 1 June 1837.

49. House of Commons, *Report from the Select Committee on Fourdrinier's Patent*.

50. David Waller, *Iron Men: How One London Factory Powered the Industrial Revolution* (London: Anthem Press, 20), p. 35.

51. Historians generally date the beginning of the Industrial Revolution in Germany to the period from around 1815 to 1835, with the main period of industrial development in the country occurring in the decades between 1830 and 1873. At the end of the Napoleonic Wars (1803-1815) and the lifting of Napoleon's Continental Blockade (1806-1814), trade barriers fell, but the German economy was then exposed to direct competition with British industry, which was increasingly becoming mechanized at a rate far faster than was occurring in Germany. The first modern factories did not emerge in Germany until near the end of the eighteenth century. In 1784, the first mechanical cotton spinning mill, Textilfabrik Cromford, went into operation in Ratingen under waterpower, and a year later, the first steam engine in the German mining industry went into operation in Hettstedt, seventy years after the first Newcomen steam engine began operation in Britain. It was not until 1796 that the first continuously producing coke oven was built at the Royal Prussian Iron Foundry in Gleiwitz, about eighty years after that technology had originated in England (coke was used as a fuel and as a reducing agent in smelting iron ore in blast furnaces—a step in iron produc-

tion). In 1798, C. F. Bernhardt's spinning mill was founded in Chemnitz-Harthau, decades after Arkwright's factories in England. Most of the early factory-like operations in Germany prior to the 1820s did not use steam power.

52. Koenig und Bauer AG, *175 Years, 1817-1992* (Würzburg: Koenig und Bauer, 1992), pp. 8-9.

53. Otto M. Lilien, *History of Industrial Gravure Printing up to 1900* (London: Autotype Co., 1957). Quoted in Moran, *Printing Presses*, p. 174.

54. Nicholson received British patent no. 1748 in 1790 for "A Machine or Instrument on a New Construction for the Purpose of Printing on Paper, Linen, Cotton Woolen and other Articles in a more Neat, Cheap, and Accurate Manner than is effected by the Machines now in use." Nicholson's specification contains several drawings: "In the first drawing, which is the outline of a hand-press A is the impression cylinder in gear with and driving the carriage HI to and fro. B is the inking cylinder, with distributing rollers; these take their ink supply from the 'ink block' (duct) at O as this advances with the carriage. In the second drawing, which shows three cylinders vertically arranged, B is an inking cylinder with distributors and an ink duct; A is a cylinder 'having the letter imposed upon it surface'; E is the impression cylinder."

55. Moran, *Printing Presses*, p. 102.

56. *Repertory of Arts and Manufactures* 5, no. XXVII (1796), pp. 145-170.

57. William Nicholson, "Observations on the Art of Printing Books and Piece Goods by the Action of Cylinders," *Journal of Natural Philosophy, Chemistry, and the Arts* (1797), pp. 18-23.

58. Moran, *Printing Presses*, p. 105.

59. Moran, *Printing Presses*, p. 105. Moran suggests the output might have been 800 sheets per hour.

60. Moran, *Printing Presses*, p. 106.

61. Friedrich Koenig and Andreas Friedrich Bauer, *Die Ersten Druckmaschinen . . .* (The first printing machines . . . (Leipzig: Druck von F. M. Brockhaus auf einer Schnellpresse von Koenig und Bauer, 1841), p. 5.

62. Moran, *Printing Presses*, p. 107.

63. The first detailed English-language historical account of Koenig's development of the steam or machine press was Samuel Smiles's "Friedrich Koenig: Inventor of the Steam-Printing Machine," published in *Macmillan's Magazine* in December 1869. Smiles expanded this article and reissued it as chapter 6 of his *Men of Invention and Industry* (London: John Murray, 1884).

64. Charles Knight, *Passages of a Working Life* (London: Bradbury and Evans, 1864), vol. I, p. 163.

65. W. H. Brock and A. J. Meadows, *The Lamp of Learning: Taylor & Francis and the Development of Science Publishing* (London and Philadelphia: Taylor and Francis, 1984), p. 60.

66. Koenig and Bauer, *175 Years, 1817-1992*, p. 12.

67. Koenig and Bauer, *175 Years, 1817-1992*, p. 13.

68. Koenig and Bauer AG History, https://www.fundinguniverse.com/company-histories/koenig-bauer-ag-history/.

69. *The London Literary Gazette and Journal of Belles-Lettres, Arts, Sciences &c.* 2, no. 50 (3 January 1818), p. 1.

70. *The London Literary Gazette and Journal of Belles-Lettres, Arts, Sciences &c.* 2, no. 51 (10 January 1818), pp. 23-24. The quotation is from p. 23.

71. J. F. Blumenbach, *The Institutions of Physiology . . . Translated from the Latin of the Third and Last Edition* (London: Bensley and Son for E. Cox and Son . . . , 1817), p. iii.

72. Knight, *Passages of a Working Life*, vol. 1 (1864), p. 163.

73. Publius Vergilius Maro, *P. Virgilii Maronis Opera* (London: Impensis F. C. et J. Rivington, 1822).

74. Gillian Cookson, *The Age of Machinery: Engineering the Industrial Revolution, 1770-1850* (Woodbridge, England: The Boydell Press, 2018), p. 32.

75. Harriet Martineau, *A Manchester Strike* (London: Charles Fox, 1832).

76. Peter Gaskell, *Artisans and Machinery: The Moral and Physical Condition of the Manufacturing Population Considered with Reference to Mechanical Substitutes for Human Labor* (London: John W. Parker, 1836), p. 361.

77. David McKitterick, *A History of Cambridge University Press. Vol. 2: Scholarship and Commerce, 1698-1872* (Cambridge: Cambridge University Press, 1998), p. 333.

78. Simon Eliot, *A History of Oxford University Press. Vol. 2: 1780-1896* (Oxford: Oxford University Press, 2013), p. 149.

79. Alexander J. Dick, "On the Financial Crisis, 1825-26." BRANCH: *Britain, Representation and Nineteenth-Century History*. Edited by Dino Franco Felluga. Extension of Romanticism and Victorianism on the Net. Web, accessed 31 July 2025.

80. The three volumes of *The Menageries* were published from 1829 to 1840.

81. Knight, *Passages of a Working Life*, vol. 2, pp. 114-115.

82. In 1832, over 75% of school-age children in England were illiterate, and another 300,000 did not attend school during the first year of publication of the magazine. Similarly, in 1841, over 30% of males and nearly 50% of females remained illiterate.

83. F. H., "The Penny Press," *The Spectator*, no. 475 (15 September 1832), p. 387.

84. Pierre Albert, "*Le Journal des connaissances utiles de Girardin* (1831-1836 . . .) *ou la première réussite de la presse à bon marché*," *Revue du Nord*, vol. 66 (1984), p. 741.

85. *Chambers's Information for the People* (London: Orr and Smith, 1835), vol. 1, p. 279.

86. J. S. Folds, *The Dublin Penny Magazine*, vol. 1, no. 28 (5 January 1833), p. 222.

87. The first supplement, "Introduction and Paper-making," appeared in issue 96 of *The Penny Magazine* (31 August-30 September 1833); the second, "Wood-cutting and Type-founding," in issue 101 (29 September-31 October 1833); the third, "Compositors' Work and Stereotyping," in issue 107 (31 October-30 November 1833); and the fourth, "Printing Presses and Machinery—Bookbinding," in issue 112 (30 November-31 December 1833).

88. Charles Knight, "The Commercial History of a Penny Magazine. No. IV.—Printing presses and machinery.—Bookbinding," *Monthly Supplement of the Penny Magazine of the Society for the Diffusion of Useful Knowledge* (30 November-31 December 1833), p. 509.

89. Charles Knight, "The Commercial History of a Penny Magazine. No. IV" (30 November-31 December 1833), p. 510.

90. Charles Knight, *William Caxton, the First English Printer: A Biography* (London: Charles Knight, 1844).

91. According to an advertisement published in *Bent's Literary Advertiser* in January 1848, the series had by then extended to 116 volumes, with each volume available in printed wrappers for 1s,, in cloth for 1s. 6d., in morocco, gilt edges for 2s. 6d. each, in double vols., g.e. for 3s. each, and in treble vols., g.e. for 4s. each.

92. Knight, *William Caxton*, pp. 237-238.

93. Francis Bond Head, "The Printer's Devil," *Descriptive Essays Contributed to the Quarterly Review* (London: John Murray, 1857), vol. 1, pp. 257-306; the above quotation is from pp. 274-277. The reprint does not indicate that Head's essay was a review of Charles Knight's publications. See Sarah Wadsworth, "Charles Knight and Sir Francis Bond Head: Two Victorian Perspectives on Printing and the Allied Trades," *Victorian Periodicals Review* 31 (1998), pp. 369-386. For an analysis of the economic writings of Charles Knight, see William F. Kennedy, "Lord Brougham, Charles Knight, and the Rights of Industry," *Economica*, New Series, 29 (1962), pp. 58-71.

94. Michael Twyman, *Lithography 1800-1850* (London: Oxford University Press, 1970) pp. 26-27.

95. Twyman, *Lithography 1800-1850*, p. 30. See also Basil Hunnisett, *Engraved on Steel: The History of Picture Production Using Steel Plates* (Aldershot: Ashgate, 1998), pp. 110-111.

96. *Analectic Magazine*, vol. 14 (1819), facing p. 67. The lithograph, signed "Bass Otis Lithographit," illustrates an article on lithography (pp. 67-73).

97. *The Illustrated London News* vol. 28, no. 797 (3 May 1856), p. 488.

98. Jean Baptiste Huguet, Specification no. 1623 for "An Improved Lithochromolitho Typographic Press" (London, 1864).

99. M. Knecht, *Nouveau manuel complet du dessinateur et de l'Im-*

primeur lithographe, nouvelle edition (Paris: Roret, 1867).

100. *Album des célébrités industrielles contemporaines . . . second volume* (Paris: E. Lacroix, 1865).

101. A discussion of this issue occurs in Peter Lanchidi, "Julius Bien and the Metamorphosis of a Kabbalistic-Masonic Lithograph (New York, 1859)," *Journal of the Printing Historical Society*, New Series, vol. 31, Winter 2019, pp. 85-119.

102. Perkins's patent was titled "Certain Machinery and Implements Applicable to Ornamental Turning and Engraving, and to the Transferring of Engraved or Other Work from the Surface of One Piece of Metal to another Piece of Metal, and to the Forming of Metallic Dies and Matrices; and also Improvements in the Construction and Method of Using Plates and Presses for Printing Bank Notes and other Papers, whereby the Producing and Combining various Species of Work is effected upon the same Plates and Surfaces, the Difficulty of Imitation increased, and the Process of Printing facilitated; and also an Improved Method of Making and Using Dies and Presses for Coining Money, Stamping Medals, and other Useful Purposes." The patent included six large, folding engineering drawings.

103. Baker Perkins Historical Society, "History of Jacob Perkins in the Printing Industry" (n.d.), http://www.bphs.net/GroupFacilities/J/JacobPerkinsPrinting.htm, accessed 06-24-2012.

104. Hunnisett, *Engraved on Steel*, p. 112.

105. Adams's work is discussed in chapter 9 of this book, "The Development of Mechanized Printing in America."

106. Nicolas Barker, "Macmillan: Or, 'tis sixty years since." In *Macmillan: A Publishing Tradition*, edited by Elizabeth James (Houndmills: Palgrave, 2002), pp. 256-265; the quote is from p. 262.

107. Geoffrey Wakeman, *Victorian Book Illustration: The Technical Revolution* (Newton Abbott: David and Charles, 1973), pp. 161-163.

108. Moran, *Printing Presses*, p. 127.

109. *Printing and the Mind of Man: Catalogue of the Exhibitions and the British Museum and Earl's Court 16-27 July 1963* [1963], no. 408. The historic *Printing and the Mind of Man* exhibition, of which this was the exhibition catalogue, included a major illustrated section on the development of printing technology, including an exhibition of handpresses and printing machines. This aspect of the history of book production was eliminated from John Carter and Percy H. Muir's *Printing and the Mind of Man* (New York: Holt, Rinehart, and Winston, 1967), by which the historic exhibition is chiefly cited and remembered.

110. The full title of Hansard's book is *Typographia: An Historical Sketch of the Origin and Progress of the Art of Printing; with Practical Directions for Conducting Every Department of an Office; with a Description of Stereotype and Lithography, illustrated by Engravings, Biographical Notices, and Portraits* (London: Printed for Baldwin, Cradock, and Joy, 1825). It consists of about 950 pages in large-octavo format, elegantly designed and produced on excellent paper, and superbly illustrated. Hansard does not say that he printed *Typographia* on the printing machine that he designed, or on any other printing machine. Therefore, it was probably printed on a handpress, much as standard editions of books continued to be printed for decades after mechanized printing was invented. It is also likely that the original printing of Hansard's book was relatively large, since the original edition remains a relatively common book on the market to this day.

111. Hansard, *Typographia*, p. 548.

112. Hansard, *Typographia*, p. 710.

113. In *The London Literary Gazette and Journal of Belles Lettres, Arts, Sciences, &c.*, no. 301 (26 October 1822), p. 671. This was relatively early for an illustration in a publication printed on a printing machine. Cowper reproduced the same image once again in a paper published in 1828.

114. Hansard, *Typographia*, pp. 665-677.

115. William Church, "An Improved Apparatus for Printing," British patent specification no. 4664.

116. Charles Babbage, *A Letter to Sir Humphry Davy, Bart . . . on the Application of Machinery for the Purpose of Calculating and Printing Mathematical Tables* (London: J. Booth; Baldwin, Cradock and Joy, 1822).

117. [Richard A. Austen-Leigh,] *The Story of a Printing House, Being a Short Account of the Strahans and Spottiswoodes* (London: Spottiswoode and Co., Ltd.,1912), p. 36.

118. Samuel Smiles, *Men of Invention and Industry* (1884), chapter 8: "William Clowes: Book-printing by Steam."

119. Smiles, *Men of Invention and Industry*, p. 218.

120. Moran, *Printing Presses, p.* 140.

121. Maurice Ernst Audouin de Geronval, *Manuel de l'imprimeur* (Paris: Crapelet, 1826). Audouin's manual, printed in duodecimo format, was an early volume in the series of pocket-sized Manuels-Roret, launched by Parisian publisher Nicolas-Edme Roret in 1825.

122. Martyn Lyons, *Le triomphe du livre: Une histoire sociologique de la lecture dans la France du XIXe siècle* (Paris: Promodis, 1987), p. 43.

123. *Proben aus der Schriftgiesserey der Andreäischen Buchhandlung im Frankfurt am Main*, 1834, facsimile edition published by the Technischen Hochschule Darmstadt, 1984.

124. Thomas Hansard must have been enthusiastic about the Cogger press, since in *Typographia* he included a spectacular folding plate of the press (facing p. 655) showing its operation by a crank, but that press was not commercially successful.

125. James Moran, *Printing Presses . . .* (Berkeley & Los Angeles: University of California Press, 1973) fig. 57 and p. 129.

126. Edward Cowper, "On the Recent Improvements in the Art of Printing," *The Quarterly Journal of Science, Literature, and Art* (January to June 1828), pp. 183-191.

127. Andrew Ure, *A Dictionary of Arts, Manufactures, and Mines: A Containing a Clear Exposition of Their Principles and Practice* (London: Longman, Orme, 1839).

128. *Institution of Civil Engineers. Minutes of Proceedings, vol. 9* (1850), pp. 409-427.

129. William Savage, *A Dictionary of the Art of Printing* (London: Longman, Brown, Green, and Longmans, 1841), pp. 448-467.

130. Savage, *A Dictionary of the Art of Printing*, p. 465.

131. Savage, *A Dictionary of the Art of Printing*, p. 467.

132. Charles H. Timperly, *The Printer's Manual* (London: H. Johnson, 1938), p. 94.

133. John Southward, *Dictionary of Typography and Its Accessory Arts*, 2nd ed. (London: Joseph M. Powell, 1875).

134. Southward, *Dictionary of Typography and Its Accessory Arts*, p. 80.

135. Robert L. Patten, *Charles Dickens and His Publishers*, 2nd ed. (Oxford: Oxford University Press, 2017), p. 56.

136. Jane E. Chadwick, *Bradbury and Evans: An Inky Tale*, 2018, www.aninkytale.co.uk.

137. Jane E. Chadwick, "Printers for Chapman and Hall." *Printers for Chapman and Hall: An Inky Tale*, 13 August 2018, www.aninkytale.co.uk/Printers_for_Chapman_and_Hall.html#Printers%20of%20the%20Novels%20of%20Dickens.

138. Robert L. Patten, *Charles Dickens and 'Boz': The Birth of the Industrial-Age Author* (Cambridge: Cambridge University Press, 2012), p. 136.

139. Patten, *Charles Dickens and 'Boz'*, pp. 135-136.

140. F. G. Kitton, *The Minor Writings of Charles Dickens: A Bibliography and Sketch* (London: Stock, 1897), p. 79.

141. R. L. Patten, *Charles Dickens and His Publishers*, pp. 57-66; J. B. Podeschi, *Dickens and Dickensiana: A Catalogue of the Richard Gimbel Collection in the Yale University Library* (New Haven, Connecticut: Yale University Press, 1980), pp. 302-303.

142. Price One Penny. A Database of Cheap Literature, 1837-1860, http://www.priceonepenny.info/database/show_periodical.php?periodical_id=27.

143. "A Christmas ghost story. Re-originated from the original by Charles Dickens, Esq., and anyulitically, [sic] condensed expressly for this work," *Parley's Illuminated Library*, vol. 1, issue 16 (1844), pp. 241-256.

144. See M. Hancher, "Grafting *A Christmas Carol*," *Studies in English Literature 1500-1900*, vol. 48 (2008), p. 814 and footnote 8.

145. M. Wills, "Pirating Charles Dickens' A Christmas Carol, in the 1840s." *JSTOR Daily*, 16 December 2019.

146. *Parley's Illuminated Library*, no. 17 (13 January 1844), p. 301.

147. Hancher, "Grafting *A Christmas Carol*," p. 816.

148. Patten, *Charles Dickens and His Publishers* (2012), p. 191.

149. The autograph letter by Dickens is in my private collection.

150. *The Holy Bible, Containing the Old and New Testaments, Translated out of the Original Tongues . . . Stereotype Edition* (Philadelphia: Stereotyped for the Bible Society . . . by T. Rutt, Shacklewell, London, 1812).

151. *The Larger Catechism, Agreed upon by the Assembly of Divines at Westminster . . . Revised by Alexander M'Leod* (New York: Stereotyped and printed by J. Watts and Co. in New York for Whiting and Watson, Theological and Classical Booksellers, June 1813).

152. Mechanized presses were sometimes called power presses in America, or just presses, as distinct from England and the Continent, where they were typically known as printing machines.

153. Morrill Wyman, "Memoir of Daniel Treadwell," *Memoirs of the American Academy of Arts and Sciences, Centennial Volume*, Vol. 11 (1888), p. 339. Wyman's "Memoir of Daniel Treadwell" contains several excerpts from Treadwell's autobiography, as well as detailed descriptions and illustrations of Treadwell's press and its operation. It is not clear whether any part of Treadwell's autobiography appeared in print prior to Wyman's memoir.

154. Wyman, "Memoir of Daniel Treadwell," pp. 344-345. Regarding the power source for Treadwell's press, James M. Wells wrote that "the power was supplied by a horse, led in a circle by a boy. When it became apparent that the job was too much for a single horse—the machine was very heavy—a second horse was added as a relay" (*American Printing: The Search for Self-Sufficiency* [Worcester, Massachusetts: American Antiquarian Society, 1985], p. 280).

155. "Constitution of the Widow's Society" (Boston: Treadwell's Power Press, 1823).

156. William Mitford, *The History of Greece* (Boston: Timothy Bedlington and Charles Ever, Cornhill, 1823).

157. *An Abstract of the American Bible Society, Containing an Account of Its Principles and Operations . . .* (New York: Printed by Daniel Fanshaw, 1830), p. 21.

158. Eugene Exman, *The Brothers Harper* (New York: Harper and Row, 1965), p. 17.

159. Quoted by Frank E. Comparato, *Chronicles of Genius and Folly: R. Hoe & Company and the Printing Press as a Service to Democracy* (Culver City: Labyrinthos, 1979), pp. 54-55.

160. Stephen O. Saxe, "Foreword," in Douglas W. Charles, *Bed & Platen Book Printing Machines: American and British Streams of Ingenious Regression in the Quest for Print Quality* (Ketchikan, Darjeeling, Kathmandu: Plane Surface Press, 2017), p. iv.

161. Jacob Abbott, *The Harper Establishment, or How the Story Books Are Made* (New York: Harper and Brothers, 1855). Abbott's work was issued as volume 4 of *Harper's Story Books. A Series of Narratives, Dialogues, Biographies, and Tales for the Instruction and Entertainment of the Young*.

162. For a discussion of the electrotyping process, see chapter 6.

163. "Electrotypes from Wood Engravings," *The American Repertory of Arts, Sciences and Manufactures*, vol. 3 (1841), pp. 161-163.

164. Daniel Davis Jr., *Davis' Manual of Magnetism, Including Also Electro-Magnetism, Magneto-Electricity, and Thermo-Electricity. With a Description of the Electrotype Process* (Boston: Daniel Davis Jr., 1842).

165. Paul C. Gutjahr, *An American Bible: A History of the Good Book in the United States 1777-1880* (Stanford, California: Stanford University Press, 1999), pp. 70-71.

166. Geo. C. Rand & Avery, *Specimens* (Boston: Rand and Avery, n.d.). (This book is not paginated.)

167. Jérôme de LaLande, *l'Art de faire le papier* (Paris: Desaint et Saillant, 1761), plates 1, 13, and 14.

168. Dominique Godineau, *The Women of Paris and Their French Revolution* (Berkeley and Los Angeles: University of California Press, 1998), p. 69.

169. Godineau, *The Women of Paris and Their French Revolution*, pp. 68-69.

170. X. Dachères, "*Les femmes typographes*," *L'Univers illustré*, vol. 8, no. 485 (15 November 1865), p. 723.

171. J. Ramsay MacDonald, *Women in the Printing Trades*, London: P. S. King and Son, 1904.

172. London: John W. Parker, 1860. The Victoria Press's imprint is on the verso of the title page.

173. Emily Faithfull, XVIII. - "Victoria Press. A Paper Read at the Glasgow Meeting of the National Association for the Promotion of Social Science, 1860," *The English Woman's Journal*, October 1860, pp. 121-126.

174. *The Victoria Regia: A Volume of Original Contributions in Poetry and Prose* (London: Printed and Published by Emily Faithfull and Co., Victoria Press, 1861).

175. "The Victoria Press," *The Illustrated London News*, vol. 38 (1861), p. 555.

176. *The Printers' Journal and Typographical Magazine*, vol. 2, no. 63 (5 August 1867), pp. 351-352.

177. In 1867, Emily Faithfull passed the management of the press to William Head, who continued to advocate for the employment of women in the printing trades until 1882. Head purchased the Victoria Press from Emily Faithfull in 1869.

178. *The Printers' Journal and Typographical Magazine*, vol. 2, no. 64 (19 August 1867), pp. 381-382.

179. Felicity Hunt, "Opportunities Lost and Gained. Mechanization and Women's Work in the London Bookbinding and Printing Trades," chapter 2 in Angela V. John, ed., *Unequal Opportunities: Women's Employment in England 1800-1918* (Oxford: Basil Blackwell, 1986), chapter 2, pp. 79-80.

180. M. L. [Mary Lloyd], *Sunny Memories. Containing Personal Recollections of Some Celebrated Characters* (London: Women's Printing Society, Limited, 1880).

181. Patricia Keats, "Women in Printing and Publishing in California, 1850-1940," *California History*, vol. 77 (1998), pp. 93-97.

182. Mary Biggs, "Neither Printer's Wife nor Widow: American Women in Typesetting, 1830-1950," *The Library Quarterly: Information, Community, Policy*, vol. 50, no. 4 (Oct. 1980), pp. 431-452.

183. Biggs, "Neither Printer's Wife nor Widow," p. 436.

184. Theodore L. De Vinne, "The Printing of 'The Century'," *The Century Illustrated Monthly Magazine*, vol. 41 (November 1890), p. 88.

185. Emmanuel Rivière, *Étude sur le salaire. Le travail de la femme dans l'industrie typographique*. [Blois] Grande Imprimerie de Blois, 1898.

186. J. Ramsay MacDonald, ed., *Women in the Printing Trades: A Sociological Study*. Investigators: Mrs. J. L. Hammond, Mrs. H. Oakeshott, Miss A. Black, Miss A. Harrison, Miss Irwin, and others (London: P. S. King and Son, 1904). Ramsay was prime minister of the United Kingdom from 1929 to 1935.

187. MacDonald, *Women in the Printing Trades*, p. 29.

188. MacDonald, *Women in the Printing Trades*, pp. 171-174.

189. MacDonald, *Women in the Printing Trades*, p. 173.

190. MacDonald, *Women in the Printing Trades*, p. 174.

191. Richard E. Huss, in *The Development of Printers' Mechanical Typesetting Methods 1822-1925* (Charlottesville: University Press of Virginia, 1973), discusses 294 methods or systems of mechanical typesetting or type distribution.

192. Etienne Robert Gaubert, *Rénovation de l'Imprimerie. Nouvelle puissance de la Mécanique. Notice sur le Gérotype ou machine à distribuer et à composer en typographie, Rapport à l'académie des sciences le 5 décembre 1842* (Paris: Chez l'Inventeur, 1843).

193. Théotiste Lefevre, *Guide pratique du compositeur d'imprimerie* (Paris: Firmin Didot, 1855).

194. C. Turner Thackrah, *The Effects of Arts, Trades, and Professions, and Civic States and Habits of Living, on Health and Longevity . . .* 2nd ed. (London: Longman, Rees . . ., 1832), pp. 42-43.

195. Louis Tanquerel des Planches, *Traité des maladies de plomb ou saturnines: Suivi de l'indication des moyens qu'on doit mettre en usage pour se préserver de l'influence délétère des préparations de plomb* (2 vols., Paris, 1839). Samuel L. Dana, *Lead Diseases: A Treatise from the French of L. Tanquerel des Planches with Notes and Additions on the Use of Lead Pipe and Its Substitutes* (Lowell, Massachusetts: Daniel Bixby and Company, 1848), p. 60.

196. William Turner Coggleshall, ed., *Five Black Arts: A Popular Account of the History, Processes of Manufacture, and Uses of Printing, Gas-Light, Pottery, Glass, Iron* (Co-

lumbus: Follett, Foster and Co., 1861). Coggeshall, a publisher and librarian, was also a self-appointed bodyguard for Abraham Lincoln. At the time *Five Black Arts* was published, Coggeshall was state librarian of Ohio. After Lincoln was assassinated, Coggeshall became the U.S. ambassador to Ecuador.

197. Lucien Alphonse Legros and John Cameron Grant, *Typographical Printing-Surfaces: The Technology and Mechanism of Their Production* (London and New York: Longmans, Green, 1916).

198. The best paper on the early history of typesetting machines, with special detail on Young and the Delcambres, is François Jarrige, "Le mauvais genre de la machine: Les ouvriers du livre et la compositions mécanique (France, Angleterre, 1840-1880)," *Revue d'histoire moderne et contemporaine* 54-1 (2007), pp. 93-221. It is also worth mentioning that many discoveries remain to be made in the identification of publications typeset by these early machines.

199. William Church, British patent specification no. 4664, p. 2.

200. Hansard, *Typographia*, pp. 665-677.

201. *Printing and the Mind of Man: Catalogue of the Exhibitions . . .* , no. 462.

202. Pierre Leroux, "D'une nouvelle typographie," *La Revue Indépendante*, vol. 6 (1843), pp. 264-291.

203. Victor Meunier, "La typographie méchanique," *L'Ami des Sciences*, vol. 3, no. 45 (8 November 1857), pp. 706-707. Meunier mentioned the machines of Young and Delcambre and of Gaubert as examples of great advances.

204. *Printing and the Mind of Man: Catalogue of the Exhibitions . . .*, no. 463. See also Legros and Grant, *Typographical Printing-Surfaces*, pp. 322-325, quoting Henry Bessemer's account of his contributions to the Pianotyp.

205. Edward Binns, *The Anatomy of Sleep; or, the Art of Procuring Sound and Refreshing Slumber at Will* (London: Churchill, 1842), pp. ix-x. Binns's book was extensively reviewed in English periodicals beginning in October and November 1842. The manner in which it was typeset was discussed in a very long review in *The Monthly Review*, vol. 3, no. 3 (November 1842): "The first thing we shall notice about this beautifully-*got-up* volume is that it has been typographically composed by machinery,—by means of an apparatus somewhat after the construction of a piano-forte, which when touched, it may be by female fingers, drops the letters into their proper places; dispensing with the usual number, and certain of the usual operations, of regular compositors" (p. 275).

206. "A New Invention," *The London Phalanx*, vol. 1 (1841-1842), p. 351.

207. "The New Composing Machine," *The London Phalanx*, vol. 1 (1841-1842), pp. 567-568. On p. 568, the editors indicated that the magazine would be set by the Pianotyp from the 28th number (18 December) "onward."

208. *Printing and the Mind of Man: Catalogue of the Exhibitions . . .* , no. 463.

209. The French printer Alphonse Alkan published a description of the machine, "*Machine propre à la composition et à la décomposition des caractères d'imprimerie*," in *Annales de la typographie française et étrangère*, vol. 30 (1840), pp. 82-87.

210. Etienne Robert Gaubert, *Rénovation de l'Imprimerie*. See R. E. Huss, *The Development of Printers' Mechanical Typesetting Methods 1822-1925* (1973), no. 5; François Jarrige, "The Genre of the Machine: Printing Workers and Mechanical Typesetting (France, England, 1840-1880)", *Revue d'histoire moderne et contemporaine* 54-1 (2007), pp. 193-221.

211. "Machine typographique de M. Gaubert. Sort des ouvriers," *La Phalange*, vol. 6, no. 106 (7 February 1843), columns 1631-1632.

212. Charles Babbage, *A Letter to Sir Humphrey Davy, Bart . . . on the Application of Machinery to the Purpose of Calculating and Printing Mathematical Tables* (London, 1822).

213. Diana H. Hook and Jeremy M. Norman, *Origins of Cyberspace* (Novato, California: HistoryofScience.com, 2002), no. 411. Jeremy M. Norman, *From Gutenberg to the Internet : A Sourcebook on the History of Information Technology* (Novato, California: HistoryofScience.com, 2005), 8.1, 9.1 etc.

214. Alexander J. Dick, "On the Financial Crisis, 1825-26." BRANCH: Britain, Representation and Nineteenth-Century History.

215. According to Paul Cheron's *Catalogue générale de la librairie française au 19e siècle* (vol. 3, col. 669), Delcambre also issued several pamphlets promoting the composing machine. Delcambre's son Isidore continued the business and is known to have exhibited his version of the machine at the Universal Exhibition in London in 1862. In 1856 and 1866, Isidore received two patents for composing machines, indicating that improved versions of the machines remained in production at least through those dates. A single scale model of the Pianotyp has survived in the Musée des Arts et Métiers, Paris. As late as 1868, Eugène Lacroix reported in his discussion of "l'imprimerie et les livres," in *Études sur l'exposition de 1867, ou les Archives de l'industrie au XIXème siècle. Nouvelle technologie des arts et métiers, des manufactures*, that the machine was being exhibited at the Paris Exposition of 1867 (pp. 150-156). Lacroix paid special attention to Young's machine for type distribution, which he illustrated on p. 156. According to Hippolyte Gautier and Adrien Desprez's *Les curiosités de l'Exposition de 1878* (Paris: Librairie Ch. Delagrave, 1878, p. 174), versions of the machine continued to be exhibited at that date. The images that Lacroix reproduced were taken from *Cassell's Illustrated Family Paper Exhibitor* (July 1862).

216. Alexander Mackie, *Italy and France: An Editor's Holiday* (London: Hamilton and Adams, 1874).

217. J. Luther Ringwalt, *American Encyclopaedia of Printing* (Philadelphia: Menamin and Ringwalt; J. B. Lippincott and Co., 1871), p. 291.

218. William H. Mitchel, "On a Type Composing and Distributing Machine," *Proceedings of the Institution of Mechanical Engineers* (1863), pp. 34-53.

219. William Winter, "Types," *The Atlantic Monthly*, vol. 13 (1864), pp. 615-621.

220. Winter, "Types," pp. 616-617.

221. According to John S. Thompson, the Mitchel machine was probably the first typesetting machine that was put to practical use for an extended period. At one point, Trow owned at least five of the Mitchel machines. See Thompson, *History of Composing Machines: A Complete Record of the Art of Composing* (Chicago: Inland Printer Co.,1904), pp. 10-11.

222. In his summary of early efforts to mechanize typesetting, Winter referenced Charles Babbage's intention to empower his Difference Engines to typeset their mathematical tables and to set this type in stereotype plates. It is possible that Winter had learned of Babbage's intentions in Babbage's autobiography, *Passages of the Life of a Philosopher*, also published in 1864. It is also possible that he read about Babbage's typesetting of numbers in Hansard's *Typographia* or some other account. Winter was also aware of the pioneering invention of William Church that had inspired later efforts to mechanize typesetting. Winter's reference to Babbage is the first that I have seen that shows that people of Babbage's time, outside the mathematics community, were aware of Babbage's work with respect to mechanizing the typesetting of mathematical tables.

223. A unit in the field of typography equal to the currently specified point size. "In metal type, the point size (and hence the em, from *em quadrat*) was equal to the line height of the metal body from which the letter rises. In metal type, the physical size of a letter could not normally exceed the em" (Wikipedia).

224. Mark Twain's relationship with developments in printing and publishing is the subject of a book by Bruce Michelson, *Printer's Devil: Mark Twain and the American Publishing Revolution* (Berkeley and Los Angeles: University of California Press, 2006).

225. In *Typographical Printing-Surfaces*, Legros and Grant state: "The Kastenbein Composing Machine invented prior to 1870 was brought into practical working form at the *Times Printing Office*, and with some modifications there introduced, is used for composing almost the whole of *The Times* and many other publications printed in the *Times Office*" (pp. 1103-1107).

226. According to The Times' *Printing in the Twentieth Century: A Survey* (London: The Times Publishing Company, 1930), p. 17.

227. Carl Schlesinger, ed., *The Biography of Ottmar Mergenthaler, Inventor of the Linotype* (New Castle, Delaware: Oak Knoll Books, 1989), pp. 18-23. Thompson, *History of Composing Machines*, p. 100.

228. In January 2015, I obtained a fine copy of this book for my collection, bound in the original red cloth, from Peter Daly in Hampshire, England. I had been

searching for a fine copy for several years; most copies are heavily worn. To my surprise, when I studied some pencil notes on the rear endpaper I noticed that a previous owner had indicated in a small, neat hand that they had bought the book from me in San Francisco on 14 August 1972! It was fortunate that the buyer, of whose name I have no recollection, kept it with great care. This copy remains the only fine copy in the original red cloth that I have seen on the market. From experience, I knew that some copies of *The Tribune Book of Open-Air Sports* were bound in blue cloth. But like the original red cloth, nearly all copies are heavily worn or ex-library. In 2021, I purchased the fine copy illustrated here from James Cummins Booksellers.

229. "The Linotype. A Revolution in Printing. The Latest Triumph of American Invention Seen in the Tribune Office," *Library of Tribune Extras*, vol. 1, no. 7 (July 1889), pp. 96-100.

230. "The Linotype," p. 97.

231. "The Linotype," p. 100.

232. William Martin, British patent no. 12,421, 1849. "Figuring fabrics, playing musical instruments, and printing, &c." (London: published at the Great Seal Patent Office, 1857).

233. Thompson, *History of Composing Machines*, pp. 120-121.

234. Tolbert Lanston, "Improvements in the Art of Printing," British patent specification no. 8183 (7 June 1887), pp. 1-2.

235. Legros and Grant, *Typographical Printing-Surfaces*.

236. William C. Barnes, Joseph W. McCann, and Alexander Duguid, *A Collation of Facts Relative to Fast Typesetting* (New York: Concord Co-Operative Printing Co., 1887), p. 7.

237. Regarding these manual typesetting speed contests, see Walter Rumble, *The Swifts: Printers in the Age of Typesetting Races* (Charlottesville and London: University of Virginia Press, 2003).

238. Regarding labor problems at hand paper mills in France, see Leonard N. Rosenband, *Papermaking in Eighteenth-Century France. Management, Labor and Revolution at the Montgolfier Mill, 1761-1805* (Baltimore, Maryland: The Johns Hopkins University Press, 2000).

239. For details, see *The British Bookmaker: A Journal for the Book Printer, the Book Illustrator, the Book Cover Designer, the Book Binder, Librarians, and Lovers of Books Generally*, vol. 5 (1892), pp. 119-121.

240. Michael Sadleir, *The Evolution of Publishers' Binding Styles 1770-1900* (London: Constable, 1930), pp. 42-43 and plate 10.

241. John Carter, *Binding Variants in English Publishing 1820-1900* (London: Constable, 1932), pp. 20 and 22.

242. John Carter, *Publisher's Cloth: An Outline History of Publisher's Bindings in England 1820-1900* (London: Constable, 1935).

243. Geoffrey Keynes, *William Pickering Publisher* (London: Galahad Press, 1969), pp. 13-14.

244. Paul W. Nash, "Two Hundred Years of Publisher's Cloth," *Journal of the Printing Historical Society*, third series, no. 1 (2020), pp. 241-301.

245. Lionel S. Darley, *Bookbinding Then and Now* (London: Faber and Faber, 1959), pp. 31-33.

246. Charles Knight, *The Results of Machinery, Namely, Cheap Production and Increased Employment* (London: Charles Knight, 1831), pp. 156-158.

247. *The Reply of the London Bookbinders to "Remarks on a Memorial addressed to their Employers, on the Effects of a Machine Introduced to Supersede Manual Labour," as Appeared in a Work Published by the Society for the Diffusion of Useful Knowledge, with Observations on the Influence of Machinery on the Working Classes in General* (London: Published for the Society by William Smith, 1831).

248. Lionel S. Darley, *Bookbinding Then and Now* (London: Faber and Faber, 1959), pp. 35-36.

249. Michael Sadleir, *The Evolution of Publishers' Binding Styles 1770-1900* (London: Constable, 1930), pp. 49-50.

250. "Technical bookbinding. Chapter XIII: Muslin or cloth binding and blocking," *The Bookbinder*, no. 30 (1889), pp. 90-92. See Bernard Warrington, "William Pickering and the Book Trade in the Early Nineteenth Century," *Bulletin of the John Rylands Library*, vol. 68 (1985), pp. 247-266; Pickering and Chatto, *Catalogue 708, William Pickering and His Successors 1820-1900* (1993).

251. Charles Babbage, *The Economy of Machines and Manufactures* (London: Charles Knight, 1832.)

252. Andrew Ure, *Dictionary of Arts, Manufactures, and Mines* (London: Longman, Orme, Brown, Greene, and Longmans, 1839), pp. 160-161.

253. Charles Tomlinson, *Cyclopedia of Useful Arts: Mechanical and Chemical, Manufactures, Mining and Engineering* (London: G. Virtue, 1852-1854). The quotation is from vol. I, pp. 158-159.

254. Graham Pollard and Esther Potter, *Early Bookbinding Manuals: An Annotated List of Technical Accounts of Bookbinding to 1840* (Oxford: Oxford Bibliographical Society, 1984).

255. John Andrews Arnett, *Bibliopegia, or, The Art of Bookbinding in All Its Branches* (London: Richard Groombridge, 1835).

256. James B. Nicholson, *A Manual of the Art of Bookbinding: Containing Full Instructions in the Different Branches of Forwarding, Gilding, and Finishing. Also, the Art of Marbling Book-Edges and Paper. The Whole Designed for the Practical Workman, the Amateur, and the Book-Collector* (Philadelphia: Henry Carey Baird, 1856).

257. Charles Tomlinson, *Cyclopaedia of Useful Arts, Mechanical and Chemical, Manufactures, Mining and Engineering* (London: G. Virtue, 1852-1854), vol. 1, p. 152.

258. Tomlinson, *Cyclopaedia of Useful Arts*, vol. 1, p. 152.

259. Works on the mechanization of bookbinding include Ernest-Peter Biesalski, *Die Mechanisierung der deutschen Buchbinderei 1850-1900* (Frankfurt am Main: Buchhänder-Vereinigung Gmbh, 1991) and Kristina Lundblad, *Bound to Be Modern: Publisher's Cloth Bindings and the Material Culture of the Book 1840-1914*, translated by Alan Crozier (New Castle, Delaware: Oak Knoll Press, 1915).

260. George Dodd, "Printing: Its Modern Varieties," in *Curiosities of Industry* (London: G. Routledge, 1852); the quotation is from p. 16. Dodd was quoting from *The Companion to the Almanac; or Year-Book of General Information for 1852* (London: Charles Knight, 1852), p. 27.

261. *Die ersten Druckmaschinen. Erbaut in London bis zu dem Jahre 1818 von Friedrich Koenig und Andreas Friedrich Bauer* [The first printing machines. Constructed in London up to the year 1818 by Friedrich Koenig and Andreas Friedrich Bauer] (Leipzig: Druck von F. A. Brockhaus auf einer Schnellpresse von Koenig und Bauer, 1851).

262. C. T. Courtney Lewis, *George Baxter (Colour Printer)* (London: Sampson, Low, 1908), p. 158.

263. Max E. Mitzman, *George Baxter and the Baxter Prints* (North Pomfret, Vermont: David and Charles, 1978), p. 50.

264. Geoffrey Wakeman, *Victorian Colour Printing* ([Leicestershire:] Plough Press, 1981), p. 9. This book was issued in only 141 numbered copies, each with numerous examples of original printing from the period bound in.

265. George Baxter (1804-1867): The 'Inventor' of Colour Printing, www.georgebaxter.com, accessed April 2025.

266. "Christmas Supplement," *The Illustrated London News*, vol. XXVII (22 December 1855), no. 776, pp. [737]-752.

267. R. M. Burch, *Colour Printing and Colour Printers* (London: Sir Isaac Pitman, 1910), p. 147.

268. Supplement to *The Illustrated London News*, 18 October 1862.

269. Burch, *Colour Printing and Colour Printers*, p. 135.

270. L'Imprimerie Berger-Levrault et Cie., *Notice historique sur le développememnt et l'organisation de la maison* (Nancy and Paris: Berger-Levrault et Cie., 1878).

271. L'Imprimerie Berger-Levrault et Cie., *Notice historique*, pp. 42-43.

272. Andreas Albert, *Der Maschinenmeister an der Schnellpresse. Ein Handbuch für Buchdruckereibesitzer, Factore, Maschinemeister und Mechaniker* (Leipzig: Dürr'sche Buchhandlung,1853).

273. Bigmore and Wyman, *A Bibliography of Printing* (1880-86; 2001 edition), vol. III, p. 91, and vol. I, p. 194.

274. Adolphe Lucien Monet, *Le conducteur de machines typographique. Études sur les différents systèmes de machines mise en trains* (Paris: Jules Claye, 1827). Monet, an employee of Claye, described himself as the firm's *prote des machines* (machine guard) on the work's title.

275. Bigmore and Wyman, Vol. II, pp. 48-49.

276. Thomas F. Adams, *Typographia, or the Printer's Instructor* (Philadelphia: Published at No. 118 Chestnut Street and No. 8 Franklin Place, 1844), pp. 270-272.

277. Adams, *Typographia, or the Printer's Instructor* (Philadelphia: L. Johnson and Co., 1857), p. 269.

278. Thomas MacKellar, *The American Printer: A Manual of Typography* (Philadelphia: L. Johnson and Co., 1866), p. 212.

279. Frederick J. F. Wilson, *Typographic Printing Machines and Machine Printing: A Practical Guide to the Selection of Bookwork, Two-Colour, Jobbing, and Rotary Machines* (London: Wyman and Sons, [1879]).

280. Frederick J. F. Wilson and Douglas Grey, *A Practical Treatise upon Modern Printing Machinery and Letterpress Printing* (London: Cassell and Co., 1888).

281. R. W. Kerr, *History of the Government Printing Office (at Washington, D.C.) with a Brief Record of the Public Printing for a Century (1789-1881)* (Lancaster, Pennsylvania: Inquirer Printing and Publishing Co., 1881).

282. Schelter and Giesecke, *Fünfundsiebzig Jahre des Hauses J. G. Schelter & Giesecke in Leizpig* (Leipzig: Schelter und Giesecke, 1891).

283. MacKellar Smiths and Jordan Foundry, *1796-1896: One Hundred Years* (Philadelphia: [MacKellar Smiths and Jordan Foundry,] 1896).

284. Vereeniging ter Bervordering van de Belangen des Boekhandels, *Internationale tentoonstelling, Juli-Augustus 1892* ([Amsterdam: Gedr. door Roeloffzen & Hübner,] 1892).

285. G. Fritz, *Die K. K. Hof- und Staatsdruckerei und deren technischen Einrichtungen* (Vienna: Kaiserlich-Königlichen Hof- und Staatsdruckerei, 1894).

286. K.K. Staatsdruckerei, *Zur Feier des einhundertjährigen Bestandes der KK Hof- und Staatsdruckerei* (Vienna: K. K. Staatsdruckerei, 1904).

287. David McKitterick, *Old Books, New Technologies* (Cambridge and New York: Cambridge University Press, 2013), p. 175).

288. McKitterick, *Old Books, New Technologies*, pp. 170-171.

289. *The Illustrated London News*, vol. LXXXI, no. 1982, p. 18.

290. Catherine M. MacSorley, *The Earl-Printer: A Tale of the Time of Caxton* (London: Shaw, 1877).

291. Henry Stevens, *The Bibles in the Caxton Exhibition MDCCLXXVII, or a bibliographical description of nearly one thousand representative Bibles in various languages chronologically arranged from the first Bible printed by Gutenberg in 1450-1456 to the last Bible printed at the Oxford University Press the 30th June 1877. With an Introduction on the History of Printing as Illustrated by the printed Bible from 1450 to 1877 in which is told for the first time the true history and mystery of the Coverdale Bible of 1535 Together with bibliographical notes and collations of many rare Bibles in various languages and divers versions printed during the last four centuries* (London: Henry Stevens . . . , 1888).

292. Henry Stevens, *The History of the Oxford Caxton Memorial Bible Printed and Bound in Twelve Consecutive Hours June 30, 1877* (London: H. Stevens, 1878), p. 23.

293. R. A. Gross and M. Kelly, eds., *A History of the Book in America*, Vol. 2, p. 165.

294. "The Napier Printing Machine," *New-York Daily American*, vol. 6, no. 604 (4 January 1826), [p. 1].

295. James L. Crouthamel, *Bennett's New York Herald and the Rise of the Popular Press* (Syracuse, New York: Syracuse University Press, 1989).

296. "New York Printing Machine, Press, and Saw Works. R. Hoe & Co.," *Graham's Magazine*, vol. 40, no. 6 (June 1852), pp. 563-576; vol. 41, no. 7 (July 1852), pp. 7-12.

297. "New York Printing Machine, Press, and Saw Works. R. Hoe & Co.," *Graham's Magazine*, vol. 41, no. 7 (July 1852), p. 8.

298. Stephen D. Tucker, "History of R. Hoe & Company," edited with an introduction by Rollo G. Silver, *Proceedings of the American Antiquarian Society*, vol. 82 (1972), p. 381, n. 31.

299. R. Hoe & Co., *Printing Machines* (New York: n.p., 1867), p. 9.

300. Bullock's press incorporated ideas for newspaper printing patented in Austria in 1858 by Alois Auer, director of Austria's state printing house, the K. K. Hof- und Staatsdruckerei. Among his many other duties, Auer was responsible for the production of the *Wiener Zeitung* newspaper.

301. R. Hoe & Co., *R. Hoe & Co. Manufacturers of Type-Revolving, Perfecting, Single and Double Cylinder and Adams' Printing Machines . . .* (New York: [R. Hoe & Co.,] 1873).

302. "Colonel Richard M. Hoe," *Harper's Weekly*, vol. 20, no. 1041 (9 December 1876), p. 997.

303. St. Bride Foundation Institute, *Catalogue of the William Blades Library*, compiled by John Southward (London, 1899), p. 16.

304. George Bullen, ed., *Caxton Celebration, 1877. Catalogue of the Loan Collection of Antiquities, Curiosities, and Appliances Connected with the Art of Printing* (London: Elzevir Press, 1877), p. xvii.

305. Éric le Ray, *Marinoni, le foundateur de la presse moderne (1823-1904)* (Paris: L'Harmattan, 2009), p. 60.

306. Éric Le Ray, "Histoire de l'imprimerie et de la presse, en marge d'un centenaire: Hippolyte August Marinoni (1823-1904)," *Cahiers Gutenberg*, no. 43 (December 2003), pp. 33-99.

307. A. L. Monet, *Les machines et appareils typographiques en France et à l'étranger suivi des procédés d'impression* (Paris: Administration du Bulletin de l'Imprimerie, 1878).

308. Theodore Low De Vinne, "The Printing of 'The Century'," *The Century Illustrated Monthly Magazine*, vol. 41 (1890), pp. 87-99.

309. De Vinne, "The Printing of 'The Century'," p. 87.

310. Robert Hoe III, *A Short History of the Printing Press and of the Improvements in Printing Machinery from the Time of Gutenberg to the Present Day* (New York: Printed and Published for Robert Hoe, 1902).

Bibliography

PATENTS

Industrial Revolution

Arkwright, Richard. No. 931. Spinning machine. 3pp. Plate. London: Printed by George E. Eyre and William Spottiswoode; published at the Great Seal Patent Office, 1856.
 Issued 1769.

Arkwright, Richard. No. 1111. Machinery for preparation of fibrous materials for spinning. 4pp. Plate. London: Printed by George E. Eyre and William Spottiswoode; published at the Great Seal Patent Office, 1856.
 Issued 1775.

Cartwright, Edmund. No. 1565. Looms for weaving. 6pp. Plate. London: Printed by George E. Eyre and William Spottiswoode; published at the Great Seal Patent Office, 1856.
 Issued 1786.

Emmons, Calvin. 1829. [Improvement in the threshing machine.] 3pp. Washington, DC: N.p., 1829.
 U.S. patent.

Hargraves, James. No. 962. Machinery for spinning, drawing, and twisting cotton. 3pp. Plate. London: Printed by George E. Eyre and William Spottiswoode; published at the Great Seal Patent Office, 1856.
 Issued 1770.

Kay, James. No. 5226. Machinery for preparing and spinning fibrous substances. 4pp. Plate. London: Printed by George E. Eyre and William Spottiswoode; published at the Great Seal Patent Office, 1856.
 Issued 1825.

Lombe, Thomas. No. 422. Engines to wind, spin, and twist silk. 4pp. London: Printed by George E. Eyre and William Spottiswoode; published at the Great Seal Patent Office, 1857.
 Issued 1718.

Maudslay, Henry. No. 3050. Steam engines. 3pp. Plate. London: Printed by George E. Eyre and William Spottiswoode; published at the Great Seal Patent Office, 1855.
 Issued 1807.

Neilson, James Beaumont. No. 5701. Hot blast for furnaces. 3pp. London: Printed by George E. Eyre and William Spottiswoode; published at the Great Seal Patent Office, 1856.
 Issued 1828.

Paul, Lewis. No. 562. Spinning machines. 2pp. London: Printed by George E. Eyre and William Spottiswoode; published at the Great Seal Patent Office, 1856.
 Issued 1738.

Roberts, Richard. No. 4726. Looms for weaving. 26pp. 6 plates. London: Printed by George E. Eyre and William Spottiswoode; published at the Great Seal Patent Office, 1856.
 Issued 1822.

Roberts, Richard. No. 5138. Spinning machines. 12pp. 2 plates. London: Printed by George E. Eyre and William Spottiswoode; published at the Great Seal Patent Office, 1856.
 Issued 1825.

Roberts, Richard. No. 5949. Spinning machines. 7pp. 1 plate. London: Printed by George E. Eyre and William Spottiswoode; published at the Great Seal Patent Office, 1856.
 Issued 1830.

Wilson, Stephen. No. 4543. Pattern machines and looms for weaving figured fabrics. 8pp. Plate. London: Printed by George E. Eyre and William Spottiswoode; published at the Great Seal Patent Office, 1857.
 Issued 1821.

Bookbinding

Cope, James, and James Bradbrook. No. 2217. Bookbinding. 8pp. 6 plates. London: Printed by George E. Eyre and William Spottiswoode; published at the Great Seal Patent Office, 1869.
 Issued 1868.

Edwards, James. No. 1462. Embellishing books. 3pp. London: Printed by George E. Eyre and William Spottiswoode; published at the Great Seal Patent Office, 1856.
 Issued 1785.

Hancock, William. No. 7247. Bookbinding. 4pp. London: Printed by George E. Eyre and William Spottiswoode; published at the Great Seal Patent Office, 1857.
 Issued 1836.

Nichol, James. No. 1125. Bookbinding. 6pp. Plate. London: Printed by George E. Eyre and William Spottiswoode; published at the Great Seal Patent Office, 1853.
 Issued 1853.

Nickels, Christopher, and Henry George Collins. No. 7515. Bookbinding and cutting paper. 4pp. Plate. London: Printed by George E. Eyre and William Spottiswoode; published at the Great Seal Patent Office, 1857.
 Issued 1837.

Starr, Charles. No. 13,166. Bookbinding. 18pp. 3 plates. London: Printed by George E. Eyre and William Spottiswoode; published at the Great Seal Patent Office, 1857.
 Issued 1850.

Papermaking

Cobb, Thomas, Jr. No. 3580. Manufacture of paper. 8pp. 3 plates. London: Printed by George E. Eyre and William Spottiswoode; published at the Great Seal Patent Office, 1856.
 Issued 1812.

Crompton, Thomas Bonsor. No. 4509. Drying and finishing paper. 3pp. Plate. London: Printed by George E. Eyre and William Spottiswoode; published at the Great Seal Patent Office, 1856.
 Issued 1820.

Dickinson, John. No. 3191. Machinery for making, cutting, and placing paper. 11pp. 2 plates. London: Printed by George E. Eyre and William Spottiswoode; published at the Great Seal Patent Office, 1855.
 Issued 1809.

Didot, Leger. No. 3568. Machinery for the manufacture of paper. 9pp. 2 plates. London: Printed by George E. Eyre and William Spot-

tiswoode; published at the Great Seal Patent Office, 1856.
Issued 1812.

Fourdrinier, Henry, Sealy Fourdrinier, and John Gamble. No. 3068*. Machinery for making paper. 18pp. Plate. London: Printed by George E. Eyre and William Spottiswoode; published at the Great Seal Patent Office, 1856.
Issued 1807.

Gamble, John. No. 2487. Machinery for making paper. 8pp. 2 plates. London: Printed by George E. Eyre and William Spottiswoode; published at the Great Seal Patent Office, 185
Issued 1801.

Koops, Matthias. No. 2392. Extracting ink from paper and converting such paper into pulp. 3pp. London: Printed by George E. Eyre and William Spottiswoode; published at the Great Seal Patent Office, 1856.
Issued 1800.

Phipps, John, and Christopher Phipps. No. 5075. Machinery for making paper. 9pp. Plate. London: Printed by George E. Eyre and William Spottiswoode; published at the Great Seal Patent Office, 1856.
Issued 1825.

Wilks, John. No. 5934. Machinery for the manufacture of paper. 16pp. Plate. London: Printed by George E. Eyre and William Spottiswoode; published at the Great Seal Patent Office, 1856.
Issued 1830.

Printing

Adams, Isaac. No. 2264. Machinery for printing. 36pp. 14 plates. London: Printed by George E. Eyre and William Spottiswoode; published at the Great Seal Patent Office, 1855.
Issued 1854.

Applegath, Augustus. No. 4249. Casting stereotype and other plates for printing. 5pp. Plate. London: Printed by George E. Eyre and William Spottiswoode; published at the Great Seal Patent Office, 1856.
Issued 1818.

Applegath, Augustus. No. 4640. Printing machines. 4pp. Plate. London: Printed by George E. Eyre and William Spottiswoode; published at the Great Seal Patent Office, 1857.
Issued 1822.

Applegath, Augustus. No. 4757. Printing machines. 7pp. 6 plates. London: Printed by George E. Eyre and William Spottiswoode; published at the Great Seal Patent Office, 1857.
Issued 1823.

Applegath, Augustus. No. 4902. Printing machines. 4pp. Plate. London: Printed by George E. Eyre and William Spottiswoode; published at the Great Seal Patent Office, 1857.
Issued 1824.

Applegath, Augustus. No. 5988. Printing machines. 9pp. 5 plates. London: Printed by George E. Eyre and William Spottiswoode; published at the Great Seal Patent Office, 1857.
Issued 1830.

Applegath, Augustus. No. 11,505. Printing machines. 13pp. 8 plates. London: George E. Eyre and William Spottiswoode; published at the Great Seal Patent Office, 1857.
Issued 1846.

Applegath, Augustus. No. 13,879. Printing machines. 5pp. 4 plates. London: Printed by George E. Eyre and William Spottiswoode; published at the Great Seal Patent Office, 1857. "Special reprint."
Issued 1851.

Applegath, Augustus. No. 2565. Printing machines. 7pp. 4 plates. Redhill: Printed for Her Majesty's Stationery Office by Malcomson & Co., 1906. Second edition.
Issued 1857.

Applegath, Augustus. No. 372. Improvements in printing machinery. 7pp. 2 plates. Redhill: Printed for His Majesty's Stationery Office by Love & Malcolmson, 1903. Second edition.
Issued 1858.

Applegath, Augustus. No. 2098. Machinery for printing and cutting paper. 9pp. 2 plates. Redhill: Printed for His Majesty's Stationery Office by Love & Malcolmson, 1905. Second edition.
Issued 1859.

Armand, Peter, Compte de Fontaine Moreau. No. 1609. Typographical printing presses. 13pp. 6 plates. London: Printed by George E. Eyre and William Spottiswoode; published at the Queen's Printing Office, 1854
Issued 1853.

Bacon, Richard Mackenzie, and Brian Donkin. No. 3757. Printing machine. 7pp. 4 plates. London: Printed by George E. Eyre and William Spottiswoode; published at the Great Seal Patent Office, 1856.
Issued 1813.

Bakewell, Frederick Collier. No. 1831. Letter printing. 15pp. 3 plates. Redhill: Printed for His Majesty's Stationery Office by Love & Malcomson, 1905. Second edition.
Issued 1869.

Banks, William. No. 6886. Machinery and pens for ruling and pressing paper. 6pp. Plate. London: Printed by George E. Eyre and William Spottiswoode; published at the Great Seal Patent Office, 1856.
Issued 1835.

Baxter, George. No. 6916. Steel and copper plate engravings. 5pp. 4 plates. London: Printed by George E. Eyre and William Spottiswoode; published at the Great Seal Patent Office, 1857.
Issued 1835.

Beach, Moses Sperry. No. 9308. Printing types, and machinery for damping, packing, printing, folding, and cutting paper. 9pp. 3 plates. Redhill: Printed for His Majesty's Stationery Office by Love & Malcolmson, 1905. Third edition.
Issued 1842.

Bell, Thomas. No. 1378. Machinery for printing textile fabrics. 3pp. 2 plates. London: Printed by George E. Eyre and William Spottiswoode; published at the Great Seal Patent Office, 1856.
Issued 1782.

Beniowski, Bartholomew. No. 11,905. Letter-press printing. 11pp. 4 plates. London: Printed by George E. Eyre and William Spottiswoode; published at the Great Seal Patent Office, 1856.
Issued 1847.

Beniowski, Bartholomew. No. 12,589. Letter-press printing. 9pp. 4 plates. London: Printed by George E. Eyre and William Spottiswoode; published at the Great Seal Patent Office, 1857.
Issued 1849.

Bond, Walter, and Joseph Foster. No. 774. Printing machines. 6pp. 3 plates. Redhill: Printed for His Majesty's Stationery Office by Love & Malcolmson, 1912. Second edition.
Issued 1871.

Bonneville, Henri Adrien. No. 1623. Printing press. 3pp. Plate. London: Printed by George E. Eyre and William Spottiswoode; published at the Great Seal Patent Office, 1864.
Issued 1864.

Bremner, Samuel. No. 2511. Improvements in the construction of printing machines, and in driving or actuating the same. Redhill: Printed for His Majesty's Stationery Office by Love & Malcolmson, 1905. Second edition.
Issued 1861.

[Bullock, William.] Bakewell, Frederick Collier. No. 955. Letter printing machines. London: Printed by George E. Eyre and William Spottiswoode; published at the Great Seal Patent Office, 1862.
Issued 1862.

Bunnett, Jacob. No. 1540. Printing paper hangings, calicoes, &c. 2pp. Plate. London: Printed by George E. Eyre and William Spottiswoode; published at the Great Seal Patent Office, 1856.
Issued 1786.

Church, Edward. No. 4565. Printing. 7pp. Plate. London: Printed by George E. Eyre and William Spottiswoode; published at the Great Seal Patent Office, 1857.
Issued 1821.

Church, William. No. 4664. Printing. 12pp. 8 plates. London: Printed by George E. Eyre and William Spottiswoode; published at the Great Seal Patent Office, 1857.
Issued 1822.

Church, William. No. 4760. Apparatus for printing. 4pp. 2 plates. Redhill: Printed for His Majesty's Stationery Office by Love & Malcolmson, 1902. Second edition.
Issued 1823.

Church, William. No. 4903. Printing machines. 8pp. 3 plates. Starr, Charles. 1850. No. 13,166. Bookbinding. 18pp. 3 plates. London: Printed by George E. Eyre and William Spottiswoode; published at the Great Seal Patent Office, 1857.
Issued 1824.

Clymer, George. No. 4174. Printing presses. 6pp. 2 plates. London: Printed by George E. Eyre and William Spottiswoode; published at the Great Seal Patent Office, 1857.
Issued 1817.

Conisbee, William. No. 582. Colour printing machines. 6pp. 2 plates. Redhill: Printed for His Majesty's Stationery Office by Love & Malcolmson, 1908. Second edition.
Issued 1862.

Conisbee, William, and James Smale. No. 739. Cutting and feeding paper to printing machines. 5pp. 2 plates. London: Printed by George E. Eyre and William Spottiswoode; published at the Great Seal Patent Office, 1874.
Issued 1874.

Congreve, William. No. 4521. Printing in colours. 11pp. 3 plates. London: Printed by George E. Eyre and William Spottiswoode; published at the Great Seal Patent Office, 1857.
Issued 1820.

Cooper, Samuel, and William Miller. No. 4570. Printing machines. 10pp. Plate. London: Printed by George E. Eyre and William Spottiswoode; published at the Great Seal Patent Office, 1857.
Issued 1821.

Cowper, Edward. No. 3974. Printing paper hangings, &c. 4pp. Plate. London: Printed by George E. Eyre and William Spottiswoode; published at the Great Seal Patent Office, 1856.
Issued 1816.

Cowper, Edward. No. 4801. Printing machines. 5pp. 2 plates. London: Printed by George E. Eyre and William Spottiswoode; published at the Great Seal Patent Office, 1857.
Issued 1823.

Davis, Benjamin Walter, and John Parsons. No. 3887. Machinery for letter-press printing. 6pp. 2 plates. London: Printed by George E. Eyre and William Spottiswoode; published at the Great Seal Patent Office, 1875.
Issued 1874.

Edwards, Ernest. No. 3543. Photo-mechanical printing, &c. 6pp. London: Printed by George E. Eyre and William Spottiswoode; published at the Great Seal Patent Office, 1870.
Issued 1869.

Edwards, Ernest. No. 3543*. Photo-mechanical printing, &c. Memorandum of alteration. 6pp. London: Printed by George E. Eyre and William Spottiswoode; published at the Great Seal Patent Office, 1870.
Issued 1869.

Foulis, Andrew, and Alexander Tillock. No. 1431. Stereotypes for printing. 3pp. London: Printed by George E. Eyre and William Spottiswoode; published at the Great Seal Patent Office, 1856.
Issued 1784.

Hansard, Thomas. No. 4176. Printing. 6pp. Plate. London: Printed by George E. Eyre and William Spottiswoode; published at the Great Seal Patent Office, 1857.
Issued 1817.

Hedderwick, Percy David. No. 1581. Printing. 14pp. Plate. Redhill: Printed for His Majesty's Stationery Office by Love &

Malcomson, 1905. Second edition.
Issued 1870.

Hill, John Reed. No. 10,284. Letter-press printing machine. 7pp. 2 plates. London: Printed by George E. Eyre and William Spottiswoode; published at the Great Seal Patent Office, 1857.
Issued 1844.

Hill, Rowland. No. 6762. Letter-press printing machines. 35pp. 9 plates. London: Printed by George E. Eyre and William Spottiswoode; published at the Great Seal Patent Office, 1857.
Issued 1835.

[Hoe, Richard March.] Newton, William Edward. No. 239. Improvements in printing machinery. 19pp. 7 plates. London: Printed by George E. Eyre and William Spottiswoode, 1862.
Issued 1862.

[Hoe, Richard March.] Newton, William Edward. No. 529. Stereotype plates. 3pp. Plate. London: Printed by George E. Eyre and William Spottiswoode, 1863.
Issued 1863.

[Hoe, Richard March.] Newton, William Edward. No. 2330. Printing machinery. 27pp. 13 plates. London: Printed by George E. Eyre and William Spottiswoode; published at the Great Seal Patent Office, 1871.
Issued 1870.

[Hoe, Richard March.] Newton, William Edward. No. 2368. Improvements in lithographic printing machines and in apparatus for grinding or preparing the surface of the stones used for lithographic purposes. 12pp. 8 plates. London: Printed by George E. Eyre and William Spottiswoode, 1871.
Issued 1870.

[Hoe, Richard March.] Newton, William Edward. No. 1762. Improved apparatus for folding printed sheets of paper. 6pp. 2 plates. London: Printed by George E. Eyre and William Spottiswoode, 1869.
Issued 1869.

[Hoe, Richard March.] Newton, William Edward. No. 3582. Improved apparatus for folding printed sheets of paper, and for cutting and folding printed sheets of paper as they issue from the printing machine. 15pp. 4 plates. London: Printed by George E. Eyre and William Spottiswoode; published at the Great Seal Patent Office, 1870.
Issued 1869.

[Hoe, Richard March.] Newton, William Edward. No. 2863.

Improvements in printing machinery. 6pp. 2 plates. London: Printed by George E. Eyre and William Spottiswoode, 1865.
Issued 1864.

[Hoe, Richard March.] Newton, William Edward. No. 1825. Improvements applicable to machinery for letter-press printing. 16pp. 7 plates. London: Printed by George E. Eyre and William Spottiswoode, 1872.
Issued 1871.

[Hoe, Richard March.] Newton, William Edward. No. 45. Improvements in printing machinery. 9pp. 4 plates. London: Printed by George E. Eyre and William Spottiswoode, 1872.
Issued 1872.

Hullmandel, Charles Joseph. No. 8683. Lithography. 7pp. London: Printed by George E. Eyre and William Spottiswoode; published at the Great Seal Patent Office, 1856.
Issued 1840.

Ingle, Henry. No. 64. Improvements in printing machines. 3pp. Plate. London: Printed by George E. Eyre and William Spottiswoode; published at the Great Seal Patent Office, 1853.
Issued 1853.

Kitchen, John. No. 6454. Printing presses. 6pp. Plate. London: Printed by George E. Eyre and William Spottiswoode; published at the Great Seal Patent Office, 1857.
Issued 1833.

Knight, Charles. No. 7673. Printing on paper, vellum, parchment and pasteboard. 11pp. 2 plates. London: Printed by George E. Eyre and William Spottiswoode; published at the Great Seal Patent Office, 1857.
Issued 1838.

Koenig, Frederick. No. 3496. Printing machines. 13pp. 6 plates. Redhill: Printed for His Majesty's Stationery Office by Love & Malcomson, 1902. Second edition.
Issued 1811.

Koenig, Frederick. No. 3725. Printing machines. 5pp. Plate. London: Printed by George E. Eyre and William Spottiswoode; published at the Great Seal Patent Office, 1856.
Issued 1813.

Koenig, Frederick. No. 3868. Printing machines. 13pp. 3 plates. London: Printed by George E. Eyre and William Spottiswoode; published at the Great Seal Patent Office, 1856.
Issued 1814.

Little, William. No. 11,203. Printing machines. 7pp. 6 plates. London: Printed by George E. Eyre and William Spottiswoode; published at the Great Seal Patent Office, 1857.
Issued 1846.

MacDonald, John Cameron, and Joseph Calverley. No. 1661. Printing apparatus. 6pp. 4 plates. Redhill: Printed for Her Majesty's Stationery Office by Malcomson & Co., 1904. Second edition.
Issued 1863.

MacDonald, John Cameron, and Joseph Calverley. No. 3222. Casting stereotype surfaces, printing and cutting paper. 13pp. 8 plates. London: Printed by George E. Eyre and William Spottiswoode; published at the Great Seal Patent Office, 1867.
Issued 1866.

Martin, William. No. 12,421. Figuring fabrics, playing musical instruments, and printing, &c. 17pp. 7 plates. London: Printed by George E. Eyre and William Spottiswoode; published at the Great Seal Patent Office, 1857.
Issued 1849.

Mennons, Marc Antoine François. No. 2088. 3pp. Plate. London: Printed by George E. Eyre and William Spottiswoode; published at the Great Seal Patent Office, 1862.
Issued 1861.

Napier, David. No. 5713. Letter-press printing machines. 5pp. Plate. London: Printed by George E. Eyre and William Spottiswoode; published at the Great Seal Patent Office, 1857.
Issued 1828.

Napier, David. No. 6010. Printing and pressing machines. 5pp. Plate. London: Printed by George E. Eyre and William Spottiswoode; published at the Great Seal Patent Office, 1857.
Issued 1830.

Napier, David. No. 7343. Letter-press printing machines. 3pp. Plate. London: Printed by George E. Eyre and William Spottiswoode; published at the Great Seal Patent Office, 1857.
Issued 1837.

Napier, James Murdoch. No. 1740. Inking apparatus for platen printing machines. 9pp. Plate. London: Printed by George E. Eyre and William Spottiswoode; published at the Queen's Printing Office, 1854.
Issued 1853.

Napier, James Murdoch. No. 3113. Letter-press printing machines. 11pp. 5 plates. London: Printed by

George E. Eyre and William Spottiswoode; published at the Great Seal Patent Office, 1858.
    Issued 1857.

Napier, James Murdoch. No. 2438. Printing machines. 7pp. Plate. Redhill: Printed for His Majesty's Stationery Office by Love & Malcolmson, 1905. Second edition.
    Issued 1859.

Newton, Alfred Vincent. No. 1603. Machinery for printing. 7pp. Plate. London: Printed by George E. Eyre and William Spottiswoode; published at the Great Seal Patent Office, 1853.
    Issued 1853.

Newton, Alfred Vincent. No. 2463. An improved construction of printing press. 6pp. Plate. London: Printed by George E. Eyre and William Spottiswoode; published at the Great Seal Patent Office, 1854.
    Issued 1853. Possibly for Stephen Ruggles's press, Boston, 1840.

Newton, William. No. 10,338. Letter-press printing. 9pp. 4 plates. London: Printed by George E. Eyre and William Spottiswoode; published at the Great Seal Patent Office, 1856.
    Issued 1844.

Nicholson, William. No. 1748. Printing machine. 11pp. Plate. London: Printed by George E. Eyre and William Spottiswoode; published at the Great Seal Patent Office, 1856.
    Issued 1790.

Palmer, William. No. 4782. Printing and staining paper for paper-hangings. 9pp. 3 plates. London: Printed by George E. Eyre and William Spottiswoode; published at the Great Seal Patent Office, 1857.
    Issued 1823.

Perkins, Jacob. No. 4400. Engine lathe for engraving surfaces, printing and coining presses, &c. 17pp. 6 plates. London: Printed by George E. Eyre and William Spottiswoode; published at the Great Seal Patent Office, 1857.
    Issued 1819.

Petter, George William, and Thomas Gilpin Dixon. 1914. Improvements in printing presses. 8pp. 3 plates. Redhill: Printed for His Majesty's Stationery Office by Love & Malcolmson, 1905. Second edition.
    Issued 1859.

Poole, Moses. No. 8159. Flexible moulds for casting printing surfaces. 3pp. London: Printed by George E. Eyre and William Spottiswoode; published at the Great Seal Patent Office, 1857.
    Issued 1839.

Ruthven, John. No. 3746. Printing presses. 7pp. Plate. London: Printed by George E. Eyre and William Spottiswoode; published at the Great Seal Patent Office, 1856.
    Issued 1813.

Rutt, William. No. 4375. Printing machines. 9pp. Plate. London: Printed by George E. Eyre and William Spottiswoode; published at the Great Seal Patent Office, 1857.
    Issued 1819.

Senefelder, John Aloysius. No. 2518. Printing textile fabrics. 18pp. Plate. London: Printed by George E. Eyre and William Spottiswoode; published at the Great Seal Patent Office, 1856.
    Issued 1801.

Treadwell, Daniel. No. 4433. Printing presses. 6pp. Folding plate. London: Printed by George E. Eyre and William Spottiswoode; published at the Great Seal Patent Office, 1857.
    Issued 1820.

[Tucker, Stephen Davis.] Newton, William Edward. No. 1304. Printing machinery. 21pp. 14 plates. Redhill: Printed for His Majesty's Stationery Office by Love & Malcomson, 1905. Second edition.
    Issued 1861.

[Tucker, Stephen Davis.] Newton, Alfred Vincent. No. 1934. An improved mode of and apparatus for stereotype plates. 7pp. 3 plates. London: Printed by George E. Eyre and William Spottiswoode, 1864.
    Issued 1863.

[Tucker, Stephen Davis.] Newton, William Edward. No. 3039. Improvements in printing machinery. Plate. London: Printed by George E. Eyre and William Spottiswoode, 1864.
    Issued 1863.

Watts, Richard. No. 4463. Printing machines. 6pp. Plate. London: Printed by George E. Eyre and William Spottiswoode; published at the Great Seal Patent Office, 1857.
    Issued 1820.

Woods, Joseph. No. 10,219. Producing and multiplying designs and copies. 13pp. 3 plates. London: Printed by George E. Eyre and William Spottiswoode; published at the Great Seal Patent Office, 1856.
    Issued 1844.

Typesetting

Abel, Charles Denton. No. 2031. Composing and distributing type. 8pp. Plate. London: Printed by George E. Eyre and William Spottiswoode; published at the Great Seal Patent Office, 1870.
    Issued 1869.

Alden, Timothy. No. 3089. Machinery for setting and distributing printing types. 42pp. 6 plates. London: Printed by George E. Eyre and William Spottiswoode; published at the Great Seal Patent Office, 1857.
    Issued 1856.

Clay, John, and Frederick Rosenborg. No. 8726. Arranging and setting up types for printing. 10pp. 6 plates. London: Printed by George E. Eyre and William Spottiswoode; published at the Great Seal Patent Office, 1856.
    Issued 1840.

Clay, John, and Frederick Rosenborg. No. 9300. Composition of type for letter-press printing. 14pp. 6 plates. London: Printed by George E. Eyre and William Spottiswoode; published at the Great Seal Patent Office, 1856.
    Issued 1842.

Delcambre, Isidore. No. 2444. Machinery for composing and distributing type. 6pp. 2 plates. London: Printed by George E. Eyre and William Spottiswoode; published at the Great Seal Patent Office, 1857.
    Issued 1856.

Delcambre, Isidore. No. 1496. Composing and distributing type. 6pp. 8 plates. London: Printed by George E. Eyre and William Spottiswoode; published at the Great Seal Patent Office, 1866.
    Issued 1866.

Felt, Charles Wilson. No. 2531. Setting, spacing, &c., printers' type. 34pp. 7 plates. London: Printed by George E. Eyre and William Spottiswoode; published at the Great Seal Patent Office, 1862.
    Issued 1861.

Gaubert, Etienne Robert. No. 8427. Machinery for distributing types. 9pp. 2 plates. London: Printed by George E. Eyre and William Spottiswoode; published at the Great Seal Patent Office, 1856.
    Issued 1840.

Hattersley, Robert. No. 1794. Machinery for distributing and setting up type. 8pp. Plate. London: Printed by George E. Eyre and William Spottiswoode; published at the Great Seal Patent Office, 1857.
    Issued 1857.

Kastenbein, Charles. No. 2864. Composing and distributing type. 16pp. 6 plates. London: Printed by George E. Eyre and William Spottiswoode; published at the Great Seal Patent Office, 1873.
    Issued 1872.

Lake, William Robert. No. 3358. Distributing type. 16pp. 2 plates. London: Printed by George E. Eyre and William Spottiswoode; published at the Great Seal Patent Office, 1870.
    Issued 1869.

Lanston, Tolbert. No. 8183. Improvements in the art of printing. 29pp. 8 plates. London: Printed for Her Majesty's Stationery Office by Darling & Son, 1887.
    Issued 1887.

Mackenzie, Duncan. No. 12,229. Jacquard machinery for figuring fabrics, transmitting designs, composing printing types, etc. 7pp. 5 plates. London: Printed by George E. Eyre and William Spottiswoode; published at the Great Seal Patent Office, 1857.
    Issued 1848.

Mackie, Alexander. No. 3396. Distributing type. 7pp. Plate. Edwards, Ernest. 1869. No. 3543. Photo-mechanical printing, &c. 6pp. London: Printed by George E. Eyre and William Spottiswoode; published at the Great Seal Patent Office, 1867.
    Issued 1866.

Mergenthaler, Ottmar. No. 11,670. Improvements in machines for producing stereotype matrices and for printing. 7pp. 4 plates. London: Eyre and Spottiswoode ... for Her Majesty's Stationery Office, 1885.
    Issued 1884.

Mergenthaler, Ottmar. No. 11,670*. Improvements in machines for producing stereotype matrices and for printing. 7pp. 4 plates. London: Printed for Her Majesty's Stationery Office by Darling & Son, 1890.
    Issued 1884.

[Mergenthaler, Ottmar.] Boult, Alfred Julius. No. 1833. Improvements in machines for producing stereotype matrices and the like. 19pp. 4 plates. London: Eyre and Spottiswoode ... for Her Majesty's Stationery Office, 1885.
    Issued 1885.

[Mergenthaler, Ottmar.] Boult, Alfred Julius. No. 1833*. Improvements in machines for producing stereotype matrices and the like. 19pp. 4 plates. London: Printed

for Her Majesty's Stationery Office by Darling & Son, 1890.
Issued 1885.

[Mergenthaler, Ottmar.] Boult, Alfred Julius. No. 2823. Improvements in machines for producing relief surfaces for letter press printing. 24pp. 5 plates. London: Eyre and Spottiswoode . . . for Her Majesty's Stationery Office, 1885.
Issued 1885.

[Mergenthaler, Ottmar.] Boult, Alfred Julius. No. 2823*. Improvements in machines for producing relief surfaces for letter press printing. 24pp. 5 plates. London: Printed for Her Majesty's Stationery Office by Darling & Son, 1890.
Issued 1885.

[Mergenthaler, Ottmar.] Boult, Alfred Julius. No. 5823. Improvements in machines for producing type bars and matrices for type bars and surfaces for letter press printing. 7pp. 2 plates. London: Eyre and Spottiswoode . . . for Her Majesty's Stationery Office, 1885.
Issued 1885.

[Mergenthaler, Ottmar.] Boult, Alfred Julius. No. 5823*. Improvements in machines for producing type bars and matrices for type bars and surfaces for letter press printing. 7pp. 2 plates. London: Printed for Her Majesty's Stationery Office by Darling & Son, 1890.
Issued 1885.

[Mergenthaler, Ottmar.] Boult, Alfred Julius. No. 8457. Improvements in method of and means for justifying matrices, types, and dies, when assembled or composed in lines. 4pp. Plate. London: Eyre and Spottiswoode . . . for Her Majesty's Stationery Office, 1885.
Issued 1885.

[Mergenthaler, Ottmar.] Boult, Alfred Julius. No. 8457*. Improvements in method of and means for justifying matrices, types, and dies, when assembled or composed in lines. 4pp. Plate. London: Printed for Her Majesty's Stationery Office by Darling & Son, 1890.
Issued 1885.

[Mergenthaler, Ottmar.] Boult, Alfred Julius. No. 9115. Improvement in machines for forming type bars and stereotype matrices. 12pp. 8 plates. London: Eyre and Spottiswoode . . . for Her Majesty's Stationery Office, 1886.
Issued 1886.

[Mergenthaler, Ottmar.] Boult, Alfred Julius. No. 9115*. Improvement in machines for forming type bars and stereotype matrices. 12pp. 8 plates. London: Printed for Her Majesty's Stationery Office by Darling & Son, 1890.
Issued 1886.

[Mergenthaler, Ottmar.] Boult, Alfred Julius. No. 10,525. Improvement in machines for casting type-bars for printing and like purposes. 8pp. 8 plates. London: Eyre and Spottiswoode . . . for Her Majesty's Stationery Office, 1886.
Issued 1886.

[Mergenthaler, Ottmar.] Boult, Alfred Julius. No. 10,525*. Improvement in machines for casting type-bars for printing and like purposes. 8pp. 8 plates. London: Printed for Her Majesty's Stationery Office by Darling & Son, 1890.
Issued 1886.

[Mergenthaler, Ottmar.] Leigh, Henry Harington. No. 5582. Improvements in Linotype machines. 21pp. 15 plates. London: Printed for Her Majesty's Stationery Office by Darling & Son, 1890.
Issued 1890.

Mitchel, William Haslett. No. 155. Machinery for composing and distributing types. 13pp. Plate. London: Printed by George E. Eyre and William Spottiswoode; published at the Great Seal Patent Office, 1857.
Issued 1857.

Young, James Hadden, and Adrien Delcambre. No. 8428. Setting up printing types. 7pp. 2 plates. London: Printed by George E. Eyre and William Spottiswoode; published at the Great Seal Patent Office, 1857.
Issued 1840.

IMAGES

Bourdelin, E. Imprimerie impériale.—La réglure. 1860. Colored wood engraving.
Extract from Julien François Turgan, L'Imprimerie impériale. Première partie: Fabrication des caractères, gravure, fonderie, etc. (Paris: A. Bourdilliat et Cie.), p. 120.

Bourdelin. La papeterie d'Essone.—Les machines Amédée et Auguste. 1861. Hand-colored wood engraving.
From Julien François Turgan, Les grandes usines de France, vol. 1 (Paris: Bourdillat).

Collings. Conflagration! Or the merry mealmongers, a new dance, as it was performed with universal applause, at the Theatre Blackfriars March 2d, 1791. 1 April 1791. Etching. N.p.: Bentley & Co.
"Etched by Barlow." From the Attic Miscellany (1791).

Deutsche Bundespost. Friedrich Koenig. 150 Jahre Druckmaschinen. 1968. Postage stamp.

Deutsche Bundespost. 125 Jahre Lette-Verein. 1991. Postage stamp.
Commemorating the founding of the Lette-Verein, a technical school for girls that included a school for compositors. The stamp's image shows women typesetting.

Exposition Marinoni. 1889. Photogravure. Paris: Motteroz; Ludovic Baschet.
Plate from F. G. Dumas and L. de Foucard, L'Exposition universelle de 1889, vol. 2 (Paris: Motteroz; Ludovic Baschet).

F. Appel. Imprimerie-lithographie. 1867 or later. Chromolithograph.
Illustrated advertisement.

[Faithfull, Emily.] Carte-de-visite portrait of Emily Faithfull. Ca. 1867. Photograph. London: London Stereoscopic & Photographic Company.
Signed by Faithfull below the photograph.

Farey, J. Printing. The Stanhope or iron press. 1815. Engraving. London: Longman, Hurst, Rees, Orme & Brown.
Engraved by Wilson Lowrie. Plate from Abraham Rees's Cyclopaedia or Useful Dictionary (London: Longman, Hurst, Rees, Orme & Brown, 1802-1820).

Flamm's typographic compositor (class 59). Delcambre's type-distributing machine (class 59). Delcambre's type-composing machine (class 59). 7 December 1867. Wood engravings. The Illustrated London News 51, no. 1458: 632.
Extract.

Fourdrinier paper machines. 1870. Wood engraving. Hartford, CT: L. Stebbins.
Extract from One Hundred Years' Progress of the United States (Hartford, CT: L. Stebbins).

Grant, C. J. Frontispiece for the Penny Magazine of the Society for the Diffusion of Useful Knowledge. N.d. [1832]. Lithograph.
Lithographed by G. Davies.

"The Illustrated London News" steam printing machines. 2 December 1843. Wood engraving. The Illustrated London News 3, no. 83.
Extract.

Interior view of our press-room, where are printed the Pictorial Drawing Room Companion, and the Flag of our Union. 29 May 1852. Wood engraving. Gleason's Pictorial Drawing Room Companion 2, no. 22: 352.
Extract.

Inventions illustrés. L'imprimerie. 2 June 1901. Chromolithograph. Le Petit Journal Supplément Illustré 12, no. 550: 173.
Shows Marinoni demonstrating one of his presses, with a frieze above illustrating printing in Gutenberg's time.

An iron foundry, Coalbrook Dale. 1 November 1799. Hand-colored mezzotint/etching. London: R. Ackermann's Repository of Arts.

J. Minot et Cie. Lithographie artistique . . . Imprimé sur la nouvelle presse lithographique Marinoni . . . 1889. Chromolithograph.
Advertising card.

Joseph Moxon. Born at Wakefeild [sic] August 8. Anno 1627. 1699. Engraving. London: W. Hawes.
Used as the frontispiece for the fifth edition of Moxon's A Tutor to Astronomy and Geography (London: W. Hawes, 1699).

[Knight, Charles.] Carte-de-visite portrait of Charles Knight. 1867. Photograph. [London: London Stereoscopic & Photographic Company.]

[Knight, Charles.] Portrait of Charles Knight by an unidentified artist. Ca. 1920. Engraving.

Manufacture of cotton. 1835. Colored wood engraving.

Extract from Andrew Ure, The Philosophy of Manufactures (London: Charles Knight).

Mackie's steam type-composing machine. 18 November 1871. Wood engraving. Every Saturday, new series, 3, no. 99: 497.
> Extract.

Mayall, [John J. E.]. Carte-de-visite portrait of Henry Brougham, 1st Baron Brougham and Vaux. 1 June 1861. Photograph.

Meyer, W. Neue Setzmaschine. Ca. 1890. Wood engraving.
> Illustration of Alexander Lagerman's "Typotheter" typesetting machine, from an unidentified periodical.

Morton, G. The new steam carriage. N.d. [1828]. Hand-colored engraving.
> Engraved by Pyall.

Mr. Gladstone at the Caxton Memorial Exhibition, South Kensington, on Saturday last. 7 July 1877. Wood engraving. Supplement to the Illustrated London News 71, no. 1982: 17.

Neumann, A. A. Auer. Ca. 1860. Lithograph.
> "X. A. v. Eduard Kretschmer sc." Portrait of Alois Auer, director of the Imperial and State Printing Office in Vienna.

Paris exhibition: The Ingram rotary machine for printing illustrated newspapers. 14 September 1878. Wood engraving. The Illustrated London News 73, no. 2046, p. 253.

The printer & the letter-founder. Der Buchdrucker & der Schriftgieszer. Ca. 1850. Hand-colored lithograph. Mainz: Joseph Scholz; Cincinnati: Eggers & Co.

Prud'homme, Antoine Daniel. Charles Stanhope, Viscount Mahon. 1 November 1775. Mezzotint. London: W. Shropshire.
> Engraved by Thomas Watson.

Pugin, Auguste Charles, and Thomas Rowlandson. Fire in London. 1 September 1808. Hand-colored aquatint. London: R. Ackermann's Repository of Arts.
> "J. Black Aquat." From the Microcosm of London (1808–1810).

Roffe. Charles Babbage Lucasian Professor of Mathematics in the University of Cambridge. 1 May 1833. Engraving. N.p.: M. Salmon, Mechanics Institute Office.

Rossin. Emile de Girardin. Paris, n.d. [ca. 1836]. Engraving.
> Published around the time Girardin began publishing La presse.

Schlegel, George. 2 chromolithographed cigar box labels for "La flor de R. M. Hoe" cigars. New York: Litho. Geo. Schlegel, n.d. [ca. 1890].
> The labels feature a portrait of Richard March Hoe. The small octagonal label includes the phrase "La flor de R. M. Hoe" in the upper banner; the large rectangular label has a blank upper banner.

Shortshanks [Seymour, Robert.] Shaving by steam. N.d. [ca. 1826]. Hand-colored engraving. [London:] Pub. by S. King, Chancery Lane.

Shortshanks [Seymour, Robert]. The march of intellect. Ca. 1828. Hand-colored engraving. N.p.: G. Humphrey.

Shortshanks [Seymour, Robert.] Locomotion. Walking by steam. Riding by steam. Flying by steam. 1835. Hand-colored engraving. London: Thos. McLean.

[Seymour, Robert.] Locomotion. Walking by steam. Riding by steam. Flying by steam. Ca. 1830. London: Thos. McLean.

A steam coach with some of the machinery going wrong. N.d. [ca. 1824–1826]. Hand-colored engraving. London: Thos. McLean.
> Caption reads: "Something wrong. My eyes Bob, if our Parson ha'nt lost his living."

Steam engine. Side view or section of one of the Albion Mill steam engines. 1830. Engraving. Edinburgh: W. Blackwood.
> Extract from The Edinburgh Encyclopaedia (Edinburgh: W. Blackwood).

Stereoscopic view of a woman and four men working at a Hoe rotary press or similar machine. Ca. 1855. Photograph.

Stradanus, Johannes. Impressio librorum. 1590. Engraving.
> From Nova reperta (ca. 1590). Framed.

Summerfield, J. Mr. Thomas Bewick, restorer of the art of engraving on wood. From an original miniature by Murphy in the possession of Mr. Bewick . . . 1 February 1816. Engraving. London: T. McLean.

Tavernier, Jules, and Paul Frénzeny. The manufacture of iron: Filling the furnace. 1873. Hand-colored wood engraving. Harper's Weekly 17, no. 879: 964.
> Extract.

Taylor's cylinder printing press. 22 November 1851. Wood engraving. Gleason's Pictorial Drawing Room Companion 1, no. 30: 480.
> Extract.

Th. Dupuy. Cette épreuve en quatorze couleurs a été tirée publiquement au Palais de l'Exposition internationale de 1867 . . . sur la presse chromolithographique de Th. Dupuy . . . 1867. Chromolithograph.

The "Times" newspaper printing machine. 12 October 1833. Wood engraving. London: Published for the proprietors, by W. Edwards.

Various processes of bookbinding. 1845. Wood engraving. The Pictorial Gallery of Arts, vol. 1, p. 356.

Visit of the Prince of Wales and Prince Alfred to Messrs. Day and Son's lithographic establishment. 3 May 1856. Wood engraving. The Illustrated London News 28, no. 797.
> Extract.

The Walter press employed for printing the "Times" newspaper. Ca. 1872. Colored wood engraving.
> Probably an extract from The Illustrated London News.

The Walter printing machine. 10 September 1877. Wood engraving. The Missouri Republican 65, no. 17,357.
> Caption reads: "Fastest press in the world—capacity 20,000 perfect copies per hour, in use by the Missouri Republican." This issue of The Missouri Republican also includes an image titled "The Bullock printing press."

MANUSCRIPT / TYPESCRIPT MATERIAL

Bryan, Elmer J. The Raffold plant of the Mead Pulp and Paper Company, Chillicothe, Ohio, its equipment, operation and control. Andover, MA, 23 February 1929.
> Carbon typescript.

Cave, Edward. Autograph letter signed to Lewis Paul. 2pp. N.p., 14 October 1740.

Cowper, Edward Shickle. Autograph letter signed to an unidentified correspondent. 3pp. Bayswater [London], 12 August 1842.

Day & Haghe, Lithographers. Invoice addressed to James Wilson, architect. 1 page. London: 20 April 1842.
> Lithographed document completed in ink.

Dickens, Charles. Autograph letter signed to Edward Chapman. 2pp. Chester Place [London], 26 April 1847.

Faithfull, Emily. Autograph letter signed to an unidentified correspondent. 1 page. London, 6 February 1860.

Peel, Robert. Autograph letter signed to an unidentified correspondent. 1 page. N.p., n.d. [November 1802].
> Re: posting printed copies of the 1802 Apprentice Act in mills and factories.

Thomas Long & Co. Invoice addressed to Messrs. Blackwood & Sons, booksellers. 3pp. Edinburgh, December 1858.
> Printed document completed in ink.

Ure, Andrew. Autograph letter signed to Davies Gilbert. 2pp. Glasgow, 9 December 1821.
> Accompanied by an engraved portrait of Ure by C. Cook after a photograph by W. H. Diamond.

PRINTED MATERIAL

1700s

Anisson, Étienne-Alexandre-Jacques. Premier mémoire sur l'impression en lettres, suivi de la description d'une nouvelle presse. Mémoires de mathématiques et de physique des Scavants Etrangers 10 (1785): 613-650.

Bewick, Thomas. A general history of quadrupeds. The figures engraved on wood ... Newcastle upon Tyne: S. Hodgson, R. Beilby, & T. Bewick, 1790.

By the King. A proclamation for suppressing riots and tumults committed by colliers, and others, in the counties of Northumberland and Durham, and for apprehending and bringing to justice the persons who have committed or shall commit the same. N.p., 1789.
Broadside.

[Deltusso, C.] Pétition à la Convention national pour l'École typographique des femmes. Paris: École typographique des femmes, [1794].

Flesselles, Jacques de. Arrêt du conseil d'état du roi, qui ordonne l'exécution de celui du 27 janvier 1739, pour règlement pour les papeteries. Lyon: Imprimerie de P. Valfray, Imprimeur du Roi, 1772.

Flesselles, Jacques de. Arrêt du conseil d'état du roi, qui condamne en des amendes, l'entrepreneur de la manufacture de papier établie à la Motte près Verberie, ainsi que quelques-uns de ses ouvriers, ci-devant employés à celle de Courtalin, près Faremoutier en Brie: Et ordonne en outre l'exécution du règlement du 27 janvier 1739, concernant les papeteries du royaume. Lyon: Imprimerie du Roi, 1777.

Great Britain, Parliament. An act for the more effectual punishment of such persons as shall demolish or pull down, burn, or otherwise destroy or spoil, any mill or mills, and for preventing the destroying or damaging of engines for draining collieries and mines; or bridges, wagon ways, or other things used in conveying coals, lead, tin, or other minerals, from mines; or fences for inclosing lands in pursuance of acts of Parliament. London: Mark Baskett, 1769.

Lacoré, Charles-André de. Arrest du conseil d'état du roi, qui condamne en des amendes l'entrepreneur de la manufacture de papier établie à la Motte, près Verberie, ainsi que quelques-uns de ses ouvriers, ci-devant employés à celle de Courtalin, près Faremoutier en Brie: Et ordonne en outre l'exécution du règlement du 27 janvier 1739, concernant les papeteries du royaume. N.p., 1777.

Lalande, Jérome de. Art de faire le papier. [Paris: Desaint et Saillant, 1761.]
Descriptions des arts et métiers, vol. 4.

London, Leeds, Wakefield, Barnsley, Sheffield and Mansfield machine ... Leeds: N.p., 2 November 1773.
Broadside advertising coach travel from Yorkshire to London.

[More, Hannah.] Black Giles the poacher; with some account of a family who had rather live by their wits than their work. Part I. London: Sold by J. Marshall ..., n.d. [1796].
A "Cheap Repository" tract.

Nicholson, William. Specification of the patent granted ... for a machine or instrument for printing on paper, linen, cotton, woollen, and other articles, in a more neat, cheap, and accurate manner, than is effected by the machines now in use. The Repertory of Arts and Manufactures 5 (1796): 145-170.
With 3 plates.

Philippeaux. Réponse ... à tous les défenseurs officieux des bourreaux de nos frères dans la Vendée, avec l'acte solemnel d'accusation, fait à la séance du 18 nivôse, suivie de trois lettres écrites à sa femme, de sa prison. Paris: De l'Imprimerie des Femmes, An III [1794-1795].

Smith, Charlotte. Elegiac sonnets and other poems. Worcester, MA: Isaiah Thomas, 1795.
First American book printed on wove paper.

Thibaudeau, Antoine-Claire. Histoire du terrorisme dans le département de la Vienne. Paris: Chez le citoyen Maret ... et chez la citoyenne Barbier; de l'Imprimerie des Femmes, sous les auspices de la Convention nationale, [1795].

The trial of a cause instituted by Richard Pepper Arden, Esq; His Majesty's attorney general, by writ of *scire facias*, to repeal a patent granted on the sixteenth of December 1775, to Mr. Richard Arkwright, for an invention of certain instruments and machines for preparing silk, cotton, flax, and wool for spinning; before the Honorable Francis Buller, one of the judges of his Majesty's court of King's Bench, at Westminster Hall, on Saturday the 25th of June 1785. London: Hughes and Watson, 1785.

1800s

100 guineas reward. Whereas between the hours of eight and ten o'clock on Saturday night last, the 31st of March, the machine engine at the Street pit ... was maliciously set on fire ... Newcastle: Akenheads, printers, 2 April 1821.
Broadside.

Abbott, Jacob. The Harper establishment; or, how the story books are made. New York: Harper & Brothers, 1855.
Harper's Story Books, no. 4.

Adams, Thomas F. Typographia; or, the printer's instructor: A brief sketch of the origin, rise and progress of the typographic art, with practical directions for conducting every department in an office, hints to authors, publishers, &c. Philadelphia: L. Johnson & Co., 1858.

Album des célébrités industrielles contemporaines, vol. 2, 1865. Paris: Direction, Administration, rue Violet, 1865.
Includes illustrations of Alauzet's printing machine factory and Th. Dupuy's steam-powered lithographic establishment.

Alden Type Setting & Distributing Machine Co. Stock certificate no. 434 (unissued). New York, n.d. [ca. 1863.]
The certificate includes an illustration of the Alden typesetting machine and a small image of a Hoe rotary printing machine.

Alfred. *See* Kydd, Samuel.

Alfred Mame et fils. Imprimerie—librairie—relieure. Notice et spécimens. Tours: Alfred Mame, 1867.

[Alkan, Alphonse aîné.] Les femmes compositrices d'imprimerie sous la revolution française en 1794. Par un ancien typographe. Paris: Dentu, 1862.

Allgemeine Zeitung, no. 185 (3 July 1824).
The first publication printed in Germany by the first printing machine built in Germany. It represents a test run of Koenig & Bauer's *Schnellpresse*, built at their factory in Oberzell.

Allgemeine Zeitung, no. 194 (12 July 1824).
3,000 copies of the 12 July 1824 issue of the Allgemeine Zeitung were printed on Koenig & Bauer's *Schnellpresse*, but the press's steam engine failed, so that the machine had to be powered by hand.

Allgemeine Zeitung, no. 267 (23 September 1824).

L'Almanach de France indiquant à tous les Français qui savent lire ... [Année 1833]. Paris: Société pour l'Émancipation Intellectuelle, 1833.
"Publié à un million trois cent mille exemplaires."

Almanach der France publié par la Société nationale. Paris: Auguste Desrez, 1838.

American Bible Society. An abstract ... containing an account of its principles and operations, and of the manner of organizing and conducting auxiliary and branch bible societies. New York: Printed by Daniel Fanshaw, 1830.

American Bible Society. The first annual report of the board of managers ... presented May 8, 1817. With an appendix containing extracts of correspondence, &c. New York: Printed for the Society by J. Seymour, 1817.

American Bible Society. Ninth report of the American Bible Society, presented May 12, 1825. New York: Printed for the Society, by Abraham Paul, 1825.

American Society for the Diffusion of Useful Knowledge. Prospectus. New York: Published by the Committee, 1837.

American Tract Society. First annual report ... with lists of auxiliaries and benefactors, addresses at the anniversary, &c. New York: Printed at the Society's house ... by Daniel Fanshaw, 1826.

Applegath's vertical printing machine, exhibited by the proprietors of the "Illustrated London News." The Illustrated London News 18, no. 489 (31 May 1851): 501-502.
Extract. Includes a wood-engraved illustration of Applegath's machine.

Appletons' railway and steam navigation guide. New York: Appletons, May 1869.

Arguments of the judge-advocate and of Mr. R. T. Merrick, private counsel for Gen. Hazen, in the Stanley trial. New York: S. W. Green's Type-Setting Machines, 1879.

Arlidge, J. T. The hygiene diseases and mortality of occupations. London: Percival & Co., 1892.

Audouin de Géronval, Maurice Ernst. Manuel de l'imprimeur, ou traité simplifié de la typographie. Paris: Imprimerie de Crapelet, 1826.
    First French printing manual to discuss machine printing.

[Auer von Welsbach, Alois.] Geschichte der K.K. Hof- und Staats-Druckerei in Wien. Vienna: Aus der Kaiserliche-Koeniglichen Hof- und Staats-Druckerei, 1851.

Babbage, Charles. On the economy of machinery and manufactures. London: Charles Knight, 1832.

[Babbage, Charles.] "Manufactures." P. Barlow, A Treatise on the Manufactures and Machinery of Great Britain (London: Baldwin and Cradock, 1836), pp. 1-84.

Baines, Edward, Jr. History of the cotton manufacture in Great Britain: With a notice of its early history in the east, and in all the quarters of the globe; a description of the great mechanical inventions which have caused its unexampled extension in Britain; and a view of the present state of the manufacture and the condition of the classes engaged in its several departments. London: H. Fisher, R. Fisher and P. Jackson, [1835.]

Bakewell, Frederick C. Great facts: A popular history and description of the most remarkable inventions during the present century. New York: D. Appleton & Co., 1860.

Barnes, William C., Joseph W. McCann, and Alexander Duguid. A collation of facts relative to fast typesetting, together with portraits and biographies... New York: Concord Co-operative, 1887.

Barthe, J.-G. Le Canada reconquis par la France. Paris: Ledoyen, 1855.
    "Paris.—Typographique mécanique d'Adrien Delcambre et Cie. 15, rue Breda." Advertisement for the Delcambre typesetting machine after p. 416.

Baxter, George. Baxter's gems of the Great Exhibition. London: Published at the offices of the patentee, [1851].

Becquey. Instruction sur la legislation relative aux brevets d'invention. Paris: Imprimerie Royale, July 1817.
    Bound in volume of pamphlets relating to the legal dispute between Léger Didot and Berte.

The beggar's complaint, against rack-rent landlords, corn factors, great farmers, monopolizers, paper money makers, and war, and many other oppressors and oppressions. Also, some observations on the conduct of the Luddites, in reference to the destruction of machinery. Sheffield: Printed for the author by J. Crome, 1812 [i.e., 1813].
    "The second edition, greatly enlarged." Front wrapper dated 22 June 1813.

Berger-Levrault et Cie. L'imprimerie Berger-Levrault et Cie. Notice historique sur le développement et l'organisation de la maison. Nancy and Paris: Berger-Levrault et Cie., 1878.

Berte, A.-J. 16 octobre 1811. Brevet d'importation de quinze ans pour une machine à fabriquer le papier de differentes longueurs et largeurs. Déscription des machines et procédés specifies dans les brevets d'invention de perfectionnement et d'importation dont la durée est expiré, vol. 5 (Paris: Mme. Huzard, 1828), pp. 180-197.
    Plate 19 in the volume includes illustrations of Berte's machine.

Beschreibung der Druckmaschine oder mechanischen Presse, worauf Friedrich König und Georg Fried. Bauer zu Nürnberg ein Privilegium erhielten. Königlich-Bayerisches Intelligenzblatt für den Isarkreis (1831): cols. 975-982.
    With 2 folding plates.

Beschreibung der Maschine zum Verfertigen und Trocknen des Papiers, worauf Friedr. König und Andr. Friedr. Bauer zu Oberzell unterm 14 März 1828 ein allergnädigstes Privilegium auf 15 Jahre erhielt. Königlich-Bayerisches Intelligenzblatt für den Isarkreis (1831): cols. 802-806.

Beurtheilungen über die k.k. Hof- und Staatsdruckerei in Wien. Vienna: Aus der kaiserl. königl. Hof- und Staatsdruckerei, 1852.

Bevan, G. Phillips, ed. British manufacturing industries. London: Edward Stanford, 1876.
    Includes chapters on paper, printing and bookbinding, engraving, photography.

Binny & Ronaldson. A specimen of metal ornaments cast at the letter foundry of Binny & Robinson. Philadelphia: Fry and Kammerer, 1809 [Meriden, CT: Meriden Gravure, 1922 or later].
    Facsimile ed.

Binns, Edward. The anatomy of sleep; or, the art of procuring sound and refreshing slumber at will. London: John Churchill, 1842.
    Note on title verso: "Printed by J. H. Young, by the Patent Composing Machine." Inscribed by the author.

Binns, Edward. The anatomy of sleep; or, the art of procuring sound and refreshing slumber at will. London: John Churchill, 1842.
    In the original cloth binding.

Binns, Edward. The anatomy of sleep, or, the art of procuring sound and refreshing slumber at will. London: John Churchill, 1845.
    Second edition.

Blades, William. The Pentateuch of printing with a chapter on Judges. With a memoir of the author, and list of his works, by Talbot B. Reed. London: Elliot Stock, 1891.

Blades, William. The Pentateuch of printing with a chapter on Judges. With a memoir of the author, and list of his works, by Talbot B. Reed. London: Elliot Stock, 1891.
    No. 35 of 100 copies printed on large paper.

Blades, William. The Pentateuch of printing with a chapter on Judges. With a memoir of the author, and list of his works, by Talbot B. Reed. Chicago: A. C. McClurg and Co., 1891.

Boulay, A. de. Précis historique sur la typographie. L'Univers illustré 2 (1859): 199-200.
    Extract. Includes an illustration of a Marinoni color-printing press, captioned "Presse Marinoni, pour les impressions de vignettes et les tirages en couleur."

British and Foreign Bible Society. The first report of the British and Foreign Binding Society. MDCCCV. With an appendix and a list of subscribers and benefactors. London: Printed by the Philanthropic Society, 1805.

Brade, Ludwig. L. Brade's illustriertes Buchbinderbuch. Leipzig: Otto Spamer, 1868.
    Second edition.

British and Foreign Bible Society. An appeal to mechanics, labourers, and others respecting bible associations. Bradford: Printed for Stump and Bubb, 1816.
    Broadside.

Brégeaut, R. L. Manuel complet théorique et pratique du dessinateur et de l'imprimeur lithographe. Paris: Roret, 1827.
    Second edition.

[Brontë, Charlotte.] Bell, Currer. Shirley. A tale. London: Smith, Elder and Co., 1849.
    3 volumes.

[Brougham, Henry.] A discourse of the objects, advantages, and pleasures of science. London: Baldwin, Cradock and Joy, 1827.

Brougham, Henry. Practical observations upon the education of the people, addressed to the working classes and their employers. London: Printed by Richard Taylor; and sold by Longman, Hurst, Rees, Orme, Brown and Green for the benefit of the London Mechanic's Institution, 1825.

Brougham, Henry. Practical observations upon the education of the people, addressed to the working classes and their employers. Manchester: Archibald Prentice, 1825.
    "For gratuitous distribution to the readers of the Manchester Guardian, 19th March, 1825."

Brougham, Henry. Speech... in the House of Lords, on Thursday, May 23, 1835. On the education of the people. To which is added a summary of the last education returns to Parliament. London: James Ridgeway and Sons, 1835.

Brougham, Henry. Taxes on knowledge. Stamps on newspapers. Extracts from the evidence of the Right Honourable Baron Brougham and Vaux... before the select committee of the House of Commons on libel law, in June 1834... London: Robert Heward, at the Office of the Westminster Review, 1834.

Brown's patent type-setting and distributing machinery. Scientific American 24, no. 15 (8 April 1871): 223-224.
    Illustrated.

Brown, Thomas. Lectures on the philosophy of the human mind. Corrected from the last London edition. Boston: S. T. Armstrong..., 1826.

Printed on Daniel Treadwell's machine press. "Treadwell's Power Press.—J. G. Rogers & Co." on the title.

Burr Printing House. Invoice addressed to H. la France. New York, 9 March 1882.
  The Burr typesetter and distributor are illustrated at the top of the document.

Bury, T. H. Six coloured views on the Liverpool and Manchester railway, with a plate of the coaches, machines, &c. from drawings made on the spot. London: R. Ackermann, 1831.

Byron, George Gordon, 6th Baron Byron. The works of Lord Byron: With his letters and journals, and his life, by Thomas Moore, Esq. London: John Murray, 1832-1833.
  17 volumes. Vol. 2 of the Murray Byron represents the first application of gilt stamping directly to a cloth binding; vol. 1 has a gilt-stamped paper label on the spine.

The cabinet of useful arts and manufactures: Designed for the perusal of young persons. Dublin: Thomas Courtney, 1821.
  Discussion of machine printing and stereotyping on pp. 123-124.

Campbell, Thomas. The pleasures of hope, with other poems. A new edition. London: Longman, Hurst, Rees, Orme, and Brown; Edinburgh: Sterling and Slade, 1822.
  First book published by Longman to contain steel engravings.

Camus, Armand-Gaston. Histoire et procédés du polytypage et de la stéréotypie. Paris: Baudouin, An X [1801].

Carey's general atlas, improved and enlarged; being a collection of maps of the world and quarters; their principal empire, kingdoms, &c. Philadelphia: M. Carey and Son, 1818.
  Third edition.

Carey, Annie. The history of a book. London, Paris and New York: Cassell, Petter & Galpin, [1873].

Chaix, Alban. Historique de l'imprimerie et de la librairie centrales des chemins de fer: Organisation industrielle et économique de cet établissement. Paris: A. Chaix et Cie, 1878.
  Inscribed by the author.

Chambers' Edinburgh journal, vols. 1-7 (only). London: Orr and Smith, 1835-1839.
  First 12 issues are later editions.

Chambers' historical newspaper. Nos. 1-26 (2 November 1832-December 1834).
  All published.

Chambers, William, and Robert Chambers. Chambers's information for the people. Edinburgh: W. and R. Chambers; London: Orr and Smith, 1835.

A Christmas ghost story [part 1]. Parley's Illuminated Library, vol. 1 (London: N.p., 1844), pp. 241-256.

Christmas supplement to The Illustrated London News 27, no. 776 (22 December 1855).
  First newspaper printed in color.

Clarkson, Thomas. Memoirs of the private and public life of William Penn, vol. 1 [of 2]. London: Richard Taylor for Longman, Hurst, Ress, Orme, and Brown, 1813.
  Sheets G (pp. 81-96) and X (pp. 305-320) in this volume were the first sheets printed on F. Koenig's cylinder flatbed press.

Colliery explosion at St. Hilda's pit, South Shields, June 28th, 1839. Gateshead: Mary Stephenson, printer, 1839.

The composing and distributing machines. The Illustrated London News 1, no. 26 (5 November 1842): 413.
  Illustrated. Describes the Rosenborg typesetting and distributing machines.

The compositor's handbook: Designed as a guide in the composing room. With the practice as to book, job, newspaper, law, and parliamentary work; the London scale of prices; appendix of terms, &c. London: Simpkin, Marshall and Co., 1854.

The costume of Yorkshire, illustrated by a series of forty engravings, being fac-similes of original drawings, with descriptions in English and French. London: T. Bensley for Longman, Hurst, Rees, Orme, and Brown; Ackermann; Robinson, Son and Holdsworth, 1814.

Cowie, [George]. Cowie's printer's pocket-book and manual, containing the compositors' and pressmen's scale of prices, agreed upon in 1810 and modified in 1816 . . . London: W. Strange, [1835].
  Second edition.

Cowper, Edward. Cowper, machinist. High Street, Newington, patentee of the improved paper cutting machine. London: W. J. White, n.d. [ca. 1813-1816].
  Engraved trade card.

Cowper, Edward. On printing machines, especially those used for printing "The Times" newspaper. Institution of Civil Engineers, Minutes of Proceedings 9 (1850): 409-427.
  Extract.

Cowper, Edward. On the recent improvements in the art of printing. The Quarterly Journal of Science, Literature, and Art (January-June 1828): 183-191.

Crichton, Andrew. Converts from infidelity: Or lives of eminent individuals who have renounced libertine principles and sceptial opinions, and embraced Christianity, vol. 1 [2]. Edinburgh: Printed for Constable and Co., 1827.
  Vols. 6 and 7 of Constable's Miscellany of Original and Selected Publications in the Various Departments of Literature, Science, & the Arts.

[Davids, Thaddeus.] The history of ink including its etymology, chemistry, and bibliography. New York: Thaddeus Davids, [1860].

Davis, Charles Thomas. The manufacture of paper: Being a description of the various processes for the fabrication, coloring, and finishing of every kind of paper . . . Philadelphia: Henry Carey Baird; London: Sampson Low, Marston, Searle & Rivington, 1886.

Davis, Daniel. Davis's manual of magnetism. Including also electro-magnetism, magneto-electricity, and thermo-electricity. With a description of the electrotype process. Boston: Daniel Davis, Jr., 1842.
  Frontispiece reproduces an engraved image and its electrotyped copy.

A day at a bookbinder's. The Penny Magazine 11, no. 673 (September 1842): 377-384.
  Extract.

De Vinne, Theodore Low. The printing of "The Century." The Century Illustrated Monthly Magazine 41, no. 1 (November 1890): 87-99.

Delaplaine, Joseph. Delaplaine's repository of the lives and portraits of distinguished Americans. Philadelphia: William Brown for Joseph Delaplaine, 1817-1818.
  Vol. I, part 2, and vol. II, part 1, only.

Deuxième procès des ouvriers typographes en première instance et en appel. Juillet 1862. Paris: Lucien Marpon, 1862.

[Dickens, Charles.] Bentley's miscellany, edited by Boz, vol. 1, nos. 1-6 (2 January-1 June 1837).
  The March 1837 number includes the "Extraordinary Gazette" announcing the coming of Oliver Twist; the masthead vignette, by George Cruikshank ("Phiz") contains the earliest drawing of Dickens recorded in Wilkins's Dickens in Cartoon and Caricature (1924).

[Dickens, Charles.] The posthumous papers of the Pickwick Club. London: Chapman and Hall, 1836-1837 [London: Piccadilly Fountain Press, 1931].
  Facsimile of the first edition in parts.

Dickens, Charles. The posthumous papers of the Pickwick Club. London: Chapman and Hall, 1847.
  Contains the first printing of Dickens' famous introduction in which he expressed his desire to make information available at a reasonable price.

Didot, contre le Sieur Berte. [Paris:] Guyot et Scribe, 1823.
  Bound in a volume of pamphlets relating to the legal dispute between Léger Didot and Berte.

Dodd, George. Days at the factories; or, the manufacturing industry of Great Britain described, and illustrated by numerous engravings of machines and process. Series I.—London. London: Charles Knight, 1843.

Dodd, George. Dodd's curiosities of industry. London: H. Lea, [1852].

Doyle, James E. A chronicle of England B.C. 55-A.D. 1485. London: Longman, Green, Longman, Roberts & Green, 1864.
  "The designs engraved and printed in color by Edmund Evans."

Dupont, Paul. Une imprimerie en 1867. Paris: Imprimerie et Librairie Administratives, 1867.
  Inscribed by the author.

Dupont, Paul. Une imprimerie en 1867. Paris: Imprimerie et Librairie Administratives, 1867.
  In original wrappers.

Dürer, Albrecht. Albert Durers designs of the Prayer Book. London: R. Ackermann's Lithographic Press, 1 September 1817.

Dusautoy, J. A. The paper-maker's ready reckoner; or, calculations to shew the prime cost of any ream of writing or printing paper. Romsey: Printed by J. S. Hollis, for, and sold by the author, 1805.

Die ersten Druckmaschinen. Erbaut in London bis zu dem Jahre 1818 von Friedrich Koenig und Andreas Friedrich Bauer. The first printing machines. Constructed in London up to the year 1818 by Friedrich Koenig and Andreas Friedrich Bauer. Leipzig: F. A. Brockhaus, 1851 [London: International Printing Machinery and Allied Trades Exhibition, 1971].
Facsimile edition.

Earhart, John F. The color printer: A treatise on the use of colors in typographic printing. Cincinnati: Earhart & Richardson, 1892.

Evans, Edmund. The art album: Sixteen facsimiles of water-colour drawings ... engraved and printed by Edmund Evans. London: W. Kent and Co., 1861.

Evelyn, Chetwood. Table-talk or books, men, and manners. New York: George P. Putnam & Co., 1853.
On the front cover: "Putnam's semi-monthly library for travellers and the fireside ... No. XXI ... Nov. 15, 1852."

Exhibition supplement to The Illustrated London Newsss 18, no. 492 (7 June 1851).
Note at the foot of p. 521: "This sheet was printed in the Great Exhibition."

Falkenstein, Karl. Geschichte der Buchdruckerkunst in ihrer Entstehung und Ausbildung. Leipzig: B. G. Teubner, 1840.

Faulmann, Karl. Illustrierte Geschichte der Buchdruckerkunst mit besonderer Berücksichtigung ihrer technischen Entwicklung bis zur Gegenwart. Vienna: A. Hartlebens Verlag, 1882.

Felkin, William. A history of the machine-wrought hosiery and lace manufactures. Cambridge: W. Metcalfe, 1867.

Fessenden, Thomas G. An essay on the law of patents for new inventions. Boston: D. Mallory & Co., 1810 [Clark, NJ: The Lawbook Exchange, Ltd., 2003].
Facsimile ed.

Finishers' Friendly Association. The book-finisher's friendly circular. London: Printed for the Association, 1845-1851 [New York: Garland Publishing, 1990].
Facsimile edition.

Five black arts. A popular account of the history, processes of manufacture, and uses of printing, gas-light, pottery, glass, iron. Columbus: Follett, Foster and Co., 1861.
Frontispiece: "Hoe's six cylinder printing press."

Folds, John S., and Richard Shorrocks. From the compositor and pressman. Proceedings of the Wisconsin Editorial Association, First, Second and Third Sessions (Madison, WI: Carpenter & Hyer, 1859), pp. 179-186.

Foucher Frères. Matériel typographique Foucher Frères ... Catalogue général pour la fonderie en caractères, la composition & l'mpression, la clicherie & la galvanoplastie, la brochure & la reliure, la zincographie & la photogravure. Paris: Foucher Frères, 1886.

Freeling, Arthur. Lacey's railway companion, and Liverpool and Manchester guide: Describing all the scenery on, and contiguous to, the railway; and pointing out to the visitor at both places all that is interesting and necessary for business and pleasure. Liverpool: Printed for Henry Lacey, [1836].

Frey, A. Manuel nouveau de typographie. Imprimerie, contenant les principes théoriques et pratiques de l'imprimeur-typographe. Paris: Librairie Encyclopédique de Roret, 1835.
2 volumes in 1.

Frey, A. Nouveau manuel complet de typographie contenant les principes théoriques et pratiques de cet art. Nouvelle édition, corrigée et augmentee par M. E. Bouchez. Paris: Librairie Encyclopédique de Roret, 1857.
2 volumes.

Freylinghausen, John Anastasius. An abstract of the whole doctrine of the Christian religion. London: Stereotyped and printed by A. Wilson for Edward Harding; sold by T. Cadell and W. Davies, 1804.
First book printed using the Stanhope stereotype process.

Fritz, Georg. Die K.K. Hof- und Staatsdruckerei und deren technischen Einrichtungen. Vienna: Kaiserlich-Königlichen Hof- und Staatsdruckerei, 1894.

Gaskell, P. Artisans and machinery: The moral and physical condition of the manufacturing population considered with reference to mechanical substitutes for human labour. London: John W. Parker, 1836.

Gaskill, Jackson. The printing-machine manager's complete practical handbook; or, the art of machine managing fully explained. London: J. Haddon & Co., 1877.

Gaubert, Étienne Robert. Notice sur le gérotype ou machine à distribuer et à composer en typographie. Paris: Chez l'inventeur, 1843.

Génie industriel. Revue des inventions françaises et étrangères ... Tome cinquième. Paris: Chez Armengaud ainé ... , 1853.
Wood-engraved vignette of "Industry" on the front wrapper.

Goebel, Theodor. Frédéric Koenig et l'invention de la presse mécanique. Paris: Paul Schmidt, 1885.

Goebel, Theodor. Friedrich Koenig und die Erfindung der Schnellpresse. Ein biographisches Denkmal. Stuttgart: Gebrüder Kröner, 1883.

Great Britain, Parliament. Returns of wages published between 1830 and 1886. London: H. M. Stationery Office, 1887.

Great Britain, Parliament, House of Commons. Report from the select committee on Fourdrinier's patent; with the minutes of evidence, and appendix. [London: N.p.,] 1 June 1837.

Greeley, Horace, et al. The great industries of the United States: Being a historical summary of the origin, growth, and perfection of the chief industrial arts of this country. Hartford, CT: J. B. Burr & Hyde; Chicago and Cincinnati: J. B. Burr, Hyde & Co., 1873.

The growth of woodcut printing, vol. 2: The modern method by machines. Scribner's Monthly 20 (1880): 34-45.

H., F. The penny press. The Mechanic's Magazine 17, no. 475 (15 September 1832): 386-389.

Hair, Thomas H. A series of views of the collieries in the counties of Northumberland and Durham . . . with descriptive sketches and an essay on coal and the coal trade, by M. Ross. London: James Madden, 1844 [Newton Abbot, Devon: David & Charles, 1969].
Facsimile edition.

Hall, Henry. The Tribune book of open-air sports prepared by the New York Tribune with the aid of acknowledged experts. New York: The Tribune Association, 1887.
First book printed by the Mergenthaler Linotype machine. Red cloth binding.

Hall, Henry. The Tribune book of open-air sports prepared by the New York Tribune with the aid of acknowledged experts. New York: The Tribune Association, 1887.
Blue cloth binding.

Handbuch der Buchdruckerkunst. Frankfurt am Main: In der Andreäischen Buchhandlung, 1827.

Hannett, John. Bibliopegia; or, bookbinding: In two parts. Part I. The books of the ancients, and history of the art of bookbinding. Part2. The practical art of bookbinding. London: Simpkin, Marchal and Co., 1865 [New York: Garland Publishing, 1990].
Facsimile edition.

Hannett, John. An inquiry into the nature and form of the books of the ancients; with a history of the art of bookbinding, from the times of the Greeks and Romans to the present day; interspersed with references to men and books of all ages and countries. A new edition. London: Simpkin, Marshall and Co., 1843.

Hansard, Thomas Curson. Treatises on printing and type-founding ... from the seventh edition of the Encyclopaedia Britannica. Edinburgh: Adam and Charles Black, 1841.

Hansard, Thomas Curson. Typographia: An historical sketch of the origin and progress of the art of printing. London: Baldwin, Cradock, and Joy, 1825.

Harland, John Whitfield. The printing arts: An epitome of the theory, practice, processes, and mutual relations of engraving, lithography, & printing in black and in colours. London: Ward, Locke, Bowden and Co., 1892.

The harmonicon, a journal of music, vol. I, parts 1 and 2. London: William Pinnock, 1823.
Printed by W. Clowes.

Harpel, Oscar H. Harpel's typograph or book of specimens containing useful information, suggestions and a collection of examples of letterpress job printing arranged for the assistance of master printers, amateurs, apprentices, and others. Cincinnati: Printed and published by the author, 1870.

[Harper & Brothers.] Complete in about 50 numbers—at 25 cents each. Harper's illuminated and new pictorial Bible embellished with sixteen hundred historical engravings ... New York: Harper Brothers, n.d. [ca. 1843].
Broadside.

[Harper & Brothers.] The illuminated Bible, containing the Old and New Testaments, translated out of the original tongues, and with the former translations diligently compared and revised... New York: Harper & Brothers, 1843-1843.
> In the original 54 parts. Title printed in blue and black. Wrapper title: Harper's illuminated and new pictorial Bible.

[Harper & Brothers.] The illuminated Bible, containing the Old and New Testaments, translated out of the original tongues, and with the former translations diligently compared and revised... New York: Harper & Brothers, 1846.
> Book-form edition in the original gilt-decorated binding. Title printed in red and black.

Harper's illustrated catalogue of valuable standard works, in the several departments of general literature. New York: Harper & Brothers, 1847.

Hasper, W. Handbuch der Buchdruckerkunst. Carlsruhe and Baden: D. R. Marx'schen Buchhandlung, 1835 [Hannover: Th. Schäfer, 1986].
> Facsimile edition.

Head, Francis B. Descriptive essays contributed to the Quarterly Review. London: John Murray, 1857.
> 2 volumes. Vol. 1 includes Head's "The printer's devil."

Herring, Richard. Paper & paper making, ancient and modern: With an introduction, by the Rev. George Croly, LL.D. London: Longman, Brown, Green, and Longmans, 1856.

Hindley, Charles. The book of ready-made speeches... with appropriate quotations, toasts, and sentiments. London: George Routledge and Sons, n.d. [1868].
> Yellowback.

The history and present state of the Post Office. Monthly Supplement of The Penny Magazine of the Society for the Diffusion of Useful Knowledge 3, no. 117 (31 December–31 January 1834).

The history of printing. London: Society for Promoting Christian Knowledge, n.d. [ca. 1855].

Hoe, R., & Co. *See* R. Hoe & Co.

Hoe, Richard M. The literature of printing. A catalogue of the library illustrative of the history and art of typography chalcography and lithography. London: Privately printed at the Chiswick Press, 1877.

Hofmann, Carl. A practical treatise on the manufacture of paper in all its branches. Philadelphia: Henry Carey Baird, 1873.

[Hollingshead, John.] The International Exhibition of 1862. The illustrated catalogue of the industrial department. British division, vol. I. London: Printed for Her Majesty's Commissioners, 1862.

The Holy Bible, containing the old and new testaments: Translated out of the original tongues: and with the former translations diligently compared and revised, by His Majesty's special command. Oxford: University Press, 30 June 1877.
> Caxton Memorial Bible, ltd. to 100 copies.

Huntley, Anna M. Poems. New York: Burr Printing House, 1883.
> Typeset using a Burr typesetter.

Imperial arming press. Mechanic's Magazine, Museum, Journal, and Gazette 17, no. 472 (25 August 1832): 337-339.
> With wood-engraved illustration.

In memoriam. William Johnston. Remarks made at bar meeting, resolutions, reminiscences, letters, and newspaper notices. Cincinnati: Heath Matrix Typograph Machine Print, [1891].

Jamieson, Alexander. A dictionary of mechanical science, arts, manufactures, and miscellaneous knowledge. London: Henry Fisher, Son, and Co., 1827.

Jerrold, Walter. The triumphs of the printing press. London: S. W. Partridge & Co., [1896].

Johnson, J. Typographia, or the printers' instructor: Including an account of the origin of printing, with biographical notices of the printers of England, from Caxton to the close of the sixteenth century; a series of ancient and modern alphabets, and Domesday characters: Together with an elucidation of every subject connected with the art. London: Longman, Hurst, Rees, Orme, Brown & Green, [1824].
> 2 volumes.

Journal des connaissances utiles, vols. 1 and 2 (1831-1833).

Journal für Buchdruckerkunst, Schriftgiesserei und die verwandten Fächer, vols. 1-6 (1834-1839).
> Edited by Johann Heinrich Meyer.

Joyner, Leo. Fine printing: Its inception, development, and practice. London: Cooper and Budd, 1895.

Judd & Glass. Counsels to authors and hints to advertisers. London: Judd & Glass, 1856.
> Includes information on the cost of typesetting, printing, illustration, etc.

Julia de Fontenelle, Jean S. E., and P. Poisson. Manuel complet du marchand papetier et du régleur, contenant la connaissance des papers divers... Paris: Roret, 1828.

Kay, James Phillips. The moral and physical condition of the working classes employed in the cotton manufacture in Manchester. London: James Ridgway, 1832 [Shannon, Ireland: Irish University Press, 1971].
> Facsimile edition.

Kerr, R. W. History of the Government Printing Office (at Washington, D.C.) with a brief record of the public printing for a century, 1789-1881. Lancaster, PA: Inquirer Printing and Publishing Co., 1881.
> Second edition.

Knecht, Edouard. Nouveau manuel complet de l'imprimeur et du dessinateur lithographe. Paris: Librairie Encyclopédique de Roret, 1867.
> 2 volumes (text and atlas). Atlas includes a plate illustrating the steam-powered printers of Voisin and Dupuy.

[Knight, Charles.] An address to the labourers, on the subject of destroying machinery. London: Charles Knight, 1830.

[Knight, Charles.] The commercial history of a penny magazine. No. 2: Wood-cutting and type founding. Monthly Supplement of The Penny Magazine of the Society for the Diffusion of Useful Knowledge 2, no. 101 (30 September-31 October 1833).

[Knight, Charles.] The commercial history of a penny magazine. No. 3: Compositors' work and stereotyping. Monthly Supplement of The Penny Magazine of the Society for the Diffusion of Useful Knowledge 2, no. 107 (31 October-30 November 1833).

Knight's Weekly Volume for all readers... London: Charles Knight & Co., 1844.
> Color-printed brochure advertising "Works to be issued from June 29 to Sept. 28, 1844."

Koenig & Bauer. Erinnerung an das Fest der Vollendung der 5000. Schnellpresse. N.p.: Koenig & Bauer, 1895.

[Koops, Matthias.] Historical account of the substances which have been used to describe events, and to convey ideas, from the earliest date to the invention of paper. Printed on the first useful paper manufactured entirely from straw. London: Printed by T. Burton, 1800.

Koops, Matthias. Historical account of the substances which have been used to describe events, and to convey ideas, from the earliest date to the invention of paper. London: Jaques and Co., 1801.
> Second edition. First book printed on recycled paper; last section (pp. 259-273) printed on wood pulp paper.

Koops, Mathias. Historical account of the substances which have been used to describe events, and to convey ideas, from the earliest date to the invention of paper. London: Jaques and Co., 1801.
> Second edition. Inscribed by the author. This version of the second edition is printed entirely on paper made from straw, except for the last section (pp. 259-273), printed on wood pulp paper.

[Kydd, Samuel.] The history of the factory movement from the year 1802, to the enactment of the Ten Hours' Bill in 1847. London: Simpkin, Marshall and Co., 1857.
> 2 volumes. Published under the pseudonym "Alfred."

L. Graham & Son. A brief description, with illustrations, of the printery of L. Graham & Son, limited, intended to convey an idea of its size and facilities. New Orleans: L. Graham & Son., October 1893.

Lacroix, Auguste. Historique de la papeterie d'Angoulême suivi d'observations sur le commerce des chiffons en France. Paris: Ad. Lainé et J. Havard, 1863.

Ladimir, Jules. Le compositeur typographe. Les français peints par eux-mêmes: Encyclopédie morale du XIXe siècle, vol. 2 (Paris: L. Curmer, 1841-1842), pp. 265-276.
> Extract. Includes colored illustration.

Lalanne, Ludovic. Curiosités bibliographiques. Paris: Paulin, 1845.

The larger catechism, agreed upon by the assembly of divines at Westminster, with the assistance

of commissioners from the church of Scotland, and received by the several Presbyterian churches in America; with the proofs from the scripture. New York: Stereotyped and printed by J. Watts & Co. for Whiting and Watson, 1813.
> First American book printed with stereotyped plates made in America.

The late Charles Knight. The Illustrated London News 62, no. 1751 (22 March 1873): 265–266.
> Extract. Obituary illustrated with a large wood-engraved portrait of Knight.

Lavoisne, M. A complete genealogical, historical, chronological, and geographical atlas; being a general guide to history, both ancient and modern . . . Philadelphia: M. Carey and Son, 1820.
> "Printed by T. H. Palmer, on the Ruthven press, and on J. & T. Gilpin's machine paper."

Lefevre, Théotiste. Guide pratique du compositeur d'imprimerie. Paris: Firmin-Didot Frères, 1873–1878.
> 2 volumes.

Leighton, Son & Hodge. Indenture of apprenticeship. 2 March 1876.
> Legal document recording the apprenticeship of Albert Charles Rudd.

The leisure hour: A family journal of instruction and recreation, vol. 16, nos. 784–835 (5 January–28 December 1867).
> Bound volume with preliminary leaves and indexes. Includes chromolithographed plates.

Lenormand, Louis-Sébastien. Nouveau manuel complet du fabricant de papiers ou l'art de papeterie contenant l'art du fabricant de carton et l'art du formaire. Atlas [only]. Paris: Librairie Encyclopédique de Roret, n.d. [ca. 1840].
> 17 folding plates.

Lenormand, Louis-Sébastien. Nouveau manuel complet du relieur en tous genres contenant les arts de l'assembler, du satineur, du brocheur, du rogneur, du cartonneur, du marbreur sur tranches et du doreur sur tranches et sur cuir. Nouvelle édition entièrement refondue et considérablement augmentee par M. Maigne. Paris: Librairie Encyclopédique de Roret, 1879.

Leroux, Pierre. D'une nouvelle typographie. La revue indépendante 6 (1843): 262–291.

Leroux, Pierre. Nouveau procédé typographique qui réunit les avantages de l'imprimerie mobile et du stéréotypage. Paris: J. Didot l'aîné, 1822.

Leschevin, Philippe-Xavier, and Pierre-Joseph Antoine. Rapports lus à l'Académie des Sciences, Arts et Belles-Lettres, de Dijon . . . sur les machines à fabriquer le papier, inventées par le sieur Ferdinand Leistenschneider . . . Dijon: Frantin, 1815.

Leuchs, Johann Carl. Darstellung der neuesten Verbesserungen in der Verfertigung des Papieres. Nuremberg: Im Contor der allgemeinen Handlungs-Zeitung, 1821.

L[loyd], M[ary]. "Sunny memories." Containing personal reflections of some celebrated characters. London: Women's Printing Society Ltd., 1880.

[Lockhart, John Gibson.] The history of Napoleon Buonaparte. London: John Murray, 1830.
> 2 volumes. The first books with printed publisher's cloth bindings.

Mackie, Alexander. Italy and France: An editor's holiday. London: Hamilton, Adams & Co., 1874.

A machine to supersede typesetting. Scientific American 60, no. 10 (9 March 1889): 143; 150.

MacKellar, Thomas. The American printer: A manual of typography. Philadelphia: MacKellar, Smiths & Jordan, 1885 [N.p.: Harold Berliner, 1976].
> Facsimile edition.

MacKellar, Thomas. The American printer: A manual of typography. Philadelphia: The MacKellar, Smiths & Jordan Co., 1889.
> Seventeenth edition.

Magistrates for the County of Northumberland. Caution. To pitmen and others. Newcastle: Edward Walker, 20 April 1831.
> Broadside.

Mairet, F. Notice sur la lithographie, deuxième édition suivie d'un essai sur la reliure et le blanchement des livres et gravures. Chatillon-sur-Seine: C. Cornillac, 1824.

Making the magazine. Harper's New Monthly Magazine 32, no. 187 (December 1865): 1–31.

Marinoni. Machines à imprimer de Marinoni. Paris: Bureaux et ateliers, 1886.

Marinoni. Marinoni . . . Exposition universelle de 1878 seul grand prix pour les machines à imprimer. N.p., n.d. [1878 or later].
> Advertising brochure and catalogue.

Marshall, James, ed. The trial of Thomas Hunter, Peter Hacket, Richard M'Neil, James Gibb, and William M'Lean, the Glasgow cotton-spinners, before the high court of justiciary, at Edinburgh, on charges of murder, hiring to commit assassinations and committing, and hiring to commit, violence to persons and property. Edinburgh: William Tait, 1838.

Martineau, Harriet. A Manchester strike. A tale. London: Charles Fox, 1832.
> Illustrations of Political Economy, no. 7.

The Mechanic's Institution. Rules and orders of the Mechanic's Institution, for the promotion of useful knowledge among the working classes. London: R. Taylor, [1823].

Mechanic's magazine, vol. 1 (1823–1824).
> Bound volume with preliminary leaves and index.

Mechanics' magazine, and register of inventions and improvements, vol. 1, nos. 1–6 (January–June 1833).
> Bound volume with preliminaries and index. Frontispiece advertisement for "Robert Hoe & Co.'s manufacturing establishment."

Die Mechanik in der Buchdruckerkunst. Die Schnellpresse und das typographische Clavier. Illustrierte Zeitung 2, no. 45 (1844): 296–298 (incomplete).
> Last page of the article (p. 299) not present. Includes wood-engraved illustrations of the Middleton steam press ("Dampffschnellpresse von Middleton") and the typesetting machines of Rosenborg and Young & Delcambre.

Melville, Herman. The paradise of bachelors and the Tartarus of maids. Harper's New Monthly Magazine 10, no. 59 (April 1855): 670–678.

Mémoire en réponse et consultation pour le Sieur A. F. Berte contre le Sieur Léger Didot. [Paris:] Le Normant fils, [1823].
> Bound in a volume of pamphlets relating to the legal dispute between Léger Didot and Berte.

Messrs. Applegath and Cowper's printing press. Mechanic's Magazine, Museum, Journal, and Gazette 6, no. 157 (26 August 1826): 257–259.
> With woodcut illustration.

Milton, John. Areopagitica: A speech of Mr. John Milton for the liberty of unlicensed printing, to the Parliament of England. With an introduction by James Russell Lowell. New York: The Grolier Club, 1890.
> Bookplate of Theodore Low De Vinne.

Minola, N. Album historique. Paris: Typographie mécanique d'Adrien Delcambre et Cie., 1855.

The Mirror of Literature, Amusement, and Instruction 1, no. 2 (9 November 1822): 17–32.
> Edited and published by John Limbird.

The Missouri Republican 65, no. 17,357. 10 September 1877.
> Includes wood-engraved images of "The Walter printing machine" ("Fastest press in the world—capacity 20,000 perfect copies per hour, in use by the Missouri Republican") and "The Bullock printing press."

Mitchell, William H. On a type composing and distributing machine. Institution of Mechanical Engineers, Proceedings. 29th January 1863 (1863): 34–56.

Moncrieff, W. T. The march of intellect; a comic poem. London: William Kidd, 1830.

Monet, A.-L. Le conducteur de machines typographiques: Guide pratique. Paris: Imprimerie de Jules Claye, 1872.

Monet, A.-L. Les machines typographiques en France et à l'étranger suivi des procédés d'impression. Paris: Administration du "Bulletin de l'Imprimerie," 1878.

Monet, A.-L. Machines typographiques et procédés d'impression: Guide pratique du conducteur. Traité complet. Paris: Gauthier-Villars, 1898.

Montgomery, James. A practical detail of the cotton manufacture of the United States of America and the state of cotton manufacture of that country contrasted and compared with that of Great Britain. Glasgow: John Niven Jr., 1840 [New York: Augustus M. Kelley, 1969].
> Facsimile edition.

Morning Herald 1, no. 1 (6 May 1835).
> The first issue of this daily penny newspaper.

The Monotype. The Sketch 19, no. 247 (20 October 1897): 548.
> Extract.

Munsell, J. The typographical miscellany. Albany, NY: Joel Munsell, 1850.

Mudie, Robert. The feathered tribes of the British islands. 2 vols. London: Whittaker & Co., 1834.
> Illustrated with color-printed wood engravings by George Baxter.

Mudie, Robert. The heavens. London: Thomas Ward & Co., 1836.
> Color frontispiece printed by George Baxter.

The new annual register, or general repository of history, politics, and literature, for the year 1810. London: John Stockdale, 1811.
> Sheet H (pp. 113-128) represents the first part of any book printed on the first Koenig steam press.

A new invention. The London Phalanx 1 (1841-1842) [New York: Greenwood Reprint Corporation, 1968]: 351.
> Facsimile edition.

New York printing machine, press, and saw works. R. Hoe & Co. Graham's American Monthly Magazine of Literature and Art 40, no. 5 (June 1852): 564-576.

New York printing machine, press, and saw works. R. Hoe & Co. Graham's American Monthly Magazine of Literature and Art 41, no. 1 (July 1852): 7-12.

The Newport Daily News as sketched in The New York Evening Telegram. N.p., 1880.

Nicholson, James B. A manual of the art of bookbinding: Containing full instructions in the different branches of forwarding, gilding, and finishing. Philadelphia: Henry Carey Baird, 1856.

Nicholson, John. The operative mechanic, and British machinist; being a practical display of the manufactures and mechanical arts of the United Kingdom. London: Knight and Lacey, 1825.

Le nouveau testament de notre seigneur Jésus-Christ . . . Édition stereotype, revue et corrigée avec soin d'après le text grec. London: Imprimé avec des planches solides par A. Wilson, 1807.

Le nouveau testament de notre seigneur Jésus-Christ. Norwich: Imprimé sur les planches stéréotypes d'A. Wilson par R. M. Bacon, 1820.

Odilon-Barrot, Camille Hyacinthe. Habitans de Paris! Paris: Vichon, 30 July 1830.
> Broadside urging the protection of private and public presses during the July 1830 revolution.

Oechelhäuser, Wilhelm. Bericht über die auf den dießjährigen Gewerbe-Ausstellungen zu Paris und Gent ausgestellten Maschinen, Metalle, Metallwaaren und Papiere. Frankfurt am Main: J. D. Sauerländer's Verlag, 1849.

Official catalogue of the Great Exhibition of the works of industry of all nations, 1851. London: Spicer Brothers; W. Clowes & Son, printers, 1851.

Olmer, Georges. Du paper mécanique et de ses apprêts dans les diverses impressions. Paris: Édouard Rouveyre, 1882.

One hundred and fifty wood cuts, selected from The Penny Magazine; worked, by the printing-machine, from the original blocks. London: Charles Knight, 1835.

Ottmar Mergenthaler's Setz- und Giess-Maschine "Linotype" ihr Siegeszug durch die Welt und ihr Erscheinen in Deutschland im Januar 1895. Berlin: Verlag der Geschäfts-Stelle des "Deutscher Buch- und Steindrucker," 1895.

P. Alauzet constructeur-mécanicien presse lithographe perfectionée . . . N.p., 1878 or later.
> Advertisement. Extract from an unidentified periodical.

Pajot des Charmes, C. The art of bleaching piece-goods, cottons, and threads of every description . . . An elementary work, composed for the use of manufacturers, bleachers, dyers, calico printers, and papermakers. London: G. G. and J. Robinson, 1799.
> Translated from the 1798 French edition.

Paris chez soi: Revue historique, monumentale et pittoresque de Paris ancien et moderne. Paris: Poul Boizard, 1855.
> Original wrappers. Advertisement for the Young & Delcambre typesetting machine after the title.

Parley's visit to the printing office; with a familiar account of the steam engine, the printing machine, and the arts of composition, engraving, and stereotyping. London: Cleave, [1841].

Parton, J. The life of Horace Greeley, editor of the New York Tribune. New York: Mason Brothers, 1855.

[Pasko, Wesley Washington.] American dictionary of printing and bookmaking. New York: H. Lockwood, 1894 [New York: Burt Franklin, 1970.]
> Facsimile edition.

Patents for inventions. Abridgements of specifications relating to books, portfolios, card-cases, &c. A.D. 1768-1866. London: George E. Eyre and William Spottiswoode, 1870.

Patents for inventions. Abridgements of specifications relating to letterpress and similar printing (excluding electrotelegraphic and photographic printing). Part 2.—A.D. 1867-1876. London: Commissioners of Patents, 1880.

Paul, Hans, and Julius Lehmann. Hülfsbuch bei Herstellung und Preis-Berechnung von Druckwerken. Breslay: Leopold Freund, 1891.
> Third edition, enlarged.

Pearson, Emily C. Gutenberg; or, the world's benefactor, the art of printing. London: Ward Lock & Co., [1870].

Pearson, Emily C. Gutenberg, and the art of printing. Boston: Noyes, Holmes and Company, 1871.

Pearson, Emily C. Gutenberg, and the art of printing. Third edition. Boston: D. Lothrop and Co., 1879.

[Pelletier, Léon.] La typographie, poëme. Geneva: Ab. Cherbuliez, 1832.

The Penny Magazine and Its Imitators

The American magazine of useful and entertaining knowledge, vol. 1, nos. 1-12 (September 1834-August 1835).
> Bound volume with preliminary leaves and indexes.

The doctor, a medical penny magazine adapted for the use of clergymen, heads of families, nurses, &c., vol. 2, nos. 121-136; vol. 3, nos. 137-215 (6 August 1834-24 February 1836).
> First page of no. 121 lacking. Bound volume with preliminary leaves.

The Dublin penny journal. 1832-1833, vol. 1, nos. 1-52 (30 June 1832-22 June 1833).
> Bound volume with preliminary leaves and indexes.

Das Heller-Magazin: Eine Zeitschrift zur Verbreitung gemeinnütziger Kenntnisse, besorgt von einer Gesellschaft Gelehrter, vol. 1, nos. 1-26 (5 October 1833-28 June 1834).
> Bound volume with preliminary leaves and indexes.

The Irish penny magazine, vol. 1, nos. 1-52 (5 January-28 December 1833).
> Bound volume with preliminary leaves and indexes.

Musée des familles, lectures du soir. Premier volume. Première et seconde année (1833-1834).
> Bound volume with preliminary leaves and indexes.

Nederlandsch magazijn ter verspreiding van algemeene en nuttige kundigheden, vol. 1 (1834).
> Bound volume with preliminary leaves and indexes.

O panorama. Jornal litterario e instructive da Sociedade Propagadora dos Conhecimentos Uteis, vol. 1 (1837).
> Bound volume with preliminary leaves and indexes.

The penny magazine of the Society for the Diffusion of Useful Knowledge, vol. 1, nos. 2, 34-48; vol. 2, nos. 49-112; vol. 3, nos. 113-176 (7 April-29 December 1832; 5 January-28 December 1833; 4 January-27 December 1834).
> Bound volume with preliminary leaves and indexes.

The penny magazine of the Society for the Diffusion of Useful Knowledge. 1832. New York: J. S. Redfield, 1845.
> "American re-issue, from the English plates."

The people's magazine of useful information, designed as a companion to the London penny magazine, of which there are more than 200,000 copies now sold, vol. 2, nos. 1-26 (22 March 1834-7 March 1835).
> Bound volume with wrappers bound in.

Das Pfennig-Magazin der Gesellschaft zur Verbreitung gemeinnütziger Kenntnisse, vol. 1, nos. 1-52; vol. 2, nos. 53-91 (4 May 1833-31 December 1834).
> Bound volume with preliminary leaves and indexes.

Le magasin pittoresque, vol. 1, nos. 1-52 (1833).
> Bound volume with preliminary leaves and indexes.

The Protestant penny magazine, vols. 1 and 2 (1834-1836).
> With general title dated 1836.

Republication of The Penny Magazine. Nos. 1, 7-9, 12-15, 17, 19, 21-24 (1832, 1834, 1835-1836, 1837, 1835, 1838-1839).
> Non-consecutive numbers, in original wrappers. "Published by J. S. Redfield New York."

The periodical press of Great Britain and Ireland: Or an inquiry into the state of the public journals, chiefly as regards their moral and political influence. London: Hurst, Robinson & Co., 1824.

The pictorial album; or, cabinet of paintings. Containing eleven designs, executed in oil colours, by G. Baxter. London: Chapman and Hall, [1837].

Piette, Louis. Traité de la fabrication du papier. Paris: F. G. Levrault, 1831.

Planche, Gabriel. De l'industrie de la papeterie. Paris: Firmin Didot Frères, 1853.

Poppe, J. H. M. Technologische Bildergallerie oder Darstellung der interessantesten und lehrreichsten Manufacturen, Fabriken, Künste und Handwerke, in getreuen Bildern und ausführlichen deutlichen Erklärungen. Reutlingen: J. C. Mäcken, jun., 1830.

Presse mécanique. Magasin pittoresque 2, no. 48 (1834): 383–384.
With two wood-engraved illustrations, the second captioned "Presse mécanique de Cowper employée pour l'impression du Magasin pittoresque."

[Prest, Thomas Peckett.] The penny Pickwick edited by "Bos." No. 33 only. Bloomsbury [London]: E. Lloyd, n.d. [ca. 1837].
Woodcut illustrations. Prest's The Penny Pickwick, a parody of Dickens's The Pickwick Papers, was published in 112 weekly parts between 1837 and 1839.

Printing and piracy: New discovery. The Athenaeum, no. 736 (4 December 1841): 932.
Anastatic printing.

Printing in the fifteenth and in the nineteenth centuries. Monthly Supplement of The Penny Magazine of the Society for the Diffusion of Useful Knowledge 6, no. 369 (30 November–31 December 1837).

Printing: Its dawn, day, & destiny. London: Bradbury & Evans, 1858.

The printing machine. Nos. 1–12 (13 February 1832–16 August 1834).
Published by Charles Knight.

1. The printer ... 2. Printing in the fifteenth and in the nineteenth centuries. Review of The Printer, by Charles Knight, and of "Printing in the fifteenth and in the nineteenth centuries." The Quarterly Review 65, no. 129 (December 1839–January 1840): 1–30.
Extract.

Proben aus der Schriftgiesserey der Andreäschen Buchhandlung in Frankfurt am Main. N.p., 1834 [Darmstadt: Technische Hochschule Darmstadt, 1984].
Facsimile edition.

Procter, Adelaide A. The Victoria Regia: A volume of original contributions in poetry and prose. London: Printed and published by Emily Faithfull and Co., Victoria Press, (for the Employment of Women), 1861.

Prospectus of The Scotsman, a new weekly political and literary journal. Edinburgh: Abernethy & Walker, 1816.

Proteaux, Albert. Practical guide for the manufacture of paper and boards ... with additions by L. S. Le Normand. Translated from the French with notes by Horatio Paine ... to which is added a chapter on the manufacture of paper from wood in the United States by Henry T. Brown. Philadelphia: Henry Carey Baird, 1866.

The Public Ledger building, Philadelphia: With an account of the proceedings connected with its opening June 20, 1867. Philadelphia: George W. Childs, 1868.

R. Hoe & Co. Printers' and binders' warehouse ... R. Hoe & Co. offer for sale, of their own manufacture, cylinder printing machines ... New York: J. Hall, steam printer, n.d. [ca. 1850].
With a wood-engraved illustration of a Hoe lightning press.

R. Hoe & Co. R. Hoe & Co. manufacturers of the revolving and single and double-cylinder printing machines, power presses, (Adams' patent), Washington and Smith hand presses, self-inking machines, &c.... New York: R. Hoe & Co., 1867.
Front cover reads: "R. Hoe & Co.'s catalogue."

R. Hoe & Co. R. Hoe & Co. manufacturers of type-revolving, perfecting, single and double cylinder and Adams' printing machines ... New York: R. Hoe & Co., 1873.

R. Hoe & Co. Second-hand printing machines at low prices. New York: R. Hoe & Co., November 1896.
Advertising brochure.

Rae, W. Fraser. The centenary of the Times. Littell's Living Age, 5th series, 49, no. 2119 (31 January 1885): 259–273.
Extract. Includes an illustration of a Linotype machine used at the New-York Tribune office.

The railway traveller's handy book of hints, suggestions, and advice, before the journey, on the journey, and after the journey. London: Lockwood & Co., 1862 [Botley: Old House Books & Maps, 2012].
Facsimile edition.

Rand & Avery. Geo. C. Rand & Avery. Printers. No. 3 Cornhill, Boston. [Boston:] Rand & Avery, [ca. 1865].
Undated book of type specimens. Front cover reads: "Rand & Avery's specimens."

[Rapp, Gottlob Heinrich.] Das Geheimniss des Steindrucks in seinem ganzen Umfange. Tübigen: J. G. Cotta'schen Buchhandlung, 1810 [Württemberg: Württembergische Landesbibliothek, 1992].
Facsimile edition.

Reach, Angus B. The comic Bradshaw: Or, bubbles from the boiler. London: D. Bogue, 1848.

Recueil de journaux [cover title]. N.p., n.d. [1863–1870].
Collection of press cuttings on the printing trade in France, especially on the question of women as typesetters.

Réponse au factotum calomnieux publié par M. Bert contre M. Didot-Saint-Léger, sous le titre: Sur quelques modèles faits à Londres en 1816, présentés en 1823 à l'exposition des produits de l'industrie française, et sur le colportage en Angleterre de la machine à fabriquer le papier continu, inventée en France. [Paris:] Imprimerie de Jules Didot aîné, [1823].
Bound in a volume of pamphlets relating to the legal dispute between Léger Didot and Berte.

Report from the committee on Dr. Cartwright's petition respecting his weaving machine; together with the minutes of evidence taken before said committee. N.p., 13 April 1808.

Report of proceedings under commissions of oyer & terminer and gaol delivery, for the county of York, held at the castle of York, before Sir Alexander Thomson ... and Sir Simon le Blanc ... from the 2d to the 12th of January 1813. London: Luke Hansard & Sons, [1813].

Reports of the inspectors of factories to Her Majesty's principal secretary of state for the Home Department, for the half-year ending 30th April 1847. London: William Clowes and Sons for Her Majesty's Stationery Office, 1847.

A revolution in printing. The latest triumph of American invention seen in the Tribune office. Library of Tribune Extras 1, no. 7 (July 1886): 96–100.

Richmond, W. D. The grammar of lithography. A practical guide for the artist and printer in commercial & artistic lithography, & chromolithography, zincography, photo-lithography, and lithographic machine printing. London: Wyman & Sons, 1878.

Ringwalt, J. Luther. American encyclopaedia of printing. Philadelphia: Menamin & Ringwalt; J. B. Lippincott & Co., 1871.

Rivière, Emmanuel. Le travail de la femme dans l'industrie typographique. Blois: Grande Imprimerie de Blois, 1898.

Robert, N.-L. 18 janvier 1799. Brevet d'invention de quinze ans, pour une machine à faire le paper d'une très-grande etendue. Déscription des machines et procédés specifies dans les brevets d'invention de perfectionnement et d'importation dont la durée est expiré, vol. 5 (Paris: Mme. Huzard, 1823), pp. 18–23.
Plates 4–6 in the volume include illustrations of Robert's papermaking machine.

Rocourt, Olivia de. Lettre d'une femme aux ouvriers typographes. Paris: E. Dentu, 1862.

Royal Society of Arts. Report of the committee of the Society of Arts, &c. together with the approved communications and evidence upon the same, relative to the mode of preventing the forgery of bank notes. London: Sold by the Housekeeper, at the Society's House, 1819.

Ruskin, John. The nature of Gothic a chapter of The Stones of Venice. Hammersmith [London]: Printed by William Morris at the Kelmscott Press, [1892].
One of 500 copies printed.

The San Francisco Chronicle and its history. San Francisco: N.p., 1879.
Includes a plate titled: "The great Hoe web perfecting presses.—33,000 copies an hour each."

Savage, William. A dictionary of the art of printing. London: Longman, Brown, Green, and Longmans, 1841.

Die Schnellpresse. Allgemeine Zeitung, no. 233 (21 August 1825): 929–931.
   Probably the first detailed article on mechanized printing published in Germany.

Die Schnellpresse. Allgemeine Zeitung, no. 34 (3 February 1827): 133–134.
   A follow-up to the 21 August 1825 article.

Senefelder, Alois. A complete course of lithography: Containing clear and explicit instructions in all the different branches and manners of that art . . . London: R. Ackermann, 1819.

Six reports from the Select Committee on Artizans and Machinery. 23 February–21 May 1824. London: H. M. Stationery Office, 1824 [London: Frank Cass and Co., Ltd.; New York: Augustus M. Kelley, 1968].
   Facsimile edition.

Sketchley, Arthur. Mrs. Brown on the Alabama claims. London: George Routledge and Sons, n.d. [1872].
   2-shilling "yellowback" novel.

Slatter, Henry. The Typographical Association: A fifty years' record, 1849–1899. Manchester: Printed for the Typographical Association by the Labour Press Ltd., [1899.]

Smiles, Samuel. Men of invention and industry. London: John Murray, 1884.

Smith, James Edward. A grammar of botany, illustrations of artificial, as well as natural classification with an explanation of Jussieu's system. New York: James V. Seaman, 1822.
   First book published in the United States to contain lithographed illustrations.

Smith, John Thomas. Antiquities of Westminster; the old palace; St. Stephen's chapel (now the House of Commons) &c. &c. London: Printed by T. Bensley for J. T. Smith . . . and sold by R. Ryan . . . and J. Manson, 9 June 1807.

Société de secours des ouvriers imprimeurs-typographiques d'Avignon. Règlement . . . en faveur des sociétaires malades; autorisée le 10 novembre 1845, et fondée, comme essai, le 6 mai 1840. Avignon: Bonnet fils, 1845.

Society for the Diffusion of Useful Knowledge. Account of Lord Bacon's Novum organon scientiarum. London: Baldwin, Cradock, and Joy, 1827.
   Library of Useful Knowledge, part 10.

Society for the Diffusion of Useful Knowledge. Animal mechanics. London: Baldwin, Cradock, and Joy, 1827.
   Library of Useful Knowledge, part 9.

Society for the Diffusion of Useful Knowledge. The British almanac of the Society for the Diffusion of Useful Knowledge, for the year of our Lord 1852, being bissextile, or Leap Year. London: Charles Knight, n.d. [1851].

Society for the Diffusion of Useful Knowledge. The companion to the almanac; or year-book of general information for 1852. London: Charles Knight, n.d. [1851].

Society for the Diffusion of Useful Knowledge. Heat. Part I [II]. London: Baldwin, Cradock, and Joy, 1827.
   Library of Useful Knowledge, parts 4 and 5. Part 4 is the second edition.

Society for the Diffusion of Useful Knowledge. Mechanics. Treatise I: On the mechanical agents, or prime movers. London: Balldwin, Cradock, and Joy, 1827.
   Library of Useful Knowledge, part 6.

Society for the Diffusion of Useful Knowledge. Mechanics. Treatise 2.—Elements of machinery.—Part 2. London: Baldwin, Cradock, and Joy, 1827.
   Library of Useful Knowledge, part 8.

Society for the Diffusion of Useful Knowledge. The menageries: Quadrupeds, described and drawn from living subjects, vol. 1 [2]. London: Charles Knight, 1829–1830.
   The first extensively illustrated book printed on a mechanized press. Library of Entertaining Knowledge, parts 1, 3, 13, and 22. In the original wrappers.

Some particulars of the conduct and of the execution of Savidge and others, who were executed at Leicester, on Thursday, April 17, 1817 . . . Nottingham: Sutton and Son, n.d. [1817].

The Spectator. In six volumes. Volume the first [only]. London: Printed by and for Andrew Wilson, 1812.
   "Stereotype" on the title page.

Speiers, William Stuart. The electrotypers' manual. Buffalo, NY: N.p., 1869.

Steinberg, Heinrich. Preis-Verzeichniss über Maschinen, Utensilen und Materialien für Lithographen, Steindrucker, Graevure etc. Berlin: Heinrich Steinberg, 1891.
   Office copy with revisions in manuscript and numerous inserts.

Stevens, Henry. The history of the Oxford Caxton memorial Bible printed and bound in twelve consecutive hours June 30 1877. London: H. Stevens, 1878.

The story of the "Scotsman": A chapter in the annals of British journalism. Edinburgh: Printed for private circulation, 1886.

Strickland, William. A description of the Hetton rail road in England. New-York American 7, no. 631 (7 April 1826): [1].
   Modern facsimile.

The Sun, [vol. 1,] no. 1 (3 September 1833).
   The first issue of the first successful penny daily newspaper.

Supplément à la réponse de M. Didot-Saint-Léger au mémoire de M. Berte. [Paris:] Imprimerie de Jules Didot aîné, [1823].
   Bound in a volume of pamphlets relating to the legal dispute between Léger Didot and Berte.

Tanquerel des Planches, L. Lead diseases: A treatise from the French . . . with notes and additions on the use of lead pope and its substitutes by Samuel L. Dana. Lowell, MA: Daniel Bixby and Company, 1848.

Taylor, Richard. On the invention and first introduction of Mr. Koenig's printing-machine. The London, Edinburgh, and Dublin Philosophical Magazine and Journal of Science 31 (July–December 1847).

The Thorne typesetting machine. Scientific American 73, no. 8 (24 August 1895): 113; 118.
   Illustrated.

Timperley, C. H. Encyclopaedia of literary and typographical anecdote. London: Henry G. Bohn, 1842.
   Second edition.

Timperley, C. H. The printers' manual; containing instructions to learners, with scales of impositions, and numerous calculations, recipes, and scales of prices in the principal towns of Great Britain. Together with practical directions for conducting every department of a printing office. London: H. Johnson; Manchester: Bancks and Co., 1838.

[Timperley, C. H.] Songs of the press and other poems, relative to the art of printing. London: Simpkin and Marshall, 1833.
   "Also epitaphs, epigrams, anecdotes, notice of early printing and printers."

Todd, Charles Burr. The history of Redding, Conn., from its first settlement to the present time. New York: John A. Gray Press and Steam Type-Setting Office, 1880.

Tomlinson, Charles. Cyclopaedia of useful arts & manufactures. London and New York: George Virtue, [1852].
   Parts 1, 4, 7, 15, 33, 35, 37–40 only, in original wrappers.

Tomlinson, Charles. Cyclopaedia of useful arts & manufactures. London and New York: George Virtue, [1852–1854].
   9 volumes.

Tomlinson, Charles. Cyclopaedia of useful arts & manufactures. London and New York: George Virtue, [1854].
   Vol. 1 (A–G) only of two.

Trades' societies and strikes. Report of the Committee on Trades' Societies, appointed by the National Association for the Promotion of Social Science, presented at the fourth annual meeting of the Association, at Glasgow, September, 1860. London: John W. Parker and Son, 1860.

Turgan, Julien. Les grandes usines de France: Tableau de l'industrie française au XIXe siècle, vol. 1. Paris: Librairie Nouvelle, 1860.

Tyng, Stephen H. Record of a life of mercy. New York: S. W. Green's Type-Setting Machinery, 1879.

United States House of Representatives. 25th Congress, 3rd session. Letter from the Secretary of the Treasury, transmitting, in obedience to a resolution of the House of the 29th of June last, information in relation to steam-engines, &c. [Washington, DC]: Thomas Allen, 1838.

Ure, Andrew. A dictionary of arts, manufactures, and mines: Containing a clear exposition of their principles and practice. London: Longman, Orme, Brown, Green, & Longmans, 1839.
   "Illustrated with twelve hundred and forty engravings on wood."

Ure, Andrew. The philosophy of manufactures: Or, an exposition of the scientific, moral, and commercial economy of the factory system of Great Britain. London: Charles Knight, 1835.

Ure, Andrew. Recent improvements in arts, manufactures, and

mines: Being a supplement to his dictionary. London: Longman, Brown, Green, and Longmans, 1844.

Vachon, Marius. Les arts et les industries du papier en France 1871-1894. Paris: Librairies-Imprimeries Réunies, 1894.

Van Winkle, C. S. The printer's guide; or, an introduction to the art of printing: Including an essay on punctuation, and remarks on orthography. New York: C. S. van Winkle, 1818 [Chicago: Lakeside Press, 1970.]
> Facsimile edition.

Vereeniging ter Bervordering van de Belangen des Boekhandels. Internationale tentoonstelling Juli Augustus 1892 bij gelegenheid van het vijf-en-zeventigjarig bestaan der Vereeniging ter Bervordering van de Belangen des Boekhandels 1817-1892. N.p., 1892.
> With 12-page "Supplement of den catalogus der Internationale tentoonstelling" laid in.

Vergil. P. Virgilii Maronis opera. Interpretatione et notis illustravit Carlus Ruaeus, Soc. Jesu. London: Impensis F. C. et J. Rivington . . . , 1822.
> Note on title verso: "Vaporante machinâ excudebat B. Bensley, Bolt Court, Fleet Street" [Printed by steam engine, B. Bensley, Bolt Court, Fleet Street].

Das vierte Säcularfest der Erfindung der Buchdruckerkunst, begangen zu Stuttgart am 24. Und 25. Juni 1840. Stuttgart: J. Kreutzer, 1840.

W., J. W. The London and Birmingham railway guide, and Birmingham and London railway companion: Containing a minute description of the railroad, and every object worthy of notice; an antiquarian and topographical account of the towns, villages, noblemen and gentleman's seats, within ten miles of the railroad. London: James Wyld, 1838.
> Second edition.

Waldow, Alexander. Die Buchdruckerkunst in ihrem technischen und kaufmännischen Betriebe. Leipzig: Druck und Verlag von Alexander Waldow, 1874-1877.
> 2 volumes plus atlas.

Waldow, Alexander. Hilfsbuch für Maschinenmeister an Buchdruck-Cylinderschnellpressen. I. Teil. Leipzig: Druck und Verlag von Alexander Waldow, 1887.

Walker, James Scott. An accurate description of the Liverpool and Manchester rail-way, the tunnel, bridges, and other works throughout the line; with an account of the opening of the railway, and the melancholy accident which occurred . . . Liverpool: J. F. Cannell, 1831.
> Third edition.

Wanamaker & Brown. Methods of business of the largest establishment in the world for the manufacture and sale of men's wear. Philadelphia: Wanamaker & Brown, 1876.
> "Printed at our own steam-powered printing office. One million copies distributed gratuitously."

White, James. A new century of inventions: Being designs & descriptions of one hundred machines, relating to arts, manufactures, & domestic life. Manchester: Printed for the author by Leech and Cheetham, 1822.

The Widows' Society. Constitution of the Widows' Society; together with some account of the institution, and of its proceedings to the present period: To which is added a list of its members. Boston: Treadwell's Power Press, 1823.

Wilkins, John. The mathematical and philosophical works of the right Rev. John Wilkins, late Lord Bishop of Chester, to which is prefixed the author's life, and an account of his works. London: C. Whittingham for Vernor and Hood; Cuthell and Martin; J. Walker, 1802.
> 2 volumes. Probably the earliest use of bleached wood-pulp paper in English book production.

Wilson, Frederick J. F. Typographic printing machines and machine printing. A practical guide to the selection of bookwork, two-colour, jobbing, and rotary machines, with remarks upon their construction, capabilities, and peculiarities. London: Wyman & Sons, [1879].
> Fourth edition.

Wilson, Frederick J. F., and Douglas Grey. A practical treatise upon modern printing machinery and letterpress printing. London: Cassell & Co., 1888.

[Wilson, John Forbes.] A few personal recollections by an old printer. London: Printed for private circulation, 1896.

Wilson, Joseph. A history of mountains, geographical and mineralogical. Accompanied by a picturesque view of the principal mountains of the world in their respective proportions of height above the level of the sea by Robert Andrew Riddell. London: T. Bensley for Messrs. Nicol . . . , 1807-1810.
> 3 volumes. Note in vol. 1, p. 17: "The print [i.e. 'picturesque view'] is already the largest that has ever been engraved on one plate of copper or printed on one sheet of paper, being 4 feet 6 inches by 3 feet, exclusive of margins . . . [made possible by] the perfection to which the enterprize [sic] and spirit of two individuals (Messrs. Henry and Sealy Fourdrinier) have brought the important invention of making paper by machinery."

Wood, John George. The common moths of England. London: George Routledge and Sons, n.d. [1878].

Wyman, Morrill. Memoir of Daniel Treadwell. Cambridge, MA: John Wilson and Son, 1888.
> Memoirs of the American Academy of Arts and Sciences, vol. 11.

1900s–2000s

100 GPO years 1861-1961: A history of United States public printing. Washington, DC: U.S. Government Printing Office, [1961].

Abbott, Jacob. The Harper establishment: How books are made. New introduction by Joel Myerson and Chris L. Nesmith. New Castle, DE: Oak Knoll Press, 2001.

Ackroyd, Peter. Dickens. New York: HarperCollins, 1990.

Allen, Robert C. The British industrial revolution in global perspective. Cambridge: Cambridge University Press, 2009.

Altick, Richard D. The English common reader: A social history of the mass reading public, 1800-1900. Second edition. With a foreword by Jonathan Rose. Columbus: Ohio State University Press, 1998.

Altick, Richard D. The shows of London. Cambridge, MA: The Belknap Press of Harvard University Press, 1978.

Amory, Hugh, and David D. Hall, eds. A history of the book in America, vol. 1: The colonial book in the Atlantic world. Cambridge: Cambridge University Press, 2000.

Anderson, Patricia. The printed image and the transformation of popular culture. Oxford: Clarendon Press, 1991.

André, Louis. Machines à papier: Innovation et transformations de l'industrie papetière en France 1798-1860. Paris: Éditions de l'École des Hautes Études en Sciences Sociales, 1996.

An anthology of Delaware papermaking. New Castle, DE: The Delaware Bibliophiles; Oak Knoll Books, 1991.

Ashton, Rosemary. Victorian Bloomsbury. New Haven, CT: Yale University Press, 2012.

Aurenche, Marie-Laure. Édouard Charton et l'invention du Magasin pittoresque (1833-1870). Paris: Honoré Champion Éditeur, 2002.

[Austen-Leigh, Richard Arthur.] The story of a printing house, being a short account of the Strahans and Spottiswoodes. London: Spottiswoode & Co., 1912.
> Second edition.

Baker, Cathleen A. From the hand to the machine: Nineteenth-century American paper and mediums: Technologies, materials, and conservation. Ann Arbor, MI: The Legacy Press, 2010.

Bakunin, Jack. Pierre Leroux and the birth of democratic socialism 1797-1848. New York: Revisionist Press, 1976.

Berg, Maxine. The age of manufactures: Industry, innovation and work in Britain 1700-1820. Oxford: Basil Blackwell, 1985.

Berg, Maxine. The machinery question and the making of political economy 1815-1848. Cambridge: Cambridge University Press, 1982.

Berry, W. Turner, and H. Edmund Poole. Annals of printing: A chronological encyclopaedia from the earliest times to 1950. London: Blandford Press, 1966.

Biagini, Cédric, and Guillaume Carnino, eds. Les Luddites en France: Résistance à l'industrialisation et à l'informisation. Montreuil: Éditions l'Échappée, 2010.

Bidwell, John. American paper mills 1690-1832: A directory of the paper trade with notes on products, watermarks, distribution methods, and manufacturing techniques. Hanover, NH: Dartmouth College Press; Worcester, MA: American Antiquarian Society, 2013.

Bidwell, John. Paper and type: Bibliographical essays. Charlottesville, VA: The Bibliographical Society of the University of Virginia, 2019.

Biesalski, Ernst-Peter. Die Mechanisierung der deutschen Buchbinderei 1850-1900. Frankfurt am Main: Buchhändler-Verein, 1991.

Bigmore, E. C., and C.W.H. Wyman. A bibliography of printing with notes and illustrations. New Castle, DE: Oak Knoll Press; London: British Library, 2001.
    Facsimile reprint of the 3-vol. 1880-1886 edition, with new introduction and index.

Binfield, Kevin, ed. Writings of the Luddites. Baltimore, MD: Johns Hopkins University Press, 2004.

Bolza, Albrecht. Friedrich Koenig: Der Erfinder der Druckmaschine, ein Pionier der deutschen Maschinenindustrie. Berlin: VDO-Verlag, 1933.

Borchers [Gebrüder Borchers]. Zum 150 jährigen Jubiläum der Lübeckischen Anzeigen und der 75 jährigen Gründung der lithographischen Anstalt und Steindruckerei. Lübeck: Gebrüder Borchers, 1901.

Bottomley, Sean. The British patent system during the Industrial Revolution 1700-1852: From privilege to property. Cambridge: Cambridge University Press, 2014.

Briggs, Asa. Iron bridge to Crystal Palace: Impact and images of the industrial revolution. London: Thames and Hudson, 1979.

Briggs, Asa. The power of steam: An illustrated history of the world's steam age. Bristol: Book Promotions Ltd., 1982.

Brightly, Charles. The method of founding stereotype. [with] An essay on the origin and progress of stereotype printing by Thomas Hodgson. With a new introduction by Michael L. Turner. New York: Garland Publishing, 1982.

Brock, W. H., and A. J. Meadows. The lamp of learning: Taylor & Francis and the development of science publishing. London: Taylor & Francis, 1984.

Brontë, Charlotte. Shirley. London: Wordsworth Editions, 1993.

Brown, Philip A. H. London publishers and printers c. 1800-1870. London: British Library, 1982.

Brown, Richard D. Knowledge is power: The diffusion of information in early America, 1700-1865. New York: Oxford University Press, 1989.

Bruce, David. History of typefounding in the United States. Edited and annotated from the holograph manuscript by James Eckman. New York: The Typophiles, 1981.

Bundock, Clement J. The story of the National Union of Printing Bookbinding and Paper Workers. Oxford: Printed at the University Press, 1959.

Burch, R. M. Colour printing and colour printers. With a chapter on modern processes by W. Gamble. London: Sir Isaac Pitman and Sons, Ltd., 1910.
    Second edition.

Burch, R. M. Colour printing and colour printers. With a chapter on modern processes by W. Gamble. Introduction by Ruari McLean. Edinburgh: Paul Harris Publishing, 1983.

Bury, T. H. Coloured views on the Liverpool and Manchester railway. A facsimile of the original edition published in 1831 by R. Ackermann with an historical introduction to the railway by George Ottley. Oldham: Hugh Broadbent, 1976.

Butt, John and Kathleen Tillotson. Dickens at work, vol. 1. London: Routledge, 2009.
    Reprint of the 1957 edition.

Cameron, Rondo, and Larry Neal. A concise economic history of the world from Paleolithic times to the present. Fourth edition. New York: Oxford University Press, 2003.

Carlson, Robert E. The Liverpool & Manchester Railway project 1821-1831. Newton Abbot: David & Charles, 1969.

Carpenter, Kenneth E., ed. Books and society in history. New York: R. R. Bowker Company, 1983.

Carter, John. Publisher's cloth: An outline history of publisher's binding in England 1820-1900. New York: R. R. Bowker; London: Constable & Co., 1935.

Cate, Phillip Dennis, and Sinclair Hamilton Hitchings. The color revolution: Color lithography in France 1890-1900. Santa Barbara, CA: Peregrine Smith, 1978.

Chauvet, Paul. Les ouvriers du livre en France de 1789 à la constitution de la fédération du livre. Paris: Librairie Marcel Rivière et Cie., 1964.

Childs, Michael J. Labour's apprentices: Working-class lads in late Victorian and Edwardian England. London: Hambledon Press, 1992.

Christie's South Kensington. The New Hall vault: Kronheim prints from the collection of the late Ernest Owen. London: Christie's, 1987.
    Auction catalogue of 19th-century prints made using the Baxter process.

Clair, Colin. A chronology of printing. New York: Frederick A. Praeger, 1969.

Clair, Colin. A history of European printing. London: Academic Press, 1976.

Clair, Colin. A history of printing in Britain. New York: Oxford University Press, 1966.

Clapperton, R. H. The paper-making machine: Its invention, evolution and development. Oxford: Pergamon Press, 1967.

Claproth, Justus. Justus Claproth's Abhandlung von 1774 über die Verwendung von Makulatur zur Papierherstellung. Leipzig: Papierfabrik Kabel A.G., 1947.

Coleman, D. C. The British paper industry 1495-1860: A study in industrial growth. Oxford: Clarendon Press, 1958.

Coleman, D. C. Myth, history and the industrial revolution. London: The Hambleton Press, 1992.

Collins, E.J.T. The diffusing of the threshing machine in Britain, 1790-1880. Tools and Tillage 2 (1972): 16-33.
    Offprint.

Collins, Irene. The government and the newspaper press in France 1814-1881. London: Oxford University Press, 1959.

Collins, Philip. Dickens and crime. Second edition. London: Macmillan & Co., 1964.

Collins, Philip, ed. Charles Dickens: The critical heritage. London: Routledge, 2009.

Comparato, Frank E. Books for the millions: A history of the men whose methods and machines packaged the printed word. Harrisburg, PA: The Stackpole Company, 1971.

Cookson, Gilliam. The age of machinery: Engineering the Industrial Revolution, 1770-1850. Woodbridge: Boydell Press, 2018.

Crouthamel, James L. Bennett's New York Herald and the rise of the popular press. Syracuse, NY: Syracuse University Press, 1989.

Cziszar, Alex. The scientific journal: Authorship and the politics of knowledge in the nineteenth century. Chicago: University of Chicago Press, 2018.

Dabundo, Laura, ed. Encyclopedia of romanticism: Culture in Britain, 1780s-1850s. New York: Garland Publishing, 1992.

Darley, Lionel S. Bookbinding then and now: A survey of the first hundred and seventy-eight years of James Burn & Company. London: Faber and Faber, 1959.

Darvall, Frank Ongley. Popular disturbances and public order in Regency England. With a new introduction by Angus Macintyre. London: Oxford University Press, 1969.

Darwin, Bernard, ed. The Dickens advertiser: A collection of the advertisements in the original parts of novels by Charles Dickens. New York: Macmillan, 1930.

David, Deirdre, ed. The Cambridge companion to the Victorian novel. Cambridge: Cambridge University Press, 2001.

Dawe, Edward A. Paper and its uses: A treatise for printer stationers and others. London: Crosby Lockwood and Son, 1914.

De Vinne, Theodore Low. The practice of typography. Correct composition: A treatise on spelling abbreviations, the compounding and division of words, the proper use of figures and numerals, italic and capital letters, notes, etc., with observations on punctuation and proof-reading. New York: The Century Company, 1901.
    Presentation copy.

De Vinne, Theodore Low. The practice of typography. Modern methods of book composition: A treatise on type-setting by hand and by machine and on the proper arrangement and imposition of pages. New York: The Century Company, 1904.
    Presentation copy.

De Vinne, Theodore Low. The practice of typography. A treatise on title-pages with numerous illustrations in facsimile and some observations on the early and recent printing of books. New York: The Century Company, 1902.
    Presentation copy.

De Vinne, Theodore Low. The printers' price list. With a new introduction by Irene Tichenor. New York: Garland Publishing, 1980.
    Reprint of the 1871 edition.

De Vinne, Theodore Low. Printing in the nineteenth century. New York: Lead Mould Electrotype Foundry, 1924.

Dickens, Charles. The letters of Charles Dickens. 12 vols. Oxford: Clarendon Press, 1965-2002.

Dickens, Charles. Hard Times: An authoritative text, contexts, criticism. Edited by Fred Kaplan. New York: W. W. Norton & Co., 2017.

Dickinson, H. W. A short history of the steam engine. Cambridge: Printed for Babcock and Wilcox, Ltd. at the University Press, 1938.

Diderot, Denis, and Jean le Rond d'Alembert. L'Encyclopédie Diderot & d'Alembert. Imprimerie-Reliure. N.p.: Bibliothèque de l'Image, 2001.

Dilke, Charles Wentworth. Catalogue of a collection of works on or having reference to the Exhibition of 1851 in the possession of C. Wentworth Dilke. Cambridge: Cambridge University Press, 2011.
    Reprint of the 1855 edition.

Dooley, Allan C. Author and printer in Victorian England. Charlottesville, VA: University Press of Virginia, 1992.

Draeger, Alain. Draeger: Les pages d'or de l'édition. Paris: Draeger, 2019.

Dudek, Louis. Literature and the press: A history of printing, printed media, and their relation to literature. Toronto: Ryerson Press; Contact Press, 1960.

Duffy, Patrick. The skilled compositor, 1850-1914: An aristocrat among working men. Aldershot: Ashgate, 2000.

Duncan, Bingham. Whitelaw Reid: Journalist, politician, diplomat. Athens, GA: University of Georgia Press, 1975.

Dyson, Anthony. Pictures to print: The nineteenth-century engraving trade. London: Farrand Press, 1984.

Eliot, Simon, ed. The history of the Oxford University Press, vol. 2: 1780-1896. Oxford: Oxford University Press, 2013.

Elton, Arthur. British railways. London: Collins, 1945.

Evans, Edmund. The reminiscences of Edmund Evans. Edited and introduced by Ruari McLean. Oxford: Clarendon Press, 1967.

Evans, Joan. The endless web: John Dickinson & Co. Ltd 1804-1954. London: Jonathan Cape, 1955.

Exman, Eugene. The brothers Harper: A unique publishing partnership and its impact on the cultural life of America from 1817 to 1852. New York: Harper & Row, 1965.

Exman, Eugene. The house of Harper: One hundred and fifty years of publishing. New York: Harper & Row, 1967.

Exman, Eugene. The house of Harper: The making of a modern publisher. With a new introduction by Jennifer B. Lee. New York: Harper Perennial, 2010.

Fae, John. The Bible cause: A history of the American Bible Society. Oxford: Oxford University Press, 2016.

Feltes, N. N. Modes of production of Victorian novels. Chicago: University of Chicago Press, 1986.

Fido, Martin. Charles Dickens: An authentic account of his life & times. London: Hamlyn Publishing Group, [1970].

Finkelstein, David. Movable types: Roving creative printers of the Victorian world. Oxford: Oxford University Press, 2018.

Freedgood, Elaine, ed. Factor production in nineteenth-century Britain. New York: Oxford University Press, 2003.

Freeman, Michael. Railways and the Victorian imagination. New Haven, CT: Yale University Press, 1999.

Friedman, Joan M. Color printing in England 1486-1870. An exhibition. New Haven, CT: Yale Center for British Art, 1978.

Fyfe, Aileen. Steam-powered knowledge: William Chambers and the business of publishing, 1820-1860. Chicago: University of Chicago Press, 2012.

Gascoigne, Bamber. Milestones in colour printing 1457-1859. Cambridge: Cambridge University Press, 1997.

Gaskell, Philip. A new introduction to bibliography. New York and Oxford: Oxford University Press, 1972.

Gebunden in der Dammpfbuchbinderei: Buchbinden im Wandel des 19. Jahrhunderts. Wiesbaden: Harrasowitz Verlag, 1994.

George, M. Dorothy. English political caricature to 1792: A study of opinion and propaganda. Oxford: Clarendon Press, 1959.

George, M. Dorothy. English political caricature 1793-1832: A study of opinion and propaganda. Oxford: Clarendon Press, 1959.

George, M. Dorothy. Hogarth to Cruikshank: Social change in graphic satire. New York: Walker and Co., 1967.

Gerry, Vance. The Ernest A. Lindner collection of antique printing machinery. Pasadena, CA: Weather Bird Press, 1971.

Giebelhausen, Michaela. Painting the Bible: Representation and belief in mid-Victorian Britain. Aldershot: Ashgate, 2006.

Gilbert, K. R. Henry Maudslay machine builder. London: H. M. Stationery Office, 1971.

Gillespie, Sarah C. A hundred years of progress: The record of the Scottish Typographical Association 1853 to 1952. Glasgow: Robert Maclehose and Co., 1953.

Gouilloud, Maurice. Essai historique 1815-1910 sur les machines à composer. Paris: Imprimerie des Cours Professionnels, 1910.

Grasselli, Margaret Morgan, and Elizabeth Savage, eds. Printing colour 1700-1830: Histories, techniques, functions, and receptions. Oxford: Published for the British Academy by Oxford University Press, 2025.

Green, James N. Mathew Carey: Publisher and patriot. Philadelphia: Library Company of Philadelphia, 1985.

Griest, Guinevere L. Mudie's circulating library and the Victorian novel. Bloomington: Indiana University Press, 1970.

Grolier Club. Lasting impressions: The Grolier Club library. New York: The Grolier Club, 2004.

Gross, Robert A,. and Mary Kelley. A history of the book in America, vol. 2: An extensive republic: Print, culture and society in the new nation, 1790-1840. Chapel Hill: University of North Carolina Press, 2010.

Gutjhar, Paul C. An American Bible: A history of the good book in the United States, 1777-1880. Stanford, CA: Stanford University Press, 1999.

Haas, Wilhelm. Beschreibung einer neuen Buchdruckerpresse von Wilhelm Hass dem Vater 1772. Wiedergegeben nach dem Druck von Wilhelm Haas dem Sohne, Basel 1790. Bern: Schweizerisches Gutenbergsmuseum, 1955.
    Includes facsimile.

Handover, P. M. Printing in London from 1476 to modern times. London: George Allen & Unwin Ltd., 1960.

Hansard, Luke. The auto-biography of Luke Hansard printer to the House 1752-1828. Edited with an introduction by Robin Myers. London: Printing Historical Society, 1991.

Hardwick, Michael, and Mollie Hardwick. The Charles Dickens encyclopedia. Reading: Osprey, 1973.

Harman, Claire. Charlotte Brontë: A fiery heart. New York: Alfred Knopf, 2016.

Harris, Elizabeth M. Patent models in the Graphic Arts collection. Washington, DC: The National Museum of American History, Smithsonian Institution, 1997.

Harsin, Jill. Barricades: The war of the streets in revolutionary Paris, 1830-1848. New York: Palgrave, 2002.

Hart, Horace. Charles Earl Stanhope and the Oxford University Press. Reprinted from Collectanea 111, 1896, of the Oxford Historical Society with notes by James Mosley. London: Printing Historical Society, 1966.

Harvey, Charles, and Jon Press. William Morris: Design and enterprise in Victorian Britain. Manchester: Manchester University Press, 1991.

Hayman, Richard. Ironmaking: The history and archaeology of the iron industry. Stroud, UK: The History Press, 2011.

Haynes, Christine. Lost illusions: The politics of publishing in nineteenth-century France. Cambridge, MA: Harvard University Press, 2010.

Henderson, W. O. The industrialization of Europe 1780-1914. London: Thames and Hudson, 1969.

Herman, Baron. The Commerciograph: A new means to study commercial geography and the trade of the world. Washington, DC: Lanston Monotype Print, 1900.
    Possibly the first commercial work set by Lanston's Monotype machine before the company's move to Philadelphia in 1900. A form letter enclosed with the pamphlet is dated 2 May 1900.

Herrmann, Carl. Geschichte der Setzmaschine und ihre Entwickelung bis auf die heutige

Zeit. Vienna: Im Selbstverlag des Verfassers, [1900].

Hewish, John. The indefatigable Mr Woodcroft: The legacy of invention. London: The British Library, 1980.

Hewitt, Martin. The dawn of the cheap press in Victorian Britain: The end of "taxes on knowledge," 1849-1869. London: Bloomsbury Academic, 2014.

Hills, Richard L. Papermaking in Britain 1488-1988: A short history. London: Athlone Press, 1988.

Hills, Richard L. Power from steam: A history of the stationary steam engine. Cambridge: Cambridge University Press, 1993.

Hilton, Boyd. A mad, bad and dangerous people? England 1783-1846. Oxford: Clarendon Press, 2006.

Hindle, Brooke, and Steven Lubar. Engines of change: The American Industrial Revolution, 1790-1860. Washington, DC: Smithsonian Institution Press, 1986.

Histoire d'un imprimeur: Berger-Levrault 1676-1976. N.p.: Berger-Levrault, 1976.

The history of The Times: "The thunderer" in the making 1785-1841. London: The Office of The Times, 1935.

Hofmann, Carl. Traité pratique de la fabrication du papier. Deuxième édition française... Tome premier. Paris: H. Everling, 1909.
　Translated from the second German edition.

Hofmann, Carl. Traité pratique de la fabrication du papier. 3e partie: Collage-charge-coloration des pâtes à papier. Grenoble: Editions de "L'Industrie papetière," 1926.

Hollis, Patricia. The pauper press: A study in working-class radicalism of the 1830s. London: Oxford University Press, 1970.

Hopkins, Richard L. Tolbert Lanston and the Monotype: The origin of digital printing. Tampa, FL: University of Tampa Press, 2012.

Hoppen, K. Theodore. The mid-Victorian generation 1846-1886. Oxford: Clarendon Press, 1998.

Horrocks, Richard. James Kay of Turton Tower: Inventor and flax spinner (1774-1857). N.p.: RHPublishing, 2020.

Howe, Ellic, ed. The London compositor: Documents relating to wages, working conditions and customs of the London printing trade 1785-1900. London: Geoffrey Cumberlege, Oxford University Press for the Bibliographical Society, 1947.

Howe, Ellic, and Harold E. Waite. The London Society of Compositors (re-established 1848): A centenary history. London: Cassell & Company, 1948.

Howe, Ellic, and John Child. The Society of London Bookbinders 1780-1951. London: Sylvan Press, 1952.

Howsam, Leslie. Cheap Bibles: Nineteenth-century publishing and the British and Foreign Bible Society. Cambridge: Cambridge University Press, 1991.

Hunnisett, Basil. Engraved on steel: The history of picture production using steel plates. Aldershot: Ashgate Publishing, 1998.

Hunnisett, Basil. Steel-engraved book illustration in England. London: Scolar Press, 1980.

Hunter, Dard. Papermaking in pioneer America. New York: Garland Publishing, 1981.

Hunter, Dard. Papermaking: The history and technique of an ancient craft. Second edition, revised and enlarged. New York: Alfred A. Knopf, 1957.

Hunter, Louis C. A history of industrial power in the United States 1780-1930, vol. 2: Steam power. Charlottesville: University Press of Virginia for the Hagley Museum and Library, 1985.

Huss, Richard E. The development of printers' mechanical typesetting methods 1822-1925. Charlottesville: Published for the Bibliographical Society of the University of Virginia by the University Press of Virginia, 1973.

Huss, Richard E. Dr. Church's "hoax": An assessment of Dr. William Church's typographical inventions in which is enunciated Church's Law. Lancaster, PA: Graphic Crafts, Inc., 1967.

International Typographical Union, Executive Council. A study of the history of the International Typographical Union 1852-1963, vol. 1. Colorado Springs, CO: International Typographical Union, 1964.

The inventions of Dr. William Church—the first patented type-casting and composing machine. Scientific American 88, no. 7 (14 February 1903): 109; 116.
　Illustrated.

Isaacs, George A. The story of the newspaper printing press. London: Co-operative Printing Society, 1931.

Jackman, Sydney. Galloping Head: The life of the Right Honourable Sir Francis Bond Head, Bart., P.C., 1793-1875, late Lieutenant-Governor of upper Canada. London: Phoenix House, 1958.

Jackman, W. T. The development of transportation in modern England. London: Frank Cass & Co., 1962.

Jackson, Mason. The pictorial press: Its origin and progress. London: N.p., 1885 [New York: Burt Franklin, 1969].
　Facsimile edition.

James, Louis. Print and the people 1898-1851. London: Allen Lane, 1976.

Jarrigue, François. Au temps des "tueues de bras": Les bris de machines à l'aube de l'ère industrielle. Rennes: Presses Universitaires de Rennes, 2009.

Jeremy, David J. Transatlantic industrial revolution: The diffusion of textile technologies between Britain and America, 1790-1830s. North Andover, MA: Merrimack Valley Textile Museum; Cambridge, MA: MIT Press, 1981.

John, Angela V., ed. Unequal opportunities: Women's employment in England 1800-1918. Oxford: Basil Blackwell, 1986.

Joyce, Patrick, ed. The historical meanings of work. Cambridge: Cambridge University Press, 1989.

K. K. Hof- und Staatsdruckerei [Vienna]. Personalstand der K. K. Hof- und Staatsdruckerei mit 5. November 1904. [Vienna:] K. K. Hof- und Staatsdruckerei, 1904.

Kahan, Basil. Ottmar Mergenthaler: The man and his machine. New Castle, DE: Oak Knoll Press, 2000.

Kander, Astrid, Paolo Malanima, and Paul Warde. Power to the people: Energy in Europe over the last five centuries. Princeton: Princeton University Press, 2013.

The Kansas state printing plant. Topeka: N.p., 1915.

Kaser, David. The cost book of Carey & Lea 1825-1838. Philadelphia: University of Pennsylvania Press, 1963.

Kasischke, Friedrich. Friedrich Koenig Erfinder der Druckmaschine und Vollender der Gutenbergschen Druckkunst. Würzburg: Koenig & Bauer, 1999.

Keeping America informed: The U.S. Government Printing Office: 150 years of service to the nation. Washington, DC: U.S. Government Printing Office, 2011.

Kimberly-Clark Corporation. Graphic communications through the ages. Neenah, WI: Kimberly-Clark Corporation, 1971.

Koenig & Bauer. Schnellpressenfabrik Koenig & Bauer G.M.B.H. Wuerzburg. Berlin: Ecksteins Biographischer Verlag, [1911].

Koenig & Bauer. Unser Werk: KB 1817 gegründet von Friedrich Koenig dem Erfinder der Schnellpresse. Würzburg: Schnellpressenfabrik Koenig & Bauer A-G, n.d. [ca. 1925].

Koenig & Bauer. A. F. Bauer. Würzburg: Schnellpressenfabrik Koenig & Bauer, 1960.

Koenig & Bauer. Jubilee publication on the 150th anniversary 1817-1967. Würzburg: Schnellpressenfabrik Koenig & Bauer AG, 1967.

Koenig & Bauer AG. 1817-1992: 175 years. N.p.: Koenig & Bauer AG, 1992.

Kosky, Jules. Mutual friends: Charles Dickens and Great Ormond Street Children's Hospital. New York: St. Martin's Press, 1989.

Kubler, George A. Historical treatises, abstracts & papers on stereotyping. New York: C.D.M.C., 1936.

Kubler, George A. A new history of stereotyping. New York: N.p., 1941.

Lalande, Jérôme de. The art of making paper taken from the Universal Magazine of Knowledge and Pleasure... Loughborough: The Plough Press, 1978.

Lalande, Jérôme de. The art of papermaking. Translated into English by Richard MacIntyre Atkinson, B.A. Mountcashel Castel, Kilmurry, Sixmilebridge: The Ashling Press, 1976.

Lanston Monotype Machine Co. The Monotype system: A book for owners & operators of Monotypes. Philadelphia: Lanston Monotype Machine Co., 1912.

Lause, Mark A. Some degree of power: From hired hand to union craftsman in the preindustrial printing trades, 1778-1815. Fayetteville: University of Arkansas Press, 1991.

Le Ray, Eric. Marinoni: Le fondateur de la presse moderne (1823-1904). Paris: L'Harmattan, 2009.

Legros, Lucien Alphonse, and John Cameron Grant. Typographical printing-surfaces: The technology and mechanism of their production. London: Longmans, Green, and Co., 1916.

Lehmann-Haupt, Hellmut. The book in America: A history of the making and selling of books in the United States. Second edition. New York: R. R. Bowker Company, 1952.

Leighton, Douglas. Modern bookbinding: A survey and a prospect (The fifth Dent Memorial Lecture). New York: Oxford University Press, 1935.

Levenson, Roger. Women in printing: Northern California, 1857-1890. Santa Barbara, CA: Capra Press, 1994.

Lewis, C. T. Courtney. George Baxter the picture printer. London: Sampson Low, Marston and Co., [1924].
With supplement: The Picture Printer Price List (London: Sampson Low, Marston and Co., 1924).

Librairies-imprimeries réunies. Bearer bond no. 8,893 [60 francs]. N.p., n.d. [1964 or later]. The bond's decorative top border illustrates various printing presses from the 15th to the 20th centuries.

London Phalanx new series 1842-1843. New York: Greenwood Reprint Corporation, 1968.

Lundblad, Kristina. Bound to be modern: Publisher's cloth bindings and the material culture of the book, 1840-1914. New Castle, DE: Oak Knoll Press, 2015.

Lycett, Phyllis, and Michael Martin. Abraham le Blond colour printer 1819-1894. Burton upon Trent: John Mackie Printers Ltd.; published by Lycett Antiques, 1994.

Lyons, Martyn. Le triomphe du livre: Une histoire sociologique de la lecture dans la France du XIXe siècle. N.p.: Promodis, 1987.

Lyons, Martyn. Readers and society in nineteenth-century France: Workers, women, peasants. Basingstoke, UK: Palgrave, 2001.

MacDonald, J. Ramsay. Women in the printing trades: A sociological study. London: P. S. King & Son, 1904.

MacLeod, Christine. Heroes of invention: Technology, liberalism and British identity, 1750-1914. Cambridge: Cambridge University Press, 2010.

MacLeod, Christine. Inventing the Industrial Revolution: The English patent system, 1660-1800. Cambridge: Cambridge University Press, 2002.

Mairet, Joseph. Les carnets de Joseph Mairet ouvrier typographe. Histoire de la Société typographique Parisienne et du Tarif (1839-1851). Montreuil: AFIG, 1995.

Makala, Jeffrey M. Publishing plates: Stereotyping and electrotyping in nineteenth-century US print culture. University Park: Pennsylvania State University Press, 2023.

Mardon, J., et al. Paper machine crew operating manual: A book for paper machine operators. New York: Lakewood Publishing Co., 1968.

Marrot, H. V. William Bulmer [and] Thomas Bensley: A study in transition. London: The Fleuron Ltd., 1930.

Marzio, Peter C. The democratic art: An exhibition on the history of chromolithography in America 1840-1900. Fort Worth, TX: Amon Carter Museum of Western Art, 1979.

Mathias, Peter. The first industrial nation: An economic history of Britain 1700-1914. London: Routledge, 2001.

Mayer & Soutter SA. Reliure industrielle. Renens, Switzerland: Mayer & Soutter, 1968.

McCalman, Iain, et al., eds. An Oxford companion to the Romantic age: British culture 1776-1832. Oxford: Oxford University Press, 1999.

McGaw, Judith A. Most wonderful machine: Mechanization and social change in Berkshire paper making, 1801-1885. Princeton: Princeton University Press, 1987.

McKitterick, David. A history of Cambridge University Press. Cambridge: Cambridge University Press, 1992-2004.

Mellottée, Paul. Histoire économique de l'imprimerie: L'imprimerie sous l'ancien régime 1439-1789. Paris: Librairie Hachette & Cie., 1905.

Mengel, Willi. Ottmar Mergenthaler and the printing revolution. Brooklyn, NY: Mergenthaler Linotype Company, 1954.

[Mergenthaler, Ottmar.] The biography of Ottmar Mergenthaler, inventor of the Linotype. A new edition, with added historical notes based on recent findings. Researched and edited by Carl Schlesinger. New Castle, DE: Oak Knoll Books, 1992.

Mergenthaler Linotype Company. The big scheme of simple operation. Brooklyn, NY: Mergenthaler Linotype Company, 1923.

Michelson, Bruce. Printer's devil: Mark Twain and the American publishing revolution. Berkeley: University of California Press, 2006.

Mitzman, Max E. George Baxter and the Baxter prints. Newton Abbot, UK: David and Charles, 1978.

Monographie de l'École Estienne école municipale professionelle des arts et industries du livre. Paris: Typographie de l'École Estienne, 1900.

The Monotype recorder. Centenary issue: One hundred years of type making, 1897-1997. New series, no. 10 (1997).

Moran, James. Farlow Wilson and the Young/Delcambre composing machine. The Black Art 1, no. 1 (Spring 1962): 20-27.

Moran, James. The composition of reading matter: A history from case to computer. London: Wace & Co. Ltd., 1965.

Moran, James. Printing presses: History and development from the fifteenth century to modern times. Berkeley: University of California Press, 1978.

Morris, Henry. Nicolas Louis Robert and his endless wire papermaking machine, with facsimiles of the inventor's original drawings of the first paper machine. Including a chapter on the papermaking historian Leonard B. Schlosser. Newtown, PA: Bird & Bull Press, 2000.

Morse, John D., ed. Prints in and of America to 1850. Charlottesville: University Press of Virginia, 1970.

Moss, Sidney P. Charles Dickens and Frederick Chapman's agreement with Ticknor & Fields. The Papers of the Bibliographical Society of America 75 (1981): 33-38.

Moxon, Joseph. Mechanick exercises on the whole art of printing (1683-4). Edited by Herbert Davis and Harry Carter. London: Oxford University Press, 1962. Second edition.

Muir, Percy. Victorian illustrated books. New York: Praeger Publishers, 1971.

Musson, A. E. Trade union and social history. London: Frank Cass, 1974 [London: Routledge, n.d.].

Musson, A. E. The Typographical Society: Origins and history up to 1949. London: Geoffrey Cumberlege, Oxford University Press, 1954.

Myers, Robin, and Michael Harris. Medicine, mortality and the book trade. London: St. Paul's Bibliographies; New Castle, DE: Oak Knoll Press, 1998.

Navickas, Katrina. Loyalism and radicalism in Lancashire 1798-1815. Oxford: Oxford University Press, 2009.

Neipp, Lucien. Les machines à imprimer depuis Gutenberg. Paris: Club Bibliophile de France, 1951.

New, Chester W. The life of Henry Brougham to 1830. Oxford: Clarendon Press, 1961.

NKOM SSSR Glavpoligrafmash. Katalog poligraficheskikh mashin [in Cyrillic]. Leningrad: State Publishing House of Standards, 1941.

Nord, David Paul. Faith in reading: Religious publishing and the birth of mass media in America. Oxford: Oxford University Press, 2004.

Obschernitzki, Doris. "Der Frau ihre Arbeit!" Lette-Verein: Zur Geschichte einer Berliner Institution 1866 bis 1986. Berlin: Edition Hentrich, 1987.

Ottley, George. A bibliography of British railway history. London: H. M. Stationery Office, 1983. Second edition.

Palmegiano, E. M. Perceptions of the press in nineteenth-century British periodicals: A bibliography. London: Anthem Press, 2012.

Papermaking: Art and craft. An account derived from the exhibition presented in the Library of Congress, Washington, D.C., and opened on April 21, 1968. Washington, DC: Library of Congress, 1968.

Patents for inventions. Abridgements of specifications. Class 100. Printing, letterpress and lithographic. Period—A.D. 1855-1866. London: Printed for His Majesty's Stationery Office by Love & Malcomson, 1905.

Patten, Robert L. Charles Dickens and his publishers. Oxford: Clarendon Press, 1978.

Patten, Robert L., ed. Dickens and Victorian print cultures. London: Routledge, 2017.

Peacock, Thomas Love. Crotchet castle. London: The Folio Society, 1964.
> Peacock's novel satirizes the SDUK as the "Steam Intellect Society"

Pellissier, Pierre. Émile de Girardin: Prince de la presse. Paris: Denoël, 1985.

Le Petit Journal 1902. Tous les jours 5 millions de lecteurs 5c le no. . . . Dijon: Imp. Jobard, 1902.
> "Imprimé sur presses Marinoni." Chromolithographed advertising calendar

Pettitt, Clare. Patent inventions: Intellectual property and the Victorian novel. Oxford: Oxford University Press, 2004.

Picturing the big shop: Photos of the U.S. Government Publishing Office, 1900-1980. Washington, DC: U.S. Government Publishing Office, 2017.

Pike, E. Royston. "Hard times": Human documents of the Industrial Revolution. New York: Praeger, 1969.

Pike, E. Royston. Human documents of the Industrial Revolution in Britain. London: Routledge, 2006.

Pleger, John J. Bookbinding. Chicago: Inland Printer Company, 1924.

Pollard, Graham, and Esther Potter. Early bookbinding manuals: An annotated list of technical accounts of bookbinding to 1840. Oxford: Oxford Bibliographical Society, 1984.

Ponot, René. La fonderie typographique de Laurent, Balzac et Barbier créee en 1827 par Honoré Balzac. Paris: Aux Éditions des Cendres, 1992.

Printing and publishing The Times. N.p., n.d. [ca. 1955].

Printing and the mind of man. Assembled at the British Museum and at Earl's Court London 16-27 July 1963. [London:] Messrs. F. W. Bridges & Sons Ltd. and the Association of British Manufacturers of Printers' Machinery, 1963.

Printing in the twentieth century: A survey. Reprinted from the special number of The Times, October 29, 1929. London: The Times Publishing Company, 1930.

Printing patents: Abridgements of patent specifications relating to printing 1617-1857. First published in 1859 and now reprinted with a prefatory note by James Harrison. London: Printing Historical Society, 1969.
> Reprints of the original editions of 1859 and 1878.

Printing The Times since 1785: Some account of the means of production and changes of dress of the newspaper. London: Printing House Square, 1953.

Proceedings at a meeting of the vat paper makers held at the Bell Hotel, Maidstone, on Tuesday 8th March 1853. North Hills, PA: Bird and Bull Press, 1970.
> Facsimile edition from a manuscript copy.

Prothero, Iorwerth. Radical artisans in England and France, 1830-1870. Cambridge: Cambridge University Press, 1997.

Pursell, Carroll W., Jr. Early stationary steam engines in America: A study in the migration of a technology. Washington, DC: Smithsonian Institution Press, 1969.

R. Hoe & Co. Catalogue of printing presses and printers' materials. With a new introduction by Frank E. Comparato. New York: Garland Publishing, 1980.
> Reprint of the 1881 edition.

Radcliffe, William. Origin of the new system of manufacture, commonly called "power-loom weaving," and the purposes for which this system was invented and brought into use . . . Stockport: Printed and sold by James Lomax, 1828 [Clifton, NJ: Augustus M. Kelley, 1974].
> Facsimile edition.

Ramazzini, Bernardino. De morbis artificium Bernardini Ramazzini diatriba. Diseases of workers. The Latin text of 1713. Revised with translation and notes by Wilmer Cave Wright. New York: Classics of Medicine Library, 1983.
> Reprint of the 1940 edition.

Ratcliffe, Eric. The Caxton of her age: The career and family background of Emily Faithfull (1835-95). Upton-upon-Severn: Images Publishing, 1993.

Raven, James. The business of books: Booksellers and the English book trade, 1450-1850. New Haven, CT: Yale University Press, 2007.

Raven, James. Publishing business in eighteenth-century England. Woodbridge: Boydell Press, 2014.

Reynolds, Siân. Britannica's typesetters: Women compositors in Edwardian Edinburgh. Edinburgh: Edinburgh University Press, 1989.

Richardson, Ruth. Dickens and the workhouse: Oliver Twist and the London poor. Oxford: Oxford University Press, 2012.

Rickards, Maurice. The encyclopedia of ephemera: A guide to the fragmentary documents of everyday life for the collector, curator, and historian. New York: Routledge, 2000.

Rieck, Eckhard. Friedrich Koenig und die Erfindung der Schnellpresse: Wege eines Pioniers der modernen Unternehmensgeschichte. Munich: AVM, 2015.

Rivard, Paul E. A new order of things: How the textile industry transformed New England. Hanover, NH: University Press of New England, 2002.

Robak, Brigitte. Vom Pianotyp zur Zeilensetzmaschine: Setzmaschinenentwickling und Geschlecterverhältnis 1840-1900. Marburg: Jonas Verlag, 1996.

Roll, Eric. A history of economic thought. Englewood Cliffs, NJ: Prentice-Hall, [1953].
> Third edition.

Romano, Frank. History of the Linotype company. Rochester, NY: RIT Press, 2014.

Rose, Jonathan. The intellectual life of the British working classes. New Haven, CT: Yale University Press, 2010.
> Second edition.

Rose, Michael E. The English Poor Law 1780-1930. Newton Abbot: David & Charles, 1971.

Rosenband, Leonard N. La fabrication du papier dans la France des lumières: Les Montgolfier et leurs ouvriers, 1761-1805. Rennes: Presses Universitaires de Rennes, 2005.

Rosenband, Leonard. Papermaking in eighteenth-century France: Management, labor and revolution at the Montgolfier mill, 1761-1805. Baltimore, MD: Johns Hopkins Univerity Press, 2000.

Rumble, Walker. The swifts: Printers in the age of typesetting races. Charlottesville: University of Virginia Press, 2003.

Rummonds, Richard-Gabriel. Nineteenth-century printing practices and the iron handpress. New Castle, DE: Oak Knoll Press; London: British Library, 2004.

St. Bride Foundation. Catalogue of the technical reference library of works on printing and the allied arts. London: Printed for the Governors, 1919 [Mansfield Centre, CT: Martino Fine Books, 1999].
> Facsimile edition.

St. Clair, William. The reading nation in the Romantic period. Cambridge: Cambridge University Press, 2004.

Sale, Kirkpatrick. Rebels against the future: The Luddites and their war on the Industrial Revolution. Reading, MA: Addison-Wesley Publishing Co., 1995.

Särkkä, Timo. The British paper industry, 1800-2000. In Lamberg, Juha-Antti, et al., The Evolution of Global Paper Industry 1800-2050 (Heidelberg: Springer, 2012), pp. 167-190.

Saxe, Stephen O. The Bruce pivotal typecaster and its influence on nineteenth-century typography. Journal of the Printing Historical Society, n. s., no. 25 (Summer 2016): 37-62.

Schirmann, Adolfhanns. Der Druckerei-Buchbinder. 2. Auflage. Frankfurt am Main: Polygraph Verlag, 1952.

Schlicke, Paul. The Oxford companion to Charles Dickens. Anniversary edition. Oxford: Oxford University Press, 2011.

Shannon, Mary L. Dickens, Reynolds, and Mayhew on Wellington Street: The print culture of a Victorian street. Farnham: Ashgate, 2015.

Shepard, Leslie. The history of street literature. Newton Abbot: David & Charles, 1973.

Slater, Michael. Charles Dickens. New Haven, CT: Yale University Press, 2009.

Slinn, Judy, Sebastian Carter, and Richard Southall. History of the Monotype Corporation. London: Printing Historical Society, 2014.

Smyth-Horne Ltd. Smyth: Standard of the work in bookbinding machinery [cover title]. London: Smyth-Horne Ltd., n.d. [ca. 1930].

Société anonyme du Petit Journal. Bearer bond no. 15,562 [500 francs]. Paris: Imprimerie Chaix, April 1900.
> The bond's decorative border includes an illustration of a Marinoni steam-powered press in the lower right corner.

Spiegel, Henry William. The growth of economic thought. Durham, NC: Duke University Press, 1971.

Stamprech, Franz. 175 Jahre österreichische Staatsdruckerei: Entwicklung und Geschichte der österreichischen Staatsdruckerei. Vienna: Verlag der Österreichischen Staatsdruckerei, 1979.

Stapleton, Darwin H. The transfer of early industrial technologies to America. Philadelphia: American Philosophical Society, 1987.

Stearns, Peter N. The Industrial Revolution in world history. Fourth edition. Boulder, CO: Westview Press, 2013.

Stephen, George A. Die moderne Grossbuchbinderei: Eine Beschreibung der Herstellung von Bucheinbänden und der dabei verwendeten Machinen. Vienna: A. Hartleben's Verlag, 1910.

Sterne, Harold E. Catalogue of nineteenth-century bindery equipment. Cincinnati, OH: Ye Olde Printery, 1978.

Stevenson, John. Popular disturbances in England, 1700-1832. Second edition. London: Longman, 1992.

Strachan, John. Advertising and satirical culture in the Romantic period. Cambridge: Cambridge University Press, 2007.

Suarez, Michael F., and H. R. Woudhuysen, eds. The book: A global history. Oxford: Oxford University Press, 2013.

Thackrah, Charles Turner. The effects of arts, trades, and professions on health and longevity. With the life, work and times of Charles Turner Thackrah by A. Meiklejohn. N.p.: WH Smith, 1989.

Thomis, Malcolm I. The Luddites: Machine-breaking in Regency England. Newton Abbot, UK: David & Charles; Hamden, CT: Archon Books, 1970.

Thompson, Alistair G. The paper industry in Scotland 1590-1861. Edinburgh: Scottish Academic Press, 1974.

Thompson, Dorothy. The Chartists: Popular politics in the Industrial Revolution. New York: Pantheon Books, 1984.

Thorp, Joseph. Printing for business: A manual of printing practice in non-technical idiom. London: Jonathan Cape, 1928.

Tresch, John. The romantic machine: Utopian science and technology after Napoleon. Chicago: University of Chicago Press, 2012.

Trewin, J. C., and E. M. King. Printer to the House: The story of Hansard. London: Methuen and Co., 1952.

Trinder, Barrie. Britain's industrial revolution: The making of a manufacturing people, 1700-1870. Lancaster: Carnegie Publishing, 2013.

Tryon, Warren S., and Willliam Charvat. The cost books of Ticknor and Fields and their predecessors, 1832-1858. New York: Bibliographical Society of America, 1949.

Twyman, Michael. Early lithographed books: A study of the design and production of improper books in the age of the hand press. Williamsburg, VA: The Book Press Ltd.; London: Farrand Press & Private Libraries Association, 1990.

Twyman, Michael. A history of chromolithography: Printed colour for all. London: British Library; New Castle, DE: Oak Knoll Press, 2013.

Twyman, Michael. Lithography 1800-1850: The techniques of drawing on stone in England and France and their application in works of topography. London: Oxford University Press, 1970.

Twyman, Michael. Printing 1770-1970: An illustrated history of its development and uses in England. London: Eyre and Spottiswoode, 1970.

Uglow, Jenny. In these times: Living in Britain through Napoleon's wars, 1793-1815. New York: Farrar, Straus and Giroux, 2014.

Uglow, Jenny. Nature's engraver: A life of Thomas Bewick. London: Faber and Faber, 2006.

Valente, A. J. Rag paper manufacture in the United States, 1801-1900: A history, with directories of mills and owners. Jefferson, NC: McFarland & Co., 2010.

Van Remoortel, Marianne. Women, work and the Victorian periodical. New York: Palgrave Macmillan, 2015.

Verneuil, Charles, and Jules Soufflet. Les presses à pédale autour de textes de Charles Verneuil & Jules Soufflet présentés par Paul-Marie Grinevald. Paris: Éditions des Cendres, 1997.

Wakeman, Geoffrey. Aspects of Victorian lithography: Anastatic printing and photozincography. Wymondham, UK: Brewhouse Press, 1970.

Wakeman, Geoffrey. Victorian book illustration: The technical revolution. Newton Abbot, UK: David & Charles, 1973.

Wakeman, Geoffrey. Victorian colour printing. Loughborough, Leicestershire, UK: The Plough Press, 1981.
No. 124 of 150 copies.

Walker, Edward. The art of book-binding, its rise and progress; including a descriptive account of the New York Book-Bindery . . . and the great New-York book-bindery. Edited with an introduction by Paul S. Koda. New Castle, DE: Oak Knoll Books, 1984.

Waller, David. Iron men: How one London factory powered the Industrial Revolution and shaped the modern world. London: Anthem Press, 2016.

Wallis, L. W. A concise chronology of typesetting developments 1886-1986. London: The Wynkyn de Worde Society in association with Lund Humphries, 1988.

Weber, Wilhelm. A history of lithography. New York: McGraw-Hill Book Company, 1966.

Weedon, Alexis. Victorian publishing: The economics of book production for a mass market, 1836-1916. Aldershot: Ashgate Publishing, 2003.

Weightman, Gavin. The industrial revolutionaries: The making of the modern world 1776-1914. New York: Grove Press, 2007.

Werner, Alex, and Tony Williams. Dickens's Victorian London 1839-1901. London: Ebury Press, 2011.

Wheelwright, William Bond. Printing papers. Chicago: University of Chicago Press, 1936.

Wilkes, Susan. Narrow windows, narrow lives: The Industrial Revolution in Lancashire. Stroud, UK: Tempus Publishing, 2008.

Wilkins, William Glyde. Charles Dickens in America. London: Chapman and Hall, 1911.

Wilson, Charles, and William Reader. Men and machines: A history of D. Napier & Son, Engineers, Ltd., 1808-1958. London: Weidenfeld and Nicolson, 1958.

Winship, Michael. Subscription publishing in America over three centuries. Charlottesville, VA: Book Arts Press, 2021.

Wolf, Hans-Jürgen. Geschichte der Druckpressen: Ein illustriertes Handbuch mit einer ausführlichen Zeittafel. Frankfurt: Interprint, 1974.

Wolf, Hans-Jürgen. Geschichte der Druckverfahren. Elchingen, Germany: Historia, 1992.

Wosh, Peter J. Spreading the word: The Bible business in nineteenth-century America. Ithaca, NY: Cornell University Press, 1994.

Young, Alan R. Steam-driven Shakespeare or making good books cheap: Five Victorian illustrated editions. New Castle, DE: Oak Knoll Books, 2017.

Zlotnick, Susan. Women, writing and the Industrial Revolution. Baltimore, MD: The Johns Hopkins University Press, 1998.

Zur Feier des einhundertjährigen Bestandes der K. K. Hof- und Staatsdruckerei. Vienna: K. K. Hof- und Staatsdruckerei, 1904.

Index

Page numbers in *italics* refer to illustrations.

Abbott, Jacob, 141, *141*, 144, 147
Abel, Charles Denton, 182
Ackermann, R., 8-9, 98
Adam, Isaac, *138*, *141*
Adams, Isaac, 135, 140-141, 203, 225
Adams, Joseph Alexander, 107, 142, *143*, 144
Adams, Thomas F., 225
Adams Power Press, 135, 140-141, *141*, *148*, 149, 159, 203, 225
Aiken, Howard, 176
Alauzet, Pierre, 224, *251*, 253
Albert, Andreas, 222
Albert, Prince, 100, *100*, 219
Albion Mills, 16-17, 219
Albion types, 225
Alden, Timothy, *179*, 180-181, 188
Alden-Type, 180-181
Alembert, d', 26, *193*, *194*
Alexander, Matilda, 158
Alfred, Prince, 100, *100*
Alfred Mames et Fils, 222, *223*
Alkan, Alphonse, 153, *153*
Allan, David, 40
American Bible Society (ABS), 138-140, *140*
Amman, Jost, *205*
Analytical Engine, 176
André, Louis, 34
Andreäschen Buchhandlung, 117, *118*
Anisson, Étienne-Alexandre-Jacques, 17, *18*, 19
Anthony, Susan B., 161
Antoine, Pierre-Joseph, 32, *33*
Apollo Press, 167
Apple, 255
Applegath, Augustus, 88, *89*, 90, 91, *92*, 94, *108*, 109, 112-113, *112*, 114, 115, *117*, *119*, 121, *121*, 126, 216, 253
Applegath Vertical Printing Machine, 217, *217*
Arago, François, 171
Arkwright Richard, xii, 10-12, 29
arming press, 194
Arming Press, 200

Armstrong, S. T., 137-138
Armstrong, Thomas, 194
Arnett, John Andrews, 203
Ashton, John, 10
Association for the promotion of the interests of booksellers, 226
Atmospheric Steam Pumping Engine, xii
Audouin de Geronval, Maurice Ernst, *115*, 116
Auer, Alois, 215-216

Babbage, Charles, 112, 176, 202, *202*
Baillou, Maturin M., 160-161
Baines, Edward, 12
Banks, Sir Joseph, 105
Barker, Nicolas, 107
Barnes, William C., 190-191, *190-191*
Barrows, Samuel J., 252
Barth, J. G., 176
Barthe, J.-G, *165*
Basset, Paul-Andre, *41*
Bastide, Madame, 151
Baudouin, François-Jean, 117
Bauer, Andreas Friedrich, 22, *43*, 45-46, *47*, *47*, *49*, *51*, *53*, *56*, *56*, *57*, *58*, *59*, 109, 117, 216, 237, 253
Baxter, George, 217-221, *218*, *219*, *220*
Bazin, A., 155
bed-and-platen power presses, *138*, *141*
Bell, Henry, 14
Bell, Thomas, *45*, 46, 78
Bennett, James Gordon, Sr., 242
Bensley, Benjamin, 59, *59*
Bensley, Thomas, *15*, 22, 45, 47, *55*, *55*, *56*, *56*, 57, 58-59, *58*, 98
Bensley's Printing Machine, 112
Bentley, Richard, 127
Bentley's Machine, 121
Berte, Antoine-François, 32, *32*, 34
Berte and Grenevich, 32-33

Bessemer, Henry, 171
Bewick, Thomas, *93*, 97-98, *97*
Bible Society of Philadelphia, 135, *136*
Biggs, George, 172
Biggs, Mary, 161
Bigmore, Edward C., 252
Bigmore, Edward M., 249
Billetes, Gilles Filleau des, 26
binder's plout, *205*
Binns, Edward, 171-172, *172*, 173
Bismarck, Otto von, 56
Blades, William, 232, 234, 252
Blake, William, 16
Bloxam and Fourdrinier, 30
Blumenbach, John Friedrich, 58, *58*
board-cutting machine, 204-205
Boizard, Paul, 176
bookbindery machines, 208-*211*, *214*
Boulton, Matthew, 13-14, 16-17
Bourdelin, Emile, *24*, *38*
Bourinot, Elias, 138
Bradbury and Evans, 87-88, 125-126
Brade, Ludwig, 206, 207
Bradshaw, George, 65-66, *65*
Brandreth, Jeremiah, 46, *54*
Brehmer, Gebruder, 207
Breitkopf and Hartel, 44
British and Foreign Bible Society, 70-72
Brothers, the, 193, *195*
Brougham, Henry, 77, 78, *78*, 79, 80, 83, 85-86, *95*
Brown, Thomas, 137-138
Browne, Hablot Knight, *126*
Bruce type-caster, xi
Bullen, George, 235
Bullock, William, 247-248
Burdon, William, *54*
Burke, Edmund, 88
Burn, William, 199, *199*
Burr, G. A., 188
Burr Printing House, *182*
Byron, George Gordon, 201-202, *201*, *202*

Cambridge University Press, 22, 63, 71, *72*, 138
Campbell, Thomas, *106*, 107
Cannell, J. F., 65
Carey, Mathew, 35-36
Carter, John, 195-196, 198
Cartwright, Edmund, 14, 14-*15*, 67
Caxton, William, 84, 94, *166*, 167, *181*, 232, 234-235
Caxton Celebration, 182
Caxton Memorial Exhibition, 232, *234*
Caxton Quadricentennial Celebration, *181*, 182, 232, 252
Chadwick, Jane E., 126
Chaix, Edmond-Albans, 222
Chaix, Napoléon, 222
Chambers, William and Robert, 87-88, *89*
Chapman, Edward, *132*, 133
Chapman and Hall, 125-126, 133
Charles X, 116-117
Chiswick Press, 250
chromolithography, *101*, *103*, 229
Church, William, 112, 169-170, *169*, 171
Churchill, John, 171, *172*, 173
Cicero, 195, *196*
City Brothers, the, 193, *195*
Clapperton, R. H., 30
Clarkson, *49*
Clarkson, Thomas, 49
Clay, John, 169, *175*, *175*
Clay and Rosenborg typesetting machines, *175*, 176, 178
Cleave, John, 127, *128*
Close, Bartholomew, 4
Clowes, George, 37
Clowes, William, 232
Cogger press, 117, *118*, 219
Coggeshall, William Turner, 168, 191
Cohen, Colin, 26
Coleman, 32
colliery engines, *44*
Columbian handpress, 225
Combes, James, *44*

293

composing machines, *170*
Constable, Archibald, 77–78
Cope and Sherwin, 200–201
Corliss steam engine, *148, 149,* 188
Cotta, Johann Friedrich, 56–57
Cowper, Edward, 56, 89, 92, *108, 111, 112*
Cowper, Edward, 216
Cowper. Edward Shickle, 88, 90, 91, 94, 109, 112, 114, 117, 120–*121, 121,* 123
Craig, William, 194
Crane, Walter, 220
Crompton, J. W., 156
Cruikshank, George, 127
cylinder flatbed press, 49
cylinder press, xi, 47, 49, *49,* 50, 109, 121, 123, *160,* 240

Daly, Benjamin H., 240
Darblay, Paul, 34
Darby, Abraham, 8
Davies, G., 85
Davis, Charles E., 182
Davis, Daniel, *142,* 144
Davis, Daniel, Jr., 142
Davis, John, 37
De Vinne, Theodore Low, x, xi, xiii, 162, 257–259, *257*
Delcambre, Adrien, *165,* 169, 171, 173–174, *175,* 176–177, 188
Delcambre, Isidore, 171, 176
Deltusso, C., 151–152, *151*
Dentu, E., 153
Derrier, Jules, *224*
Désauguliers, Jean-Théophile, *8*
Develly, Jean-Charles, *108*
Dickens, Charles, 125–127, *126, 129,* 130, *130, 131, 132,* 133
Dickinson, John, 32–33, 35, 36
Diderot, 26, 193, *194*
Didot, Firmin, 115
Didot, Henri, 171
Didot, Pierre-François, 25, 28
Didot, Saint-Leger, 28, 30, 32–33, 34
Didot family, 19, 170
Difference Engine No. 1, 112, 176
Difference Engine No. 2, 176
Dodd, George, 213
Dolet, viii
Donkin, Bryan, 30
Donkin machines, 32
double cylinder "Nay-Peer" printing machines, *113*
Double Imperial printing machines, 237
double-cylinder flatbed press, 50, *51, 52*
double-cylinder perfecting machine, 88, *89,* 90, 92, 94, 109, *112,* 238
double-cylinder printing machines, 53, 120–*121,* 237–238, 240
Duguid, Alexander, 190–191, *190–191*
Dumas, Alexandre, 176
Dunning, Thomas Joseph, 156
Dupont, Paul, 155, 176
Dupuy, Théodore, *101,* 102, *102, 103, 104, 251*
Dusautoy, Jean Abbot, 30–31, *31*

Edward IV, King, 232, 234
Eggers and Co., *110*
eight-cylinder vertical printing machines, 121, *121*
Eisenmann, Andreas, 222
Elliotson, John, 58, *58*
Elzevier, viii
embossing machines, 207, 208–209
embossing press, *205*
Estienne, viii
Evans, Edmund, *219,* 220, 221
Evans, Oliver, 13
Exposition Universelle, 104, 253
Eyre and Spottiswoode, 232

Fain, Louis-Armand-Jean, 117
Fairbairn, Thomas, 194
Fairman, Gideon, *106*
Faithful, Emily, *154,* 155–158, *156,* 163
Fanshaw, Daniel, 138
Felkin, William, 10
Fischer, C. F., 222
flatbed press, *49, 51*
Folds, J. S., 89–90
Forster, John, 232
Foulis, Andrew, 19–21
four-cylinder horizontal machines, 121
Four-Cylinder Type-Revolving Printing Machine, *147*
four-feeder cylinder press, *119*
four-feeder printing machine, *126*
Fournier, 115
frame knitting machine, 10
Franklin, Benjamin, 4, *5,* 180
Franz I, Emperor, 215
Fraser, 188
Fraser, James, 193, *195*
French Academy of Sciences, 174
Freylinghausen, John Anastasius, 20–21, 21–22, 36, *36,* 252
Fritz, G., 226, *230*
Fröbel, Herr, 234
Frowde, Henry, 235
Fulton, Robert, *13,* 14

Gamble, John, *28,* 29–30
Gaskell, Peter, 67, 70, *71*
Gaskill, Jackson, 225, *226*
Gaubert, Etienne Robert, 167, 169, 171, *173,* 174–175
Gaveaux, Alexandre Yves, 253
Ged, William, 4–5, 6, 19
Geo. C. Rand and Avery Printers, *148,* 149
George III, 6
Gérotype, 174
Gilpin, Thomas, 34–36
Girardin, Émile de, 85, 87, *87,* 88, *250,* 252–253, 255
Gladstone, William, 232, 234, *234, 235*
gold blocking press, *205*
Golden Type, 260
Gough, Richard, 36

Government Printing Office (GPO), 225, *227*
Graham, L., 208
Grant, C. J., 85
Grant, John Cameron, 168–169
Great Exhibition, *122,* 213, 215, *215,* 216–217, *216, 217, 218*
Greeley, Horace, 188
Greenaway, Kate, 220
Greenfield, William, 198
Grey, Douglas, 225
Grolier Club, The, xi, *258,* 259
Günther, C. F. H., *39*
Gurney, Goldsworthy, 61
Gutenberg, Johannes, viii, xi, 1, 2, 4, 98, 123, *256,* 259, 260, 261

Haas, Wilhelm, 17
Haas, Wilhelm, (the younger), xiv, 17, *17*
Haddock, John, 130
Hale, Nathan, 237
Hall, Henry, *184*
Hall, Mr., 234
Hancock, William, 202
handpress, 74, 117
handpresses, xi, 1, *1, 3,* 4, 17, *18,* 19, 22, 94, *110,* 218, *223,* 249, 261
Hansard, Thomas Curson, xii, 109, 111–112, *113,* 115–116, 170
Hansard, Thomas Curson, Jr., 168, 191
Hardy, Philip Dixon, 90
Hargreaves, James, xii, 10
Harper, J. and J., 140
Harper and Brothers, 107, 141, 142, 144–145, 147, 159, 161, 244
Harper's, 12, *143, 144, 145,* 146–*147*
Harvard Mark 1 electronic computer, 176
Havell, R. & D., 22
Head, Sir Francis Bond, 94
Head, William Wilfred, 158
Heath, Charles, 105, *106,* 107
Heath, Henry, 60
Heferstein, A., *39*
Herzog, J. R., 206
Hewitt, Henry, 127, *129,* 130
Highs, Thomas, 12
Hill, Mr., 83
Hilliard, Gray, Little, 138
Hodson, Mr., 234
Hoe, Richard, 135, *258*
Hoe, Richard March, 135, 232, 242, 244, 247, *247,* 249–250, 252, 259
Hoe, Robert, 135, 138, 225, 240, 242, 259
Hoe, Robert, III, *138,* 259
Hoe Octuple Press, *256*
Hoe Type-Revolving Printing Machine, *179*
Hoe's Four-Cylinder Type-Revolving Press, 244
Hollander beater, 26, *27,* 29
Hopkinson and Cope Improved Albion handpress, 259–260
Hostrup, Mr., 57
Hot-Metal Machine, 190
Huguet, Jean Baptiste, 100
Hunter, Dard, 34–35

Imperial and Royal Court and State Printing Office, 230–*231*
Imperial Arming Press, 200–201, *200,* 202, 203, *205*
Imperial Printing Press, 200, *200*
Imprimerie des Femmes at the École Typographique des Femmes, 152
Imprimerie Royale, 19, 117, 193
ink-distributing table, 109
iron handpress, *18,* 22, 98, 225, *261*
Iron Press, 21

J. G. Rogers & Co., 138
J. G. Schelter & Giesecke, 225–226, *228*
J. Minot & Cie., *104*
J. Watts and Co., 135
Jacobi, Moritz von, 107, 141
Blaeu, Willem Janszoon, 3
Jensen, Nicolas, 260
jobbing presses, *160*
Johnson, John, *166,* 167
Jones Hydraulic Signature Press, 208
Jugenstil (Art Nouveau), 226

K. K. Hof- und Staatsdruckerei, 215, 226, 228
Kastenbein, Charles, 182
Kastenbein's Composing and Distributing Machines, *181*
Kay, John, 12
Kelmscott Press, 259, *261*
Kerr, R. W., 225, *227*
Keynes, Sir Geoffrey, 195, 198
Knapp, Wilhelm, 206
Knecht, M., *101,* 102
Knight, Charles, 37, 50, 59, 67, 79, 80, *81,* 83–84, *83,* 85, 88, 90–91, *92, 93,* 94, *95,* 196, 198, 199–200, *222,* 252, 257
Koenig, Friedrich, 15, 22, 43, *43,* 44–46, 47–51, *47, 48, 49,* 50, *51, 52,* 53, 55–58, *55, 56,* 59, 98, 109, 111, 116, 117, 121, 123, 135, 216, 237, 253
Koenig and Bauer, *118,* 222
Krebs, Benjamin, 117
Kronheim, Joseph Martin, 220–221

L. Graham & Sons, *192,* 206, *207, 208*
L. Graham and Son, 208
La Grande Imprimerie de Blois, 162
la Rue, Charles de, 59
Ladvocat, Pierre-François, 117
Lalande, Jérôme de, 26, *27*
Lanston, Tolbert, 178, *187,* 188–190, *189*
Lanston Monotype Company, 190
Lanston Monotype Machine Company, 190
Lanston Type Machine Company, 188
Lavoisne, M., 35–36, *35*
Lawson, R. E., 198

Index

le Blond, Abraham and Robert, 220
Lee, Richard, 130
Lee, William, 10
Lefevre, Théotiste, 167-168, *167*
Legros, Lucien Alphonse, 168-169
Leighton, Archibald, 198, 199, 201, *201*
Leighton, Charles Blair, 220
Leighton, George C., *212*, 220, *220, 221*
Leighton, Steve, 220
Leighton Brothers, 220
Leistenschneider, Ferdinand, 32
Lemercier, Joseph-Rose, 104-105, 222
Leroux, Pierre, 170-171, *170*
Leschevin, Philippe-Xavier, 32, *33*
Lette, Wilhelm Adolf, *163*
Leuchs, Johann Carl, *35*, 36
Lewis, C. T. Courtney, 219
Librairies-Imprimeries Réunies, 255
Lilien, Otto M., 46
Limbird, John, 77, 84, 114, *114*
L'Imprimerie Berger-Levrault et Cie., 222
Linotype machines, xi, 162, 163, 167-168, 169, *180*, 182, *182*, 183-184, *183, 184*, 186-187, *186*, 188, 190-*191*, 191, 259
Linton, H., *24, 38*
lithographic presses, 98, 99, 100, *100, 101*, 102, *102*, 104, *104*
Livingston, Robert, *13*
Lloyd, Edward, *130*, 133
Lloyd, Mary, 158
Loch, Mr., 83
London Society of Bookbinders, 199
London Society of Machine Managers, 225
London Trades Council, 156
London Union of Compositors, 172
Longman, George, 32, 107
looms, power, 14, 14-15, 67, 70, *70, 71*
Louis Philippe, 117
Loutherbourg, P. J., *10*
Ludlam, Isaac, 46
Ludwig I, King, 57
Lytton, Lord, 232

MacDonald, J. Ramsay, *162*, 163
Machine à Journaux, 255
MacKellar, Smiths, and Jordan, 226
MacKellar, Thomas, 225
Mackie, Alexander, 177, 178, 188
Mackie Steam Composing Machine, *177*, 178
Macmillan and Co., 107
MacSorley, Mary, 234-235
Mame, Alfred, 222, *223*
Marinoni, Hippolyte, viii, 104, *104*, *224, 252*, 253, 254-255, 255
Martin, William, 177-178, 188
Martineau, Harriet, 67
Maudslay, Henry, 44, *44*
Maunder, Samuel, 133

McCann, Joseph W., 190-191, *190-191*
McKitterick, David, 232
McLeod, Alexander, *136*
Mergenthaler, Ottmar, 183-184, 186, 188
Mergenthaler's Blower Linotype, 186-187
Mergenthaler's Linotype, *182*, *183, 184*
Metz, Robert, 206
Meunier, Victor, 171
Middleton, Thomas, *119*
Millaud, Moise Polydore, 255
Miller, Mr., *115*
Milton, John, *257*, 259
Mitchel, William H., 178-179, 180, 188
Mitford, William, 137, *139*
Mitzman, Max, 219
Monet, Adolphe-Lucien, 222, 224, *224*, 255
Monotype machines, 162, 167-168, 169, *180*, 182, *186*, 187, 188, *189*, 190-*191*, 191, 259
Moran, James, 17, 115
More, Hannah, 1, *1*
Morris, Henry, *25*
Morris, William, 259-260, *260, 261*
Morton, G., *61*
Moxon, Joseph, 2, *2, 3*, 4, 22, 109, 257
"Mr. Rutt's Printing Machine," xii
Mudie, Charles Edward, 125
Munsell, Joel, 30
Murray, John, 78, *79*, 201-202, *201*

Napier, David, 111, *113*, 115, 116, 123, 135, 216, 237
Napier drum cylinder machines, 115
Napier Imperial printing machines, *134, 238*
Napier Printing Machine, 237-238, 240
Napoleon, 45
Nash, Paul W., 198
National Association for the Promotion of Social Science, 156
Nay-Peer machines, 111, 116
New Patent Composing Machine, 171
Newcomen, Thomas, xii, 8
Newcomen engines, 5-6, 8, *8*, 13, 22, *22*, 44
Newly Invented Weaving Machine, 14
Nicholson, James B., 203
Nicholson, William, 46-47, *46*, 117, 123
Nouvelle Machine Rotative, 255

Octuple Press, 259
Odilon-Barrot, Camille Hyacinthe, 117
Odilon-Barrot, Camille Hyacinthe, *116*
Opie, Iona and Peter, 130
Otis, Bass, 100
Oxford University Press, 23, 70, 71-72, 138, 234, 235

Paige, James W., 181
Paige Compositor, *180*, 181-182
Palmer, Samuel, 4
Palmer, T. H., 36
paper cutting machines, *207, 211*
papermaking machines, 25-26, 28, 29-30, *29*, 32-34, *32, 33, 34, 35*, 37, *38, 39*
papermills, 27, *31*
Parker, Joseph, 234
Parrish, Douglas M., *29*, 50
Paterson, Emma, 158
Paul Dupont and Cie., 176
Peacock, Thomas Love, 79
perfecting machines, 49, 56, 58, 59
Perkins, Jacob, 99-100, 105, *105*, 106, 107
Petersen, Agnes B., 161, *161*
Philippeaux, Pierre, *152*
Philo of Byzantium, 10
Pianotyp, 171, 176-177
Pickering, William, 194-196, 196-*197*, 198-199, 201
"Pickpocket," *177*, 178
Piette, Louis, 32-33, *34*
platen press, *47, 48*, 140-141
Pollard, Graham, 202-203
Potter, Esther, 202-203
power bed-and-platen presses, 160
Presse à Reaction, 253
Presse Chromo-Lithographique-Mécanique, *103*, 104
Prest, Thomas Peckett, *130*, 133
Prince of Wales, 100
Printers' Pension Corporation, 234
printing machines, 58, 83, 93
Procter, Adelaide A., *157*
Prud'homme, Antoine, 18
Pugin, Augustus Charles, 15
Pynson, Richard, 166

R. Hoe & Co., 104, 140, 141, 147, 236, 240, *240*, 241, 242, *242*, 244, *245*, 247, 248-249, *248*, 256, 259
R. Hoe & Co. Four Cylinder Type-Revolving Printing Machine, *146*, 242, *243*, 245, 246-247
Ramazzini, Bernardino, 4
Rand, Avery & Frye, 149
Rand and Avery, 149
Reed, Sir Charles, 232, 234
Reid Whitelaw, 186
Religious Tract Society, 37, 221
Remnant and Edmonds bindery, 202
Richard Taylor and Co. Printers, 47-48
Richmond, L. P., 161
Richmond, Mrs. L. G., 161
Ridell, Robert Andrew, 37
Ringwalt, J. Luther, 178
Rivière, Emmanuel, 162-163, *162*
Robert, Louis-Nicolas, 25-26, *25*, 28-29, 30, 32, 34, *38*, 193
Roberts, Richard, 66-67, 70
Roberts power looms, *71*
Robinson, Thomas, 67, *70*
Rocourt, Olivia de, 153, *153*
rolling press/machine, 194, 199, *199, 200*, 203
Rosenband, Leonard, 30

Rosenborg, Frederick, 169, 175, *175*
rotary carding machines, 12
rotary drum printing press, 242, 244
rotary press, 49, *118*, 120-121, 253, 255
rotary presses, 45, 46, 97, 101
rotary printing machine, 46
Rowlandson, Thomas, 15
Royal Society of London for Improving Natural Knowledge, 2
Ruskin, John, 259, *261*
Ruthven, John, *35*
Ruthven press, 36, 117, *118*
Rutt, Thomas, 111-112, 135

Sachs, Hans, *205*
Sadleir, Michael, 195, 201
Sadler, James, 44
Sallust, 4, *6*
Sampson, Henry, 67-69
San Jacopo di Ripoli convent, 151
Savage, William, 121, 123
Schelter & Giesecke, 225-226, *228*
Scholtz, Joseph, *110*
Senefelder, Alois, 97, *98, 99*
sewing machines, 205-206, *207, 208*
Seymour, Robert, 40, 76, 95
Shakespeare, William, 196-*197*
shaving machine, 40, *41*
single-cylinder flatbed press, 50
single-cylinder printing machines, *115*
single-cylinder rotary printing machine, 120-121
Smiles, Samuel, 115
Smith, Hoe & Company, 240
Smith, John Raphael, 10-*11*
Smith, John Thomas, 98, *99*
Smith, Matthew and Peter, 240
Smith, W. H., 63, 66, 67-69
Smits, Édouard, 176
Smythe, David, 207
Société pour l'Émancipation Intellectuelle, *88*
Society for Promoting the Employment of Women, 155
Society for the Diffusion of Useful Knowledge (SDUK), 63, 78-80, *78, 80, 81*, 82-83, 84, 85, 85-86, 95
Society of Arts, 199, *199*
Southward, John, 123
Spamer, Otto, 206
spinning frames, 10, 12
spinning jenny, xii, 10, 12, *12*, 28-29
spinning machines, xii
Spottiswoode, Andrew and Robert, 113
Spottiswoode, G., 234
Spottiswoode, William, 232
Staatsdruckerei, K. K., *230*
standing press, 204-205
Stanhope, Charles, 17, *18*, 20-21, 21-23, 36, 135
Stanhope Press, *18*, 21, 22, 48, 71, 108, 110, *115*, 116, 117, *118*, 219, 225, 252, 259
Stanhope process, *136*
Stanton, Elizabeth Cady, 161
staple sewing machines, *210*

295

steam engines, 13–14, 16–17, 40, 43–44, *44*, 120–*121*, *128*, *148*, *160*, *214*, *230*, *245*, *252*
steamboats, *13*, 14
steam-powered carriages, *61*
steam-powered presses, 48–51, 53, 55–56, 59, 63, 87, 90, 94, *102*, 213
steam-powered printing machines, *15*, *47*, *52*, 71–72, 90, 98, *108*, *159*, *160*, *237*
steam-powered railroad, *22*, *60*, *62*, 63, *64*, 65–66, *67*
steamships, *67*
steel engraving presses, *105*, *106*
Stephenson, George, 63
stereotype plates, 4, 19–23, *19*, 71, 72, *72*, 91, 94, *107*, 135, *136*, *138*, *247*
Stevens, Henry, 235
stocking frame, 10
Strahan, 112–113
Suhl press, 44
Sully, Charles, 198

T. H. Carter & Co., 138
Tanquerel des Planches, Louis, 168
Taylor, Alva Burr, *160*
Taylor, Richard, 45, 47, *48*, 49, 55, *55*
Taylor Cylinder Press, 147, 159, *160*

Tennyson, 157
Thackeray, 157
Thackrah, Charles Turner, 168
Thibaudeau, Antoine-Claire, 152, *152*
Thom, Robert, *5*, *189*, *260*
Thompson, 188
Tilloch, Alexander, 19–21
Timperley, Charles H., 123
Tomlinson, Charles, 200, 202, 203, *203*, 204–*205*, 206
Towgood, Matthew, 30
Treadwell, Daniel, 59, 135–140, *137*, *138*, *159*, *237*
Treadwell Power Press, 59, 135–140, *137*, *138*, *139*, 141, *159*, *237*
Treuttel & Würtz, 117
Trevithick, Richard, 63
Trow, John F., 180
Tucker, Steven, 244
Turgan, Julien François, *38*
Turner, William, 46
Twain, Mark, *180*, 181
type-distribution machines, *165*
typesetting machines, *169*

*U*nited Friends, the, 193, *195*
Ure, Andrew, 67, *70*, 120–*121*, *121*, 202
U.S. Centennial Exposition, 249

*V*achon, Marius, 208, *210*, *253*
van der Straet, Jan, xii
van Hove, Frederick Henrik, *2*
Vaughan, Benjamin, 4
Verteneuil, F., *176*
Victoria, Queen, *105*, 157, 219, 235
Victoria Press, *154*, 155–158
Vienna Secession, 226
Virgil, *59*, *59*, 195

*W*akeman, Geoffrey, 26, 219
Waldow, Alexander, 222
Walker, George, 22
Walker, James Scott, *64*, 65
Walter, J., 49
Walter, John, II, 22, 49–50, *50*, *55*, 56, 121
Washington handpresses, *160*
water-frame spinning machine, 10–11
Watson, T., *18*
Watt, James 16–17
Watt, James, 13–14
web press, xi, *224*, *236*, *248*
web printing machine, 249–250
Wellcome Collection, 85
Westall, R., *106*
Widows' Metropolitan Typographical Fund, 156
Wilkins, William Glyde, 127
William Clowes and Sons, 37
William Clowes Ltd., 80, 83, 90, 94, 114–115, 213, *215*
Wilson, Andrew, 20–*21*, *21*–23, 71
Wilson, Frederick J. F., 225

Wilson, Joseph, 37
Winckler, Emil, 206, *207*
Winter, William, 179–180
Wittig, C. F., 222
Women's Co-Operative Printing Unions, 161, *161*
Women's Printing Society (WPS), 158
Women's Protective and Provident League, 158
Wood, William, 194
Woodfall, George, 45, 47, 55
Worde, Wynkyn de, *166*
Workmen's Club and Institute Union, 158
Wright, Joseph, 10–*11*
Würtz, 117
Wyld, James, 72–74, *75*
Wyman, Charles W. H., 252
Wyman, Morrill, *137*

*Y*oung, J. Z., 171
Young, James Hadden, 169, 171, 173–174, *175*, 176
Young and Delcambre Pianotyp, *172*
Young and Delcambre typesetting machine, *165*, *170*, 171–173, *175*, 176, 178

*Z*aehnsdorf, *257*
Zocher, Max and Rudolf, 213, *214*

The engraved images on pp. i–iv show men operating different printing presses that were developed from the 15th through the 19th centuries, including the Stanhope iron press, Gutenberg's handpress, Hippolyte Marinoni's web press, and a hand-driven engraving press. A hand typesetter is working to the right of the web press. These images were adapted and enlarged from graphic design details of bonds issued by Librairies-Imprimeries-Réunies.

HistoryofScience.com
P.O. Box 867, Novato, CA 94948

Published by The Grolier Club
47 East 60th Street
New York, NY 10022
www.grolierclub.org

Distributed by The University
of Chicago Press
1427 East 60th Street
Chicago, IL 60637
www.press.uchicago.edu

Printed and bound in the USA.

Copyright © 2026 by Jeremy M. Norman

All rights reserved. No part of this publication may be reproduced or transmitted in any form or by any means, electronic or mechanical, including photocopy, recording, or any information storage or retrieval system, without permission in writing from the publisher.

ISBN: 978-1-60583-126-8